Energy

Energy
Perspectives, Problems, and Prospects

Michael B. McElroy

UNIVERSITY PRESS
2010

OXFORD

UNIVERSITY PRESS

Oxford University Press, Inc., publishes works that further
Oxford University's objective of excellence
in research, scholarship, and education.

Oxford   New York
Auckland   Cape Town   Dar es Salaam   Hong Kong   Karachi
Kuala Lumpur   Madrid   Melbourne   Mexico City   Nairobi
New Delhi   Shanghai   Taipei   Toronto

With offices in
Argentina   Austria   Brazil   Chile   Czech Republic   France   Greece
Guatemala   Hungary   Italy   Japan   Poland   Portugal   Singapore
South Korea   Switzerland   Thailand   Turkey   Ukraine   Vietnam

Published by Oxford University Press, Inc.
198 Madison Avenue, New York, New York 10016
www.oup.com

Oxford is a registered trademark of Oxford University Press

Library of Congress Cataloging-in-Publication Data
McElroy, Michael B.
Energy : perspectives, problems, and prospects / Michael B. McElroy.
p.   cm.
Includes bibliographical references and index.
ISBN 978-0-19-538611-0
1. Power resources. I. Title.
TJ163.2.M3843 2010
333.79—dc22   2009007390

9 8 7 6 5 4 3

Printed in the United States of America
on acid-free paper

For Veronica

Preface

As this book goes to press, the world is confronting a baffling array of interconnected, in many respects unprecedented, crises. Financial markets worldwide are in a state of disarray. Governments have moved aggressively to shore up the balance sheets of banks, to guarantee deposits, and provide liquidity. Major financial institutions have failed overnight. Currencies have fluctuated wildly and unpredictably. Oil prices over less than a few months have moved from a high of close to $150 a barrel to less than $40 a barrel recovering more recently (May, 2009) to about $60 a barrel. Other globally traded commodities—copper, gold, silver, and even corn—have gyrated over equally extreme ranges. What happens to markets overnight in Asia influences markets in Europe a few hours later and then on to New York impacting lives of peoples everywhere. Never before has it been so clear that we live in an instantaneously interconnected world.

The focus of this book is energy, the lifeblood of the human enterprise. For the most part our present world runs on fossil fuels—coal, oil, and natural gas—exploiting energy harnessed from the sun by plants—land and sea-based—many millions of years ago. As the late Roger Revelle famously remarked more than fifty years ago, we are conducting a great global geological experiment with uncertain foreknowledge as to the consequences. The concentration of carbon dioxide in the atmosphere is greater now than it has been at any time over the past several million years. If we continue our present course it will be higher a few decades from now than it was 65 million years ago when dinosaurs last roamed the Earth. Global climate is changing. Earth is absorbing increasingly more energy from the sun than it is returning to space. The globe is getting warmer. Weather patterns are shifting. The Arctic Ocean is rapidly losing its ice cover. Even the stability of the great ice sheets covering Greenland and Antarctica is threatened. And all of this is taking place at a time when the ability of peoples to adjust to deteriorating local conditions by moving elsewhere is restricted by increasingly impenetrable international borders, at a time when world population is greater than it has ever been, destined to increase by several additional billions over the next few decades.

There is growing conviction that we need to radically alter the structure of our global energy economy. Carbon-based fossil fuels have served us well for more than a century but we are forced now to confront the consequences. We need to change course: to switch from an unsustainable fossil fuel based energy economy to one based on more sustainable resources. Wind, sun, and geothermal heat offer, we will argue, attractive options. And with appropriate precautions, nuclear power can also make an important contribution.

I was prompted at the outset of the project that led to this book by a debate at my university as to what should constitute the elements of a modern liberal arts education. There was consensus that undergraduate students should be required to pursue at least one subject in depth: they should graduate with specific disciplinary skills in at least one area of important intellectual activity. But more should be required. The last great debate on general education at Harvard University (in 1979) mandated that undergraduates should be introduced to "approaches to knowledge in seven areas considered indispensable to the contemporary student: foreign cultures, historical study, literature and the arts, moral reasoning, quantitative reasoning, science, and social analysis." They would fulfill their requirements by selecting from a menu of more or less specialized courses offered in each of these seven areas. The current debate recognizes, however, that to an increasing extent the challenges we face in modern life require multidisciplinary skills and our curriculum should be structured accordingly.

This book is an attempt to respond to this challenge with a particular focus on energy but with a recognition that the choices we make now for our energy future can have long-range consequences. We need quantitative skills to select among the multiple options available to address it. We must appreciate that our human society, like nature itself, is dynamic. We need to recognize that there is a unity to life on Earth, that we are part of nature, not independent, and that we must live within its bounds. We need to appreciate that we have the potential to alter our environment globally but that we must exercise this power with discretion. We need to absorb the lessons of history, to draw not just on insights from science and technology but also from the expansive intellectual heritage codified in the world's great philosophical and cultural traditions. In terms of the seven core areas identified for general education at Harvard University 30 years ago, lessons from all seven are relevant to the critical choices we must make today to ensure the continuing success of our civilization. My hope is that this book can help delineate these choices, that it can provide at least context for the debate that must take place to ensure a sustainable future for our interdependent global society.

Acknowledgments

A few years ago, I responded to a request from my colleague Professor Frederick Abernathy to join him in offering a new course on Energy, Technology, and the Environment directed at non-technically oriented Harvard undergraduates. Fred is a man for all seasons with expertise not only in technology but also in how our modern industrial world evolved. I learned a lot working with Fred. The experience provided the initial stimulus for the project that led to this book. I am grateful to Fred for his encouragement and for our many stimulating conversations.

I am indebted to a number of other colleagues—notably Professors James G. Anderson, Paul C. Martin, Ruth A.Reck, Daniel Schrag, Frans Spaepen and Richard Wilson—who gave generously of their time to read portions of the text and for the many contributions they made to my education. A special tip to Dan who, as Director of the Harvard University Center for the Environment, sponsored a University-wide energy initiative which hosted a distinguished cadre of energy experts drawn from government, academia, and industry and for the wide-ranging discussions prompted by these visits. Thanks also to a number of my non-scientist friends—George Cuomo, Donald Forte, Morton Myerson, and the Honorable Simon Upton—who committed time to read earlier versions of portions of the manuscript and for the questions they raised that forced me to focus. A special thanks to my son Stephen McElroy for numerous informative discussions on the practical aspects of energy and for his insights on how we could do more with less.

It would not have been possible to complete this project without the invaluable contributions I received from my immediate colleagues and students. A special thanks to Lu Xi, Chris Nielsen, and Yuxuan Wang. Yuxuan was unfailingly responsive to my frequent calls for help even after she left Harvard to take up her present position as Professor of Environmental Engineering at Tsinghua University in Beijing. I am indebted also to Hugo Beekman who stepped into the breach late in the project to prepare the index and to help with some other essential materials, to Laurence Tai who served as my research assistant during the early stages of this project and to Cecilia McCormack for her unfailing facility to keep the entire enterprise under control. Thanks also to the multi-talented undergraduates who signed up for my junior seminar and who read and critiqued the entire manuscript. Last though not least, my thanks to the National Science Foundation for uninterrupted support of my research at Harvard University over the past 40 years, most recently under grant ATM-0635548.

Contents

Energy

1

Introduction

1.1 BACKGROUND

I began the project that led to this book with a question: is the time in which we live now unique in the history of our planet? Global population has risen to an unprecedented level of 6.4 billion and is projected to climb to close to 9 billion over the next 40 years. I was conscious of the good news: that an increasing fraction of the world population is enjoying a measure of unprecedented economic success, fruits of technological developments that ensued over the past several centuries. According to some estimates, three hundred million people have been lifted from extreme poverty in China since 1990 with an additional two hundred million in India, a monumental achievement by any development measure. Globally, people are living longer. Rates of infant mortality are on the decline. More people have access to adequate facilities for health care and education. And, in many parts of the world, but surely not all, women are assuming their rightful role as coequal partners in society. The bad news is that progress is confined to less than half of the world's population.

More than a billion of our fellow citizens are trapped today in unspeakable poverty, forced to survive on less than a dollar a day (Sachs, 2005). In parts of the modern world, notably in sub-Saharan Africa, deadly diseases such as AIDS, tuberculosis, cholera, and malaria are on the rise. The quality of physical environments is in many instances on a path to ruin reflecting unsustainable demands on soils, waters, and the biota imposed by peoples driven to survive in the present without the luxury of planning for the future. It is a sad fact that aspirations for poverty alleviation and environmental protection are often antithetical. Added to this, the toll from disasters, natural and man-made, is in many cases catastrophic and the situation is getting worse rather than better.

Unanticipated variability in climate—droughts, floods, and violent storms—pose challenges for those least equipped to cope, a problem experienced increasingly in many different parts of the world. Those of us who live in the so-called developed world owe much of our affluence to the relatively inexpensive sources of fossil energy—coal, oil, and gas—we employ to drive our economic engine. But are the unanticipated consequences of our reliance on energy resources bequeathed from the past destined to bring recent progress to a halt? I refer here to the possibility of unpredictable changes in global climate triggered by the build up of greenhouse gases emitted as byproducts of the combustion of fossil fuels, gases that trap heat emitted from the surface of our planet, a circumstance that may be expected to result in an increase in global average temperature with potentially disruptive effects on local climate stressing to the breaking point the ability of local communities, particularly poor communities, to cope.

How did we arrive at this state of affairs and what are the challenges we must confront if we are to survive and prosper as a species? Can we continue to rely on fossil sources of energy or do we have alternatives that might be environmentally more benign? Can we use what we have more efficiently? Are the resources of the Earth up to the challenge of supporting the aspirations and demands of a future population of close to 10 billion? Can we survive in a world where the gap between rich and poor continues to widen? Are there strategies we can employ to ensure a more equitable distribution of resources? These are the challenges we must confront and knowledge of whence we have come is essential if we are to develop strategies that can be successful in the future. Access to economically viable and environmentally sustainable sources of energy is essential if we are to survive and prosper. The question motivating this book is: how did our present energy system evolve and what are our options for the future? It is instructive to begin with a sense of history. How did our planet arrive at its present state? How do we humans fit into the larger geological picture? We are not supernatural and we need to understand our place in nature.

We begin in Section 1.2, with a brief history of the forces that shaped the evolution of the Earth, continuing in Section 1.3 with an account of what we know of the history of life on our planet. Section 1.4 is devoted to the story of the emergence of humans and how we arrived at the point where we now have the capacity to alter the metabolism of our planet on a global scale. The organization of the book is outlined in Section 1.5.

1.2 A BRIEF HISTORY OF THE EARTH

The Earth is approximately 4.6 billion years old. Together with the other planets and the sun, it formed by condensation from a spinning mass of gas and dust that composed the original solar nebula. As matter accumulated in the proto Earth, the planet began to heat up, responding in part to heat released as a result of gravitational accretion (conversion of potential to kinetic energy), in part to energy released by decay of radioactive elements such as uranium, thorium, and potassium. Heating of the interior resulted in an instability, with warmer, and consequently lighter, material at depth rising in some regions to be replaced by colder, and thus heavier, material sinking elsewhere. Weightier elements such as iron settled to the core while more volatile species such as hydrogen, carbon, and nitrogen accumulated at the surface forming the primitive atmosphere, ocean, and crust.

Chemical differentiation driven by changes in pressure and temperature as a function of depth and by associated vertical motions led to the formation of the distinct zones we associate with the Earth today—the core, inner and outer mantle, crust, ocean, and atmosphere. In particular, it resulted in the formation of the discrete elements of crustal material geologists refer to as plates. As fresh material was added to crustal plates by upward motion, old material was removed by downward motion elsewhere. Segregation of light from heavy material led to the formation of the continents. The crustal plates floated like rafts on the heavier material that composed the underlying mantle. The pattern of vertical motion and related horizontal motions prompted the plates to move around. The configuration of the continents was altered accordingly. At times, the continental plates found themselves close together forming super continents. At other times, they were more geographically dispersed. Mountains appeared where continental plates converged (came together) and were eroded subsequently by weathering (chemical changes induced by contact with water and corrosive atmospheric gases). The Himalayas and the Tibetan Plateau arose, for example, as a consequence of a relatively recent (15 million years ago) collision of India with Asia. The juncture of North and South America was even more recent—it took place only a few million years ago. Previously, North and South America were separated by open water and the Pacific and Atlantic Oceans were connected through what we identify now as the Panama Straits. The changing configuration of continents through time is illustrated in Fig. 1.1.

The vertical overturning of the Earth—we refer to it as mantle convection— and the associated movement of the crustal plates—referred to as plate tectonics—have had a significant influence on global and regional climate. Weathering of continental rocks provides a sink for atmospheric carbon dioxide. Conversely, carbon dioxide is returned to the atmosphere as a component of volcanic eruptions when carbon-rich sedimentary material is drawn down into the mantle as a consequence of the convergence (collision) of crustal plates. As we shall discuss later, carbon dioxide is an important greenhouse gas—it is effective in limiting transmission of Earth's heat—infrared radiation—to space. Concentrations of CO_2 have varied significantly over geologic time and climate has responded accordingly. There were times in the past when climate was unusually warm—temperatures were moderate even at high latitude in winter. And there were times when the Earth was freezing cold even at the equator. Changes in CO_2 are thought to have been at least partially responsible for these large-scale fluctuations in climate.

50 Ma (Cenozoic/Tertiary/Paleogene/Eocene/Lutetian)

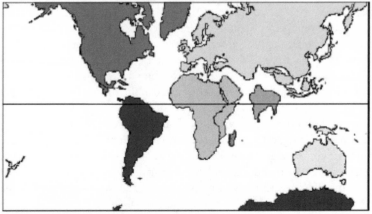

100 Ma (Mesozoic/Cretaceous/Gallic/Albian)

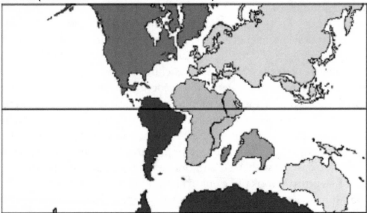

260 Ma (Paleozoic/Permian/Rotliegendes/Artinskian)

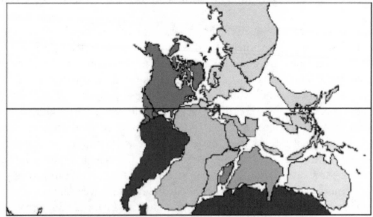

Figure 1.1 The changing configuration of continents (http://www.dinosauria.com/dml/ maps.htm, read 17 September, 2008. Picture source: Windley, 1995).

6

320 Ma (Paleozoic/Carboniferous/Pennsylvanian/Bashkirian/Yeadonian)

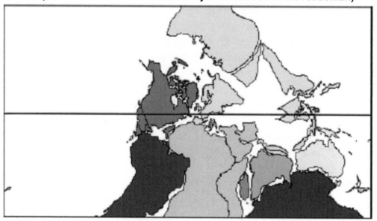

410 Ma (Paleozoic/Silurian/Pridoli)

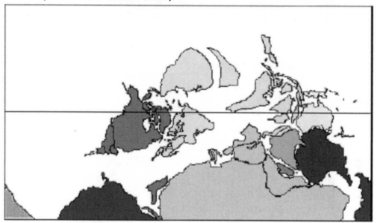

510 Ma (Paleozoic/Ordovician/Canadian/Tremadoc)

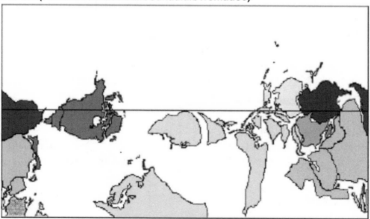

Figure 1.1 *continued*

The configuration of the continents also has an impact on climate. It affects the albedo of the Earth (the color of the planet that determines the fraction of incident sunlight absorbed or reflected by the Earth) and has an influence also on the circulation and consequently on the redistribution of heat by the ocean. Climate has been generally cold for the past several million years with large quantities of water withdrawn from the ocean sequestered in thick sheets of ice on continental landmasses at higher latitudes. Twenty thousand years ago, the Earth was in the depths of the most recent ice age with ice covering large regions of North America and Northwestern Europe. Sea level was 110 m lower than it is today and, as we shall see, it was possible for humans then to make their way from Asia to North America taking advantage of a land bridge that linked the two continents over what are now the open waters of the Bering Straits. The relatively cold climate of the past several million years (interrupted by brief periods of comparative warmth) has been attributed to a change in ocean circulation linked to the closure of the sea gap between North and South America, in combination with regular, astronomically related, changes in the seasonal distribution of sunlight incident at different latitudes (for a more detailed discussion, see McElroy, 2002).

The salt content of surface waters in the North Atlantic is significantly higher today than that in the Pacific reflecting net transfer of moisture through the atmosphere from the former to the latter. The presence of a high north to south mountain chain in the western Americas (the Rockies and the Andes) contributes to this distinction by restricting transfer of moisture westward through the atmosphere from the Pacific to the Atlantic at mid and high latitudes where winds blow generally from the west. In contrast, the absence of a high mountain barrier in Central America where the prevailing trade winds blow from the east allows moisture to move freely from the Atlantic to the Pacific. When salt-rich waters in the high latitude Atlantic cool in winter they become denser than the underlying water and sink to the deep. Circulation of deep waters in the world oceans today moves from the North Atlantic to the Pacific passing around the Antarctic continent where they receive an additional influx of cold dense water. This so-called deep-water conveyor belt, illustrated in Fig. 1.2 , has a significant influence on global climate drawing warm surface water from south to north in the Atlantic contributing to the relatively mild climate of Northwestern Europe. It depends ultimately on the contrast in salinity between the Atlantic and Pacific a distinction that was surely much less when the two oceans were linked through what is now Central America and what was earlier open water. Changes in the sense and strength of the deep ocean conveyor belt linked to changes in the spatial extent of land-based ice are thought to have played an important role in determining the climate of the past several hundred thousand years of Earth history.

1.3 HISTORY OF LIFE

It is clear that life has been present from an early stage of Earth history, for at least 3.5 billion years and arguably longer. Precisely how it came about

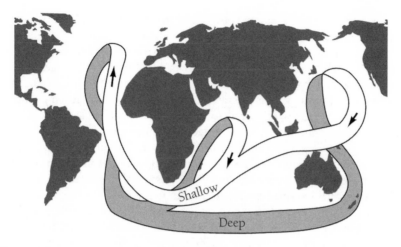

Figure 1.2 The Conveyer Belt: Broecker (1987)

is unclear. Some believe that Earth was seeded at the outset from space with primitive forms of life delivered in a rain of comets—manna from the primordial heavens. Others contend that the molecules that constituted the building blocks of early living organisms were formed in the primitive atmosphere as a consequence of chemical reactions triggered by either lightning or solar ultraviolet radiation. More likely, in my opinion, it developed in the ocean, in the vicinity of hot springs emanating from regions where fresh mantle material enters the ocean. Deep-sea vents distributed along regions of sea floor spreading (regions where the ocean plates are moving apart) support a remarkable ecological system in the ocean today. Bacteria feeding on energy supplied by the oxidation of sulfur contained in the hot (300°C) spring water represent the bottom of a contemporary food chain supporting a dense population of worms and clams in the immediate vicinity of these present-day vents. Oxygen would have been a minor constituent of the early atmosphere. Nitrite or nitrate formed in the primitive atmosphere, however, could have supplied the oxidant needed to oxidize sulfur emitted in the primitive vents and thus, as a consequence, the energy source required for synthesis of the organic molecules essential for early life.

The earliest forms of life consisted of organisms known as prokaryotes. Bacteria and blue green algae are examples of these simple life forms that developed over the first few billion years of Earth history. They have been remarkably resilient and continue to play an important role in the complex web that characterizes life on our planet today. Bacteria function as nature's garbage disposers transforming all kinds of organic wastes to more useful compounds. There are bacteria that rely on chemical energy for their metabolism and bacteria that have evolved the capacity to utilize sunlight. The blue green algae, for example, have the ability, using sunlight, to convert chemically inert diatomic nitrogen (the dominant component of the atmosphere) to forms of nitrogen such as ammonia that can be readily incorporated in the structural material of plants and animals. To transform inert molecular nitrogen to biologically useful nitrogen fertilizer today we rely on chemical factories consuming large quantities of fossil energy to accommodate the high temperatures (thousands of degrees) required to fracture the tight triple bond linking the nitrogen atoms in molecular nitrogen. As another example of the diverse function of bacteria consider the role they play in allowing animals such as cattle, sheep, and goats (so-called ruminants) to feed on plant material (grass for example) that we would find totally indigestible. The bacteria in this case play an essential intermediary role in the ruminant food chain. They convert low-grade organic matter in the second stomach of these animals to higher-grade materials that can be ingested directly by the animals. In the process they evolve waste products such as methane that are released to the atmosphere with implications, as we shall see later, for global climate (the concentration of methane in the atmosphere has been increasing steadily over the past several centuries). Bacteria are ubiquitous, multifunctional, and essential for life. And all of this remarkable diversity arose over the first few billion years of Earth history.

The next stage of evolution involved the appearance of eukaryotes, unicellular organisms significantly more complex than their prokaryotic antecedents. Lynn Margulis (1970) suggested that the eukaryotes evolved as a result of the fusion of cells of preexisting prokaryotes. Think of one species of prokaryote eating another. The ingested cell is incorporated by the scavenging cell to form what is known as a mitochondrion, one or more of which are present in all eukaryotes. The new organism combines, and in many cases extends, the functionality of its progenitors. The evolution of the eukaryotic cell paved the way for the development of more complex multicellular organisms, eventually for plants and animals and ultimately for humans. This monumental step—the evolution of the eukaryotic cell—took place roughly 1.5 billion years ago. Along the way, eukaryotes developed the capacity to survive in the presence of free oxygen, the inevitable by-product of photosynthesis. Until organisms evolved this facility, the influence of photosynthetic organisms was limited by the requirement the organisms had to dispose of their oxygen waste. They did so by attaching the oxygen they produced to conveniently available, chemically reduced, elements such as iron. The development of an oxygen-mediating enzyme eliminated the need for reduced iron allowing oxygen to be released to the external environment. This resulted in a build up of oxygen in the atmosphere and ocean triggering, roughly 2 billion years ago, the first known consequential change in the global environment. Organisms without the capacity to tolerate free oxygen were banished to anoxic microenvironments or refugia. The appearance of free oxygen in the atmosphere led also to the production of ozone in the atmosphere in concentrations sufficient to shield the surface of the Earth from the biologically lethal doses of ultraviolet solar radiation that prevailed previously, paving the way later, about 440 million years ago, for the migration of life from the ocean to land.

The pace of evolution has been distinctively sporadic. There were periods when large numbers of new species developed over relatively brief episodes of geologic time and times characterized by massive extinctions. Progenitors of all of the modern phyla are present, for example, in the Burgess Shale, a remarkable assemblage of fossils, dated at about 440 million years before present (BP), discovered by C.D. Walcott high in the Canadian Rockies in 1909. The explosion of new life forms that arose during the Cambrian Period (543 million to 510 million years BP), recorded in the Burgess Shale, was followed a few hundred million years later, at about 225 million year BP, by what the late Steven Gould (1989) termed the "granddaddy of all extinctions," responsible for elimination of as much as 95% of all marine species alive at that time. A second major extinction, triggered 65 million years ago by the impact of a giant meteorite, led to the demise of the dinosaurs paving the way for the later ascendancy of large mammals and eventually our human ancestors. As Gould points out, mammals spent their first 100 million years (they appeared first about 160 million years BP) as "small creatures living in the nooks and crannies of a dinosaur's world." The demise of the dinosaurs was a happy event for the mammalian world and indeed for us humans. As Gould drolly remarked: "in an entirely literal sense, we owe our existence, as large and reasoning animals, to our lucky stars."

1.4 EMERGENCE OF HUMANS

Our closest relatives in the animal kingdom are the giant apes (including the gorilla and the chimpanzee). Much of the development that led to the appearance of modern humans is thought to have taken place in Africa. The details of the evolutionary sequence are unclear. It is thought, however, to have proceeded along a trajectory that involved sequentially the appearance of Australopithecus africanus, Homo habilis, and Homo erectus. Homo erectus arrived on the scene about 1.7 million years ago and evolved later into Homo sapiens. Interpretations of human mitochondrial DNA suggest that we share a common maternal ancestor, a primitive Eve, and that she may have lived in Africa about 150 thousand years ago. As Diamond (1997) tells the story, our early ancestors were nomadic—hunter-gatherers—constantly on the move in the search for new sources of plant and animal species for food. Changes in climate may have had an important influence on their migration. The Earth was relatively warm 120 thousand years ago and we might expect that food and fiber would have been relatively available at that time in tropical and sub-tropical Africa. Climate then deteriorated as the world entered what developed as the last great ice age, which persisted with intermittent warm periods for close to a hundred thousand years ending only about 20,000 years ago. It is reasonable to expect that our ancestors in search of food would have made their way first to the Middle East and to southern Africa (perhaps at the time when climate began first to deteriorate) populating subsequently much of the more temperate regions of Europe and Asia. They had arrived in New Guinea and in Australia by 40,000 years BP and had made their way north to Siberia and eventually to the Americas by the time the ice age had reached its peak 20,000 years BP. As noted earlier, sea level at that time was about 120 m lower than it is today. Asia was connected to the Americas by way of a land bridge across the Bering Straights. Within a few thousand years, the human presence had expanded to encompass most of the habitable portions of both North and South America (the exception, the region occupied by the receding ice sheet that covered most of Canada for a further six or seven thousand years). By 20,000 years BP, humans had extended their range to encompass all of the world's continental landmasses with the exception of Antarctica.

The total worldwide population of humans in the hunter/gatherer age was relatively small, a few million or less (McEvedy et al., 1978). To put this in context, there were probably more bison (buffalo) roaming the plains and grasslands of the Americas a few tens of thousands of years ago than there were humans on the Earth as a whole: the population of bison when the Europeans first arrived in the Americas has been estimated at between 20 and 30 million (Lott, 2002). Despite their small population, however, early humans had an impressive, and generally negative, impact on the environment. A short time after their arrival in Australia, most of the native large mammals disappeared. A similar fate befell the bulk of the mega fauna in the Americas. Diamond (1997) suggests that 15,000 years ago, the American west would have looked much like the Serengeti Plain in Africa today with "herds of elephants and horses pursued by lions and cheetahs, and joined by members of such exotic species as camels and giant ground sloths." In a blink of geologic time this spectacular diversity was no more.

There can be little doubt that humans were responsible for this collapse. Diamond (1997) suggests that the large animals that populated Africa and Eurasia survived because they had coevolved for a long time with humans before humans had refined their hunting skills. As a consequence they had learned to take appropriate evasive action. The large mammals that populated Australia and the Americas had no such experience and had little chance to survive their first encounter with our environmentally insensitive, voracious, ancestors. We have only recently begun to appreciate the extent to which our future depends on a harmonious coexistence with nature. E.O. Wilson (1992) has argued that we are in the midst today of an extinction event that rivals any the world has witnessed over at least the past 500 million years, greater even than the momentous event that terminated the era of the dinosaurs. The weapons we employ now are not the spears or bows and arrows of our hunter/gatherer ancestors but the fire and bulldozers we use to transform landscapes and to eliminate whole ecosystems in the quest for what we perceive as more productive uses of land. Nowhere is this problem more serious than in the tropics where the diversity of life is greatest and where community structures

Table 1.1 Chronology of important events in Earth history

Earth events	Time (year BP)	1-year perspective
Earth forms	4.50E+09	January 1st
Eukaryotes	1.50E+09	September 1st
Burgess Shale	5.00E+08	November 20th
Vascular plants	4.45E+08	November 24th, 9 pm
Life expands to land	4.40E+08	November 25th, 7 am
Amphibians	3.00E+08	December 6th, 4 pm
Massive extinction	2.25E+08	December 12th, 6 pm
Mammals	1.60E+08	December 18th, 4 pm
Dinosaurs eliminated	6.50E+07	December 25th, 5 pm
Homo erectus	1.70E+06	December 31st, 8 pm, 41min
Homo sapiens	1.50E+05	December 31st, 11 pm, 42 min, 29 s
Human migration	5.00E+04	December 31st, 11 pm, 54 min, 10 s
Humans reach Americas	3.00E+04	December 31st, 11 pm, 56 min, 30 s
Agriculture	1.00E+04	December 31st, 11 pm, 58 min, 50 s
Industrial Revolution	250	December 31st, 11 pm, 59 min, 58 s

are most complex. A recent study by the United Nations suggests that 25% of the species living in tropical forests today may be doomed to extinction over the next few decades if current trends in deforestation are not reversed. Deforestation in the tropics is responsible today also for significant emissions of CO_2 contributing to the problem of global climate change (the source is estimated to amount to as much as 30% of the source from global combustion of fossil fuels).

The chronology of the important events in Earth history discussed here is summarized in Table 1.1. With times ranging from billions to hundreds of years before present, it is easy to lose perspective. To provide this perspective, we have opted to present the chronology not only on the real time line (years before present, BP) but also to indicate how it would apply if the entire history of the earth were compressed into a single year. Only in the last few seconds of this metaphorical year have humans developed the capacity to alter the environment of the Earth on a global scale. The question is whether we can develop the wisdom and foresight to ensure a harmonious co-existence of our species with nature over the first few seconds of the year that is about to begin. A specific challenge relates to the strategy we select to satisfy our requirements for energy.

Chapter 2 presents a lightning quick tour of history, from the development of agriculture in the Middle East 12,000 years or so ago to the emergence of the modern nation states of Europe, the opening up of the Americas to European colonists, and the beginning of what is referred to, somewhat misleadingly, as the industrial revolution. The word revolution normally implies a singular transforming occurrence. It is difficult, however, to identify any such unique event that could be highlighted as marking a revolutionary transition from the pre-industrial to the industrial age. James Watt is often identified as the father of the industrial revolution and, inaccurately, as the inventor of the steam engine. He was certainly a giant figure. But his singular achievement, as we shall see, was not the invention of the steam engine but rather his success in refining an earlier invention by Thomas Newcomen and his ingenuity in establishing a myriad of useful applications for this new technology.

There are a number of recurrent themes to the story in Chapter 2. The success of civilizations depended invariably on a favorable local climate together with access to reliable sources of energy. All too often, the two factors were inextricably interlinked. For much of human history, wood represented the primary energy source. John Perlin (1989) makes a persuasive case that access to adequate supplies of wood played a critical role in the development of early civilizations. When the wood ran out, civilizations collapsed. It was a pattern repeated many times over the course of history, not just for the first great civilization of recorded history, the Sumerian, but for many of the civilizations that succeeded it—in Egypt, the Indus region, Babylon, Crete, Greece, Cyprus, and later for the case of Rome. Wood was required for the construction of buildings, for furniture (tables and chairs), for heating and cooking, and for the fabrication of wheeled vehicles (carts and chariots for example), for weapons, and ships. It was the critical component in the manufacture of tools. Arguably, though, the greatest single demand for wood in the ancient world, and indeed for the modern world up to the age of mechanization, involved the production of charcoal, a carbon-rich compound formed from incomplete combustion of wood. Vast quantities of charcoal were consumed in smelting copper, tin, and iron. As late as 1900 wood was the dominant energy source in the United States. When access to adequate supplies of timber in homelands ran out, civilizations resorted frequently to military action to secure these supplies from their neighbors.

Changes in regional climate led to the demise of a number of important civilizations over the course of history, opening opportunities for new civilizations to evolve and to prosper elsewhere. Demand for resources unavailable locally provided motivation often for territorial expansion in addition to incentives for long distance trade leading inexorably to the interconnected global society that prevails today. Religious belief played a critical role in determining the cohesion of individual civilizations. It exercised an influence, which sometimes positive, was often negative and on occasions destructive—a motivation for conflict most commonly local but often in recent years more global. All of these influences we would argue are ubiquitously present in our modern world. Developing a viable strategy for future energy supplies must account from the outset for the implications of particular choices for international security and for the integrity of the environment, not only on a local but also on a global scale. We have already demonstrated our willingness to go to war to secure access to energy sources judged essential for the orderly function of our society—oil the most recent case in point. And religious differences are at least partially responsible for the threat we perceive today from militant Islam. A sense of history, as elaborated in Chapter 2, provides an essential context for the formulation of policy options with promise to enhance the prospects for a socially equitable, sustainable, future for global society.

Chapter 3 addresses the question of what exactly we mean by energy and how we measure it. Chapters 4–9 are devoted to the energy sources that fueled the development of the modern industrial economy—wood, wind, water, coal, oil, natural gas, and nuclear.

Chapters 5–7 recount the story of the emergence of the fossil fuels—coal, oil, and natural gas—as the dominant sources of energy for the modern industrial economy, including accounts of their origin. Coal replaced wood as the fuel of choice in England as early as 1800, a transition facilitated, as we shall see, by the development of the steam engine. Coal is by far the most abundant of the fossil fuels but from an environmental point of view the most problematic. Burning untreated coal adds large concentrations of sooty materials to the atmosphere in addition to gaseous compounds of nitrogen and sulfur oxides (contributing to the problem of acid rain) and a variety of toxic elements including mercury and even radioactive elements such as uranium. The impacts of coal combustion on the local and regional environment and for public health are serious and are discussed in Chapter 5. Local and regional effects of burning oil and natural gas are less severe but still consequential, especially for the case of oil as indicated in Chapter 6. The environmental impacts of fossil fuel use, moreover, are not confined locally. All three of these fuels are responsible for large-scale emissions of CO_2, with implications for global climate change—for global warming—as discussed in Chapter 13 (see also Chapter 4).

The concentration of CO_2 is higher in the atmosphere today than at any time over the past 650,000 years. It is rising steadily and, given our current reliance on fossil fuels, is likely to climb to as much as twice its pre-industrial level (or more) by the end of this century. Coal is the largest single contributor to the rise in CO_2. The energy liberated in burning coal is associated primarily with oxidation of carbon to CO_2. Production of CO_2 per unit of energy released in burning oil is consequently less than for coal—a fraction of the energy released in this case is associated with oxidation of hydrogen to H_2O. From the point of view of CO_2, natural gas (CH_4) with its higher content of hydrogen, is the least problematic of the fossil fuels but still significant.

The United States has been responsible until recently for the largest single national source of CO_2: China has now supplanted the U.S. With 5% of the world's population, the U.S. accounts for 22% of global emissions. Electricity, generated largely by combustion of coal, accounts for 40% of U.S. emissions. Transportation, fueled primarily by oil, is responsible for an additional 32%, with the balance due to a combination of home/office heating and cooling (11%) and various industrial processes (18%). China, with more than a fifth of the world's population, accounts for more than 20% of global CO_2 emissions (Netherlands Environmental Assessment Agency). If we are to reduce the climate altering potential of fossil fuels we need policy options that can work not only for large developed economies such as the U.S. but also for large developing economies such as China.

The potential for wind and water sources of energy is discussed in Chapter 8. Wind power, we argue, can provide an important, environmentally benign, economically competitive, alternative to fossil fuels as an energy resource in the future.

The history, potential, and problems of nuclear power are addressed in Chapter 9. The fuel for nuclear power is the isotope of uranium with atomic mass number 235, written as U^{235}. (The superscript here indicates the total number of protons plus neutrons contained in the nucleus of a given isotope. Different isotopes of a specific element (uranium for example) contain different numbers of neutrons). The nucleus of U^{235} includes 92 protons and 143 neutrons. The nucleus of U^{235} is stable but it may be induced to break apart—to fission—upon collision with a neutron. Two distinct elements are formed as a consequence of such a fission reaction with production of additional neutrons (2.5 on average) that can collide with additional U^{235} resulting in what is known as a chain reaction (once you start it, it can keep running on its own). The fission products, the atoms formed by the reaction between the neutron and U^{235}, are produced with extremely high (kinetic) energy, as much as a million times greater than the energy released when an atom of carbon in coal, oil, or gas is converted to CO_2. This accounts for the origin of the energy involved in nuclear power.

The nuclear age began on October 12, 1939 when President Roosevelt authorized government funding for nuclear research responding to a letter from Albert Einstein. Nuclear power was developed at the outset for military purposes, to make the bombs dropped later on Hiroshima and Nagasaki. The first steps to apply nuclear energy in the civilian sector—to generate electricity—occurred later, on August 1, 1946 when President Truman signed the Atomic Energy Act authorizing formation of the U.S. Atomic Energy Commission. Civilian use of nuclear power expanded extensively in the 1970s in response to the rise in oil prices triggered by the Arab oil crisis. Enthusiasm waned, however, following accidents at Three Mile Island in Pennsylvania on March 28, 1979 and at Chernobyl in the Soviet Union on April 25, 1986. The former was relatively inconsequential: the reactor containment vessel was not breached and minimal radiation was released to the outside. The problem at Chernobyl was more serious. The reactor in this case was totally destroyed. An intense fire lifted radiation to high altitude and the plume spread over a large area of northern and central Europe. Precisely how many people died either promptly or on a detailed basis is unclear: estimates range from as low as 30 to numbers in the thousands. There is no doubt, however, that these accidents seriously undermined the confidence of the public in the safety and future security of nuclear power.

There are a number of issues that must be addressed if use of nuclear power is to expand significantly in the future. First, there is the need for a better-educated public, one that understands the distinction between nuclear power and nuclear bombs. Second, there is need for a strategy to deal with the radioactive waste produced by nuclear industry. Third, we need to take steps

to ensure that facilities servicing a civilian nuclear industry are not subverted to produce materials that could be used to make nuclear bombs. And fourth, without compromising safety, we need to develop a new generation of economically more competitive nuclear power plants. If we can address these challenges, we can reduce our reliance on energy supplies from politically unstable sources—the Middle East for example. And we can take an important step to reduce the threat to the global climate system posed by increasing emissions of CO_2.

The immediate product of the combustion of wood, coal, oil and natural gas, and the fission of uranium is heat. Chapters 10 and 11 outline the important ways in which this heat can be channeled to perform useful work. The original purpose of Newcomen's steam engine was to pump water from mines. Watt refined and extended the power of steam to a myriad of additional industrial applications. Early uses of steam are discussed in Chapter 10. Later applications in the production of electricity, the backbone of our modern economy, are treated in Chapter 11 while the history of the internal combustion engine and the automobile is discussed in Chapter 12.

The nature of the challenge to the climate system posed by increasing human-related emissions of greenhouse gases is treated in Chapter 13. The possibility of capturing CO_2 prior to emission to the atmosphere and sequestering it in a long-lived geological reservoir, either subsurface on land or in the ocean, is explored in Chapter 14.

Capture and sequestration of carbon, were it to prove feasible economically and environmentally, could have far-reaching consequences for the future of the global energy economy. It could encourage continuing and most likely increasing, reliance on coal (as supplies of conventional oil and gas begin to run down), switching the onus from dealing with the problem of CO_2 to the potentially simpler task of addressing the other coal-related pollutants (particulates, sulfur, nitrogen, mercury etc.), investing in what is referred to as clean coal technology.

In its most advanced form, clean coal technology relies on conversion of the carbon contained in coal to syngas (CO and H_2) prior to combustion (by reacting the coal with hot steam). The CO reacts subsequently with hot steam to form additional H_2 and CO_2. This provides the opportunity not only to capture CO_2 prior to release to the atmosphere but also to eliminate (or at least significantly reduce) emissions of a large fraction of the other offending coal-related pollutants. The downside is that additional CO_2 is produced inevitably in conjunction with the transformation of the coal to syngas and the subsequent conversion of syngas to other desirable energy end products (both liquid and gaseous). The need to sequester carbon is increased accordingly. Given the large quantities of CO_2 that would have to be disposed of to fully resolve the climate problem—more than 20 billion tons (1 metric ton is equal to a mass of 1000 kg) of CO_2 each year for the foreseeable future—and the questions that will arise inevitably concerning the costs and environmental consequences of specific carbon sequestration strategies, we view it as unlikely that capture and

sequestration of carbon, at least in the near term, can make more than a modest contribution to the challenge of arresting the future accumulation of greenhouse gases in the atmosphere.

Current U.S. policy envisages an important role for ethanol produced initially from corn but potentially eventually from cellulose (extracted from grasses and trees) to reduce consumption of gasoline. Prospects for ethanol are addressed in Chapter 15. We shall argue that the advantage of corn-based ethanol as a substitute for gasoline is limited, not only in terms of its potential to reduce emissions of greenhouses gases but also in terms of its potential to significantly reduce the demand for oil. Production of ethanol from sugar cane, as practiced for example in Brazil, is more promising. Ultimately, technology may be developed that may allow for economically viable conversion of cellulose in grasses and wood chips to useful fuels. It is difficult, however, to assess at present the potential of this technology though clearly it should be pursued.

Current patterns of energy use in the U.S., Canada, the UK, and China are discussed in Chapter 16. We conclude in Chapter 17 with suggestions for steps that could contribute, we suggest, to a more sustainable, more secure, energy future, initiatives with promise also to protect the integrity of the global climate system.

REFERENCES

Diamond, J. (1997). *Guns, Germs, and Steel*. New York: W.W. Norton and Company Inc.

Gould, S.J. (1989). *Wonderful Life: The Burgess Shale and the Nature of History*. New York: W. W. Norton and Company, Inc.

Lott, D. (2002). *American Bison: A Natural History*. Berkeley: University of California Press.

Margulis, L. (1970). *Origin of Eukaryotic Cells*. New Haven: Yale University Press.

McElroy, M.B. (2002). *The Atmospheric Environment: Effects of Human Activity*. Princeton: Princeton University Press.

McEvedy, C. and Jones, R. (1978). *Atlas of World Population*. New York: Viking Press.

Netherlands Environmental Assessment Agency, http://www.mnp.nl/en/dossiers/Climatechange/moreinfo/Chinanowno1inCO2emissionsUSAinsecondposition.html, read 16 September 2008.

Perlin, J. (1989). *A Forest Journey: The Role of Wood in the Development of Civilization*. Cambridge, MA: Harvard University Press.

Sachs, J.D. (2005). *The End of Poverty: Economic Possibilities for Our Time*. New York: Penguin.

Windley, Brian F. (1995). *The Evolving Continents*. West Sussex: John Wiley & Sons Ltd.

2

From Hunter-Gatherers to English Factories

2.1 INTRODUCTION

This chapter is intended to provide an account of the historical events that set the stage for the emergence of the modern industrial age. The industrial age took off first, improbable as it might appear, in Britain, on the northwestern periphery of the Eurasian landmass. Why there, and not elsewhere—in the larger, more coherent, society of China for example? It is a story that begins with the development of agriculture and animal husbandry in the Middle East, in a region endowed with unique natural resources in terms of both plants and animals. It is a story of the progression from rain-fed agriculture (Section 2.2) to irrigation-based agriculture in the river basins of Mesopotamia, Egypt, northwestern India, and northern China (Sections 2.3–2.6), transitioning eventually to the previously densely forested regions of northwestern Europe. It is a tale of the emergence of great civilizations that flourished for a time only to succumb and be replaced by others. Depletion of natural resources, wood for example, played a crucial role in the demise of a number of these civilizations. Changes in climate, and in the environment more generally, were important for others.

The river-based civilizations were unusually vulnerable to changes in climate, subject to the twin vagaries of either too much or too little water—floods or droughts. Build-up of salt in soils subjected to excessive irrigation in a hot dry climate was also a problem. The climate of northwestern Europe, downwind of an abundant source of year-round moisture (neither too much nor too little) supplied by evaporation from the relatively warm Atlantic, was more reliable and more conducive to sustainable agriculture than that of either China or of the more southerly river basin-based societies of the Middle East.

China's climate for example is monsoonal, subject to large seasonal variations in wind strength and direction. Prevailing winds blow from land to sea in winter: winters are typically cold and dry. The prevailing winds switch direction in summer blowing in that season from sea to land. Summers are hot and wet as a consequence. Rainfall, however, is unevenly distributed both in space and time. Northern regions of China have suffered through extended droughts in recent years while precipitation in the south has often been excessive, to the point where Chinese authorities are currently contemplating major investments in projects designed to move water from south to north. It is likely that rainfall patterns were similarly variable in the past, with important implications both for food security and national governance. A strong central government in China could facilitate transfer of essential food stocks from regions enjoying temporary surpluses to others suffering through deficits, minimizing thus the impact of regional adversity on society as a whole. This provided a measure of legitimacy to national authority. In contrast to China, northwestern Europe enjoys a relatively moderate climate. Precipitation is distributed more or less uniformly throughout the year and there is no need for a central authority to implement and manage large-scale irrigation works. This accounts, at least to some extent, for the diversity of European societies in contrast to the relative cohesion of China, a distinction that has applied essentially from the beginning. The development paths followed by societies in the two regions were quite distinct.

When humans first ventured into the northwestern margins of Europe they encountered an environment that was heavily forested (a consequence of its seasonally moist, mild climate). Soils built up over long years since the end of the last ice age were heavy and difficult to cultivate. The invention of the iron plow and the introduction of draft animals—first oxen and then horses—to pull these heavy plows and turn over the dense soils of the region made all the difference in the development of northwestern European agriculture in the Middle Ages (Section 2.12). Agriculture evolved as a capital-intensive rather than a labor-intensive activity. Success depended on access to good equipment and strong healthy oxen and/or horses. In contrast, the success of agriculture in China was determined to a greater extent by the supply of abundant, inexpensive labor. The important crops there were millet (native to the region) and later wheat and barley (introduced from the Middle East) and rice (from Southeast Asia). Girls married young in China and family sizes were typically large. Boys were important both as a source of labor and to contribute to the support of parents in their old age. Large family sizes provided a measure of security in a society visited all too frequently by devastating famines (if a few children died, others might be expected to survive). European society developed with different mores with respect to family structure reflecting differences in the structure of its agricultural economy and the relatively greater security of its food supply. Men married later, waiting often until they could inherit land and capital equipment when their parents either died or retired. Not surprisingly given the differences in value attached to capital and labor, family sizes in

Europe were typically smaller than in China: population densities (people per unit land area) were correspondingly lower. The relative shortage of labor in Europe as compared to China played an important role, we shall argue, in promoting mechanized means for production of goods in the former as compared to the continuing emphasis on labor in the latter.

The path from the initial development of agriculture in the Fertile Crescent to the emergence of northwestern Europe as a major economic power center proceeded along a slow if inexorable trajectory. Milestones included the rise and fall of important civilizations clustered in and around the Mediterranean Basin, associated most notably with Crete, Greece, and Rome (Sections 2.7–2.11). Trade and territorial conquest played major roles in the evolution of these civilizations. Success, as we shall see, was dictated as often as not by military prowess. To succeed, a civilization could ill afford to stand still. It had to expand to acquire the physical and human resources needed to maintain its prosperity. And with success came threats from without, from the east in the south and from the north in the west. Such is the story of the great empires established by Alexander the Great (Section 2.10) and later by Rome (Section 2.11). All too often, empires expanded to the point where they were unmanageable, unable to maintain the economic surpluses and critical resources needed to sustain the military forces required for their defense or the bureaucracies on which they depended for imperial administration. Rome's empire in the west fell to Germanic tribes that invaded from the north. France emerged as a major power center reaching the peak of its imperial influence in the ninth century under Charlemagne (Section 2.12) but again the empire proved unsustainable. The Byzantine Empire centered in Constantinople (the Roman Empire transplanted east) proved more durable but eventually it also would fall, in this case to the Ottoman Turks, who emerged as the new imperial power in the east. Europe evolved as a series of competing states. Conflict was more common than cooperation. Often, as we shall see, religion was invoked as a reason for nations to go to war. The underlying motivational force, however, was usually more venal: the thirst for power and advantage that could be secured best on the battlefield.

Christianity and Islam, discussed in Sections 2.12 and 2.13, were the two great religious traditions that influenced the development of modern Europe. Not too long after the death of Mohammad, Arabic Muslims were established as the dominant presence in the Mediterranean Basin. They controlled Spain and, but for a defeat at the hands of the Franks, might have extended their reach much further into Europe. They emerged not only as a powerful religious and military force but also as a critical bridge to the earlier intellectual traditions of the civilizations that preceded them in the Mediterranean region and even further to the east. Europe was the beneficiary of this legacy when the Arabs were forced to retreat in the tenth and eleventh century, when Spain was overrun eventually by Christian forces from the north.

The Crusades (Section 2.14) played an important role in the resurgence of Europe following the Dark Ages that ensued following the death of

Charlemagne in 814 AD. Paradoxically, an unprecedented decrease in European population that occurred as a consequence of twin insults—a climate-driven failure of agriculture and the ravages of the Plague—contributed to this resurgence. The population of Europe fell by as much as a third in the middle years of the fourteenth century triggering a large-scale reorganization of European society. The value of labor increased challenging the influence of entrenched landowners and the aristocracy. At the same time, Europe began to turn outward and the balance of power shifted to the west. Countries on the Atlantic seaboard replaced countries and city-states centered on the Mediterranean as the new European power centers. Trade along the Atlantic seaboard supplanted trade within the Mediterranean Basin as the new engine of economic growth. The Netherlands was pivotally located to take advantage of this opportunity. Timber from the Baltic moved south in Dutch ships while goods from Spain, Portugal, and France moved north. Dutch agriculture emerged as the most efficient in Europe at the same time as Dutch ships and Dutch ingenuity came to dominate fishing in the North Sea. Portugal and Spain led the way in opening up Europe to the world at large extending the European presence to the Far East and to the Americas (Section 2.15). Prosperity, however, would fail to bring peace. The authority of the Catholic Church would be severely challenged by forces unleashed by the Reformation. Spain as the most vigorous defender of the Catholic status quo, would lose out in the struggle that ensued (Section 2.16).

The ultimate beneficiaries of Europe's opening to the world and the contemporaneous challenges to ecclesiastical and secular authority were the Dutch and eventually the English. The perfection of the movable printing press by Johann Gutenberg in the mid-fifteenth century and the subsequent widespread dissemination of written material contributed to a more educated, more assertive populace. Intellectual horizons expanded. It was a time to question old myths and experiment with new ideas, not only in religion and philosophy but also in science and technology. The Catholic south largely missed out on this social revolution. The death penalty was introduced in 1558 in Spain for printing or importing unauthorized books. To be published, books had to be vetted and could be printed only with an official imprimatur. So much for intellectual freedom: Spain's recidivism would provide the opportunity for Holland to emerge as Europe's intellectual and economic powerhouse. England would take off more slowly but eventually the British would overtake and surpass the Dutch (England and Scotland were united under the Act of Union in 1707: the name Great Britain refers to the combined entity formed under the terms of this Act). The factors responsible for this transition in power, the ground swell that led eventually to the dawn of the modern industrial age in Britain, are discussed in Section 2.17 and further in the wrap-up to the Chapter, Section 2.18.

2.2 ORIGINS OF AGRICULTURE

As discussed in Chapter 1, for most of human history our ancestors were nomads—hunter/gatherers. They survived by hunting wild animals and by

foraging for edible plants. Through time they made their way north from Africa to the Middle East, to the region known as the Fertile Crescent extending from the northern part of the Red Sea through portions of modern Jordan, Israel, Lebanon, Syria, Turkey, Iraq, and Iran. Those that arrived there after about 14,000 years BP must have had a sense that they were entering Nirvana. The Fertile Crescent would have evolved by that time into a veritable Garden of Eden with abundant sources of edible plants—wheat, barley, peas, and olives—and animals—sheep, goats, and the ancestors of modern cattle—that could be exploited for meat and milk, for skins and wool, and for bones that could be used to make primitive tools. And there would have been dogs to help with hunting. Living would have been easy and it is likely that many who arrived in the region during this propitious time would have chosen to settle, at least temporarily. Earlier travelers would have found the area less hospitable and would have opted most likely to keep moving.

Recovery of the Earth from the last ice age began about 20,000 years ago, accelerating at around 14,000 years BP. Sea level rose by close to 20 m between 20,000 years BP and 14,000 years BP responding to melting and the steady retreat of the continental ice sheets (present mainly in the northern hemisphere over North America and Northwest Europe). It increased by almost 40 m between 14,000 years BP and 12,000 years BP (Fig. 2.1) during the period of rapid warming that would likely have been most conducive to human settlement in the Fertile Crescent. But the good life came to an end, or at least a pause, at the onset of the climatic epoch known as the Younger Dryas, at about 11,800 years BP, when the Earth plunged back into cold, near glacial, conditions (McElroy, 2002).

The Younger Dryas climatic reversal (transition from warm to cold in this case) is thought to have been triggered by a change in the circulation of waters

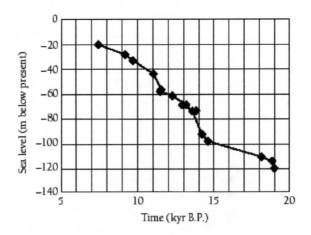

Figure 2.1 Rise in sea level during the last deglaciation.

in the North Atlantic Ocean. As discussed earlier, deep waters in the world's oceans are supplied today mainly by water sinking from the surface at high latitudes in the North Atlantic driven by a build-up of surface salinity (the increase in salinity results in an increase in the density of the water increasing the likelihood that it will sink as it cools in winter). The waters that sink at high latitude are supplied primarily by a northward extension of the Gulf Stream carrying warm waters to the high latitude regime, contributing to the relatively temperate contemporary climate of northwestern Europe. The circulation of the ocean was very different during glacial time. Deep waters were replenished at that time mainly by water sinking around Antarctica. The high latitude northern Atlantic was generally ice covered during the ice age and the Gulf Stream followed a more southerly track across the ocean. The circulation of the ocean switched from its glacial mode at the end of the ice age to a mode similar to that prevalent today. It reverted however to its glacial configuration at the onset of the Younger Dryas.

The change in circulation that led to the Younger Dryas was triggered, it is thought, by a flood of fresh water discharged to the North Atlantic Ocean through the Saint Lawrence Seaway. This water originated in a large neoglacial lake, Lake Agassiz, located to the north and west of the present Great Lakes. In the early stages of deglaciation, Lake Agassiz drained mainly to the south through the Mississippi River. Retreat of the ice sheet, however, opened up a new channel for discharge to the east through the Saint Lawrence. As the ice dam that had blocked this channel earlier finally broke, a vast flood of fresh water was released to the Atlantic. The salt content of surface waters dropped precipitously and formation of deep water in the North Atlantic came to an abrupt end. The circulation of the ocean switched back to its glacial mode and in short order, climate reverted globally to the cold, dry conditions that had prevailed up to a few thousand years earlier (in the latter stages of the ice age). Life was no longer easy in the Fertile Crescent. But the people who lived there were able to adapt. They learned to harvest their water resources and, rather than foraging for plants in the wild, they switched to cultivating the varieties they favored. And they began to feed, water, and care for the animals they had come to depend on. In short, they made the transition from a hunter/gatherer lifestyle to one of agriculture and animal husbandry.

The transition was probably neither that abrupt nor that immediate. It represented more likely an orderly shift from one lifestyle to another—an increased reliance on farming as resources in the wild became less reliable. Over time, the residents of the Fertile Crescent would have made a gradual transition from an entirely nomadic existence to one in which they opted to establish semi-permanent abodes. They could not have failed to notice then that the plants they favored for eating tended to grow more readily in the vicinity of the places where they chose to set up their tents or huts. It would not have required much in the way of ingenuity to appreciate the relevance of this observation—the connection between seeds dispersed from plants harvested in 1 year and the growth of new plants in the year that followed. Through time, they would have

learned to identify the best seeds to plant and cultivate—those derived from plants with the most desirable qualities. And they would have been quick also to see the advantage of herding and managing, and eventually selectively breeding, the animals they had come to rely on, especially as populations in the wild began to drop off in response to a deterioration in local environmental conditions (the animals would presumably have opted otherwise to move on in search of a more favorable habitat). The shift to agriculture was a transformative event in human history (Bar-Yosef and Belfer-Cohen, 1992; Moore and Hillman, 1992; Bar-Yosef, 1998; Weiss, 2001). The yield of wheat or barley or peas from a planted field would have exceeded that that could have been gathered from a comparable area in the wild even under the most favorable conditions by at least several orders of magnitude. And access to reliable sources of meat, milk, and wool from increasingly domesticated herds of cattle, goats, and sheep would have had comparable advantages. Life would never be the same after the development of agriculture and animal husbandry and it would only get better as climate resumed its more temperate course when the Younger Dryas cold snap ended, as abruptly as it had begun, at about 11,400 years BP.

2.3 IRRIGATION AND THE DEVELOPMENT OF CIVILIZATION IN MESOPOTAMIA

Climate was relatively mild for about three thousand years following the end of the Younger Dryas. Agriculture at this time was largely rain fed. A further cold snap set in, however, at about 8400 years BP, prompted again, it is thought, by a change in the circulation of the North Atlantic induced as before by an increase in the supply of fresh water to the ocean from the receding North American ice cap. Weiss (2001) points out that many if not all of the village settlements in the Levant were abandoned at this time. People migrated elsewhere in search of a more hospitable environment. A significant number found their way down the Euphrates River to the river delta where, without too much difficulty, they could tap the slowly moving waters of the river before they entered the Persian Gulf. This adjustment may have resulted in the first exploitation in human history of river waters for purposes of agriculture. A further cold snap and associated drought that developed at about 5200 years BP and persisted again for several centuries (3200–3000 BC), may have led to the demise of the deltaic society. This latter drought is thought however to have provided the stimulus for the development of the first great civilization for which we have an historical record, upstream in Sumeria where the supply of waters in the rivers would have been more reliable.

The flow of waters in the Tigris and Euphrates Rivers is generally erratic. Rainfall in the region is distinctly seasonal, confined mainly to winter. This is true as much today as it was 5000 years ago. River flows are greatest in April and May, fed by a combination of winter rain and spring snow melt in the headlands to the north—in the mountains of Turkey and Iran. Flows slow to a trickle in summer. The rivers frequently change course as they pass through the

flat alluvial plains of Mesopotamia on their way to the sea. To ensure a reliable year-round supply of water and to facilitate waterborne transportation of supplies, the Sumerians developed an elaborate system of canals. Construction and management of the canals required a complex social infrastructure. Through time, responsibilities in Sumerian society became increasingly specialized and distributed. Farmers farmed; artisans fabricated tools and constructed drainage ditches; bureaucrats managed the supply and distribution of waters in the canals; merchants engaged in trade; soldiers fought to defend property or to steal from others; priests interceded with the gods. It was a matter of simple economics: the output of goods and services was enhanced by allocating responsibilities for their provision to those best capable of providing them at least cost and managing the collective enterprise to optimize the interests of the society as a whole. The inevitable result was the evolution of a hierarchical social structure, a segregation of rich and poor, powerful and subordinate, instituting a social pattern that was to become the norm for most if not all subsequent organized societies.

The Sumerians had a remarkable record of achievement over a period of several hundred years. They built great cities with monumental temples erected to honor and care for their gods. Different cities were dedicated to different gods. The city of Uruk, for example, one of the first of the great city-states, was committed to Anu, the sky god, and to E-Anna, the goddess of love and war (Lamberg-Karlovsky and Sabloff, 1995). More than 5000 years ago, the Sumerians pioneered in the production of impressive forms of pottery, in the manufacture of textiles (including linen woven from fibers of locally available flax), in science and engineering, in metallurgy, and in the development of a system of standards for measures and weights. They introduced the first known written language, based initially on simple pictographs impressed on clay tablets, refined later into a form of writing known as cuneiform (wedge-shaped characters imprinted on clay: cuneus, Latin for wedge), incorporating as many as several thousand distinct characters. And they developed a system of counting based on a sexagesimal standard (60 as the basic unit), the legacy of which remains with us today (Kagan et al., 2002) as we divide the hour into minutes (60), minutes into seconds (60), and the circle into degrees (360). All of these achievements were necessary to support their increasingly complex society.

Since resources, other than rich agricultural soil, were in short supply in the flood plain of the rivers where they lived, the Sumerians were forced of necessity to develop a facility for trade. They reached out, south through the Persian Gulf, and north following the course of the Tigris and Euphrates, to satisfy demands for resources such as timber, stone, and metals (notably copper). In this manner they exported critical aspects of their civilization. There can be little doubt that the outreach of the Sumerians for purposes of trade had a significant influence on the development of the great civilizations that evolved later, or perhaps contemporaneously, in the Indus Valley and in the valley of the Nile. Skills once learned are readily transferred and by such transfer occurs progress.

The belief system of the ancient Sumerians held that humanity was created by the gods to serve their every day needs. The priests, as intermediaries to the gods, had positions of particular initial status in Sumerian society. Life, however, was not particularly peaceful and through time power shifted from the priests to a new class of warrior kings. The different city-states, Uruk, Kish, Ur, and Lagash, were in a constant battle for supremacy. And there were threats not only from within but also from the outside. Eventually the Early Dynastic Period of Sumeria, which had lasted from about 2900 to 2371 BC, came to an end when the city-states were conquered by Sargon of Akkad who established the first great dynastic empire in the Middle East. Over a 150-year period, the Akkadian dynasty (there were five successor kings) extended its reach to control not only the original Sumerian city-states but also all of northern Mesopotamia and large parts of Syria and western portions of Iran.

While the city-states of Sumeria were competing for influence, a second great civilization was evolving in Egypt along the banks and flood plain of the Nile. As with Sumeria, this civilization depended on a surplus of agricultural produce for its power and influence. The Nile extends some 4000 miles north from its headwaters in the Ethiopian highlands to the Mediterranean. The bulk of the precipitation that feeds the river falls over a period of about 3 months between June and August linked physically to the circulation pattern in the atmosphere that is responsible for the Indian monsoon. The flow of water in the Nile peaks predictably in late summer and early fall depositing a blanket of rich alluvial soil on the flood plain at precisely the optimal time to accommodate the fall planting. In contrast, as we have seen, the flows of waters in the Euphrates and Tigris rivers peak in late winter or early spring and are at a minimum when water is in greatest demand for agriculture in late summer. The Sumerians had to develop an elaborate system of canals to respond to this mismatch and they were constantly vulnerable to a build-up of salt in their fields as a consequence of the intense evaporation from soils that ensued during the hot dry summers. Egypt offered in many respects a more hospitable environment for agriculture than the Sumerians had to confront in Mesopotamia.

2.4 EMERGENCE OF CIVILIZATION IN EGYPT

Egypt divides naturally into two distinct geographic environments. Upper Egypt consists of a narrow valley reaching from Aswan in the south to the rich alluvial soils of the delta extending about 100 miles inland from the Mediterranean (Lower Egypt). The entire region was unified as early as 3100 BC by Menes, the first of a long line of pharaohs who ruled the country—with a brief hiatus as we shall discuss—for close to 3000 years until Egypt was absorbed into the Roman empire on the death of Cleopatra in 30 BC. The early pharaohs had the status of gods—god-kings. They ruled with absolute power supported by an elaborate bureaucracy responsible for oversight of the state granaries, for maintenance of public records and order, for recruitment to the army, for

collection of taxes and for monumental building projects as exemplified most notably by the pyramids, the elaborate tombs designed to house and support the pharaohs in the afterlife. The pyramids were built during a relatively brief interval, a period of about 100 years during the late Third and Early Fourth Dynasties of the Old Kingdom—by about 2500 BC. The Great Pyramid, commissioned by Khufu, son of Snefru, founder of the Fourth Dynasty, was constructed on a desert plateau near Memphis, the capital of the Old Kingdom on the border between Upper and Lower Egypt. It covered an area of 13.1 acres, rose to a height of 481 ft, and was composed of some 2.3 million blocks of stone weighing each an average of about 2.5 tons (Kagan et al., 2002). It was a remarkable achievement of engineering accomplished without benefit of modern technology. An estimate of the energy expended in its construction is developed in Box 2.1. Box 2.2 outlines a computation of the requirements in terms of human labor.

2.5 CLIMATE CHANGE AND THE COLLAPSE OF CIVILIZATIONS

The Old Kingdom came to an abrupt end at about the same time that the Akkadian Dynasty collapsed in Mesopotamia—at around 2200 BC. A period of prolonged political instability ensued in Egypt during the First Intermediate

---

Box 2.1

As noted in the text, the base of the Great Pyramid covered an area of 13.1 acres. It rose to a height of 481 ft and was composed of 2.3 million blocks of stone each weighing 2.5 tons. We calculate here the minimum energy required to construct the pyramid. Specifically, we compute the energy required to lift the blocks of stone from the ground and to put them in place. Assume that the average block is raised to a height of 120.25 ft (the center of mass or centroid of a pyramid is located at an elevation equal to 25% of its total height). Express the answer in units of J.

Solution:

It is a good idea from the outset to convert all quantities into a common set of units. For most of the computations in this book we choose to use the SI or MKS system (see treatment of units in Chapter 3).

$$\text{Mass of the Great Pyramid } (M) = 2.5 \text{ tons} \times 2.3 \times 10^6$$
$$= 5.75 \times 10^6 \text{ tons}$$
$$= 5.75 \times 10^9 \text{ kg}$$

$$\text{Energy required to lift the blocks } (E) = Mgh$$
$$= 5.75 \times 10^9 \text{ kg} \times 9.8 \text{ m/s}^2 \times 120.25 \text{ ft}$$
$$= 5.75 \times 10^9 \text{ kg} \times 9.8 \text{ m/s}^2 \times 36.6 \text{ m}$$
$$= 2.06 \times 10^{12} \text{ J}$$

Box 2.2

Assume for the moment that the pyramid was erected over a 20-year period by a workforce engaged for an average of 8 h a day. Assume further that a typical Egyptian worker could expend energy at an average rate of 100 W (about the equivalent of a typical light bulb). Before the pyramid could have been assembled, the blocks would have had to have been quarried and shaped and transported to the site. To allow for the additional labor this required, assume that it required an expenditure of energy equal to at least the energy involved in assembling the pyramid (this assumption is probably conservative). Using the solution developed in Box 2.1, calculate the total size of the workforce required to construct the pyramid and discuss the implications of the result.

Solution:

The energy expended by a typical Egyptian worker over a 20-year period
$$= 100 \text{ W} \times 8 \text{ h/day} \times 3600 \text{ s/h} \times 365 \text{ day/year} \times 20 \text{ year}$$
$$= 2.1 \times 10^{10} \text{ J}$$

Energy expended in constructing the pyramid
$$= 2 \times \text{energy involved in assembling the pyramid}$$
$$= 2 \times 2.06 \times 10^{12} \text{ J}$$
$$= 4.12 \times 10^{12} \text{ J}$$

Size of the workforce over a 20-year period
$$= 4.12 \times 10^{12} \text{ J} / 2.1 \times 10^{10} \text{ J per person}$$
$$= 196 \text{ persons}$$

Assuming that workforce was employed for only 2 months a year—that they were busy the rest of the year with other activities, with agriculture for example—we should increase this number by a factor of 6. The workforce required in this case would climb to 1176. To lift the stones and put them in place, the Egyptians are thought to have constructed ramps up which the blocks of stone could be pushed or pulled. The advantage of the ramp is that a lesser force (fewer men pushing or pulling at any given time) could be used to raise the blocks of stone. Ignoring effects of friction, the work required to push the blocks up the ramp to a given height would be the same as the work involved if the weight was lifted directly. Work equals force times distance. The distance the weight had to be moved up the inclined plane would be greater than the distance over which it would be moved if lifted vertically. As a consequence, the force required to push or pull the blocks up the inclined plane would be less than that required to lift the blocks vertically. The equal work condition applies of course only if the contact between the blocks and the inclined plane is frictionless. The Egyptians would have taken steps to reduce the frictional resistance, by pouring water on the surface over which the blocks had to be moved for example or by using logs to roll the weights up the ramp. It is certain though that they would have had to pay some cost in terms of extra work relative to the ideal frictionless limiting case. Even if we were to double the workforce to allow for this extra labor, we would still fall far short of the figure for the workforce of 100,000 attributed to the Greek historian, Ptolemy. Ptolemy wrote about the construction of the Great Pyramid many years after it was built and his estimate of the size of the workforce was based largely on anecdotal information picked up on his visits to Egypt. Surely it took a large number of people to build the pyramids: but probably less than one might have thought—less than historical accounts might suggest—at least according to the analysis presented here.

Period that followed (2200–2025 BC) and there is convincing evidence that a change in climate was the proximate cause for this reversal of fortune. Lake levels near the headwaters of the Nile, in Ethiopia, dropped precipitously (Gasse, 2000; Gasse and van Campo, 1994; Johnson and Odada, 1996) and the previously reliable summer/fall floods of the river failed. The agricultural surpluses that had provided the lifeblood of the Old Kingdom disappeared and a devastating famine took hold that persisted for more than a hundred years. People were reduced to "commit unheard of atrocities such as eating their own children and violating the sacred sanctity of the dead," according to Fekri Hassan (2001). And, again quoting Hassan, "within the span of 20 years, fragmentary records indicate that no less than 18 kings and possibly one queen ascended the throne with nominal control over the country." In short order, the pharaohs lost their god-like aura—a status that had depended to no small extent on the belief that they had power over the life giving sustenance of the river. Several hundred years would elapse before climate would recover and Egypt would resume its path to progress in the Middle Kingdom (2025–1630 BC).

The climate change that prompted the demise of the Old Kingdom in Egypt was widespread—arguably global. Its impact was experienced as far away as India and Pakistan (no great surprise since this region is coupled climatically to the Ethiopian Highlands through the Indian monsoon) where it contributed most probably to the collapse of the Harappan civilization that had flourished for centuries previously on the alluvial plain and upstream reaches of the Indus River. The associated changes in the strength and frequency of the Indian monsoon are discussed by Bentaleb et al. (1997). There can be little doubt that the climate event that set in at 2200 BC was implicated directly also in the problems that developed in Mesopotamia. Conclusive evidence for a decrease in precipitation over the Mesopotamian region is indicated by data from a deep-sea core drilled in the Gulf of Oman (Cullen et al., 2000; deMenocal, 2001) and by measurements of changes in the composition of sediments sampled in Lake Van, located near the headwaters of the Tigris River (Lemcke and Sturm, 1997). Careful dating of the Gulf of Oman core indicates that the aridification event that resulted in such dire consequences for Middle Eastern civilizations persisted for about 300 years. Elsewhere in the region, the water level in the Dead Sea dropped by close to 100 m (Frumkin et al., 1994). And, further afield, sea surface temperatures in the North Atlantic decreased by between 1 and 2°C (Bond et al., 1997; deMenocal et al., 2000). deMenocal (2001) suggests that the decrease in precipitation observed in the Middle East was linked causally to the change in temperatures inferred for the North Atlantic. He attributes the decrease in precipitation to a change in the strength and spatial pattern of the winter westerlies that are normally responsible for carrying moisture to the region. The concurrent changes in precipitation observed over the Ethiopian Highlands (a summer rather than a winter phenomenon in this case) suggests, however, the impact of the 2200 BC climatic anomaly must have been significantly more extensive, probably global.

Archeologists refer to the first stage of human development as the Paleolithic (old stone) Age. It persisted until about 10,000 BC when the transition from hunter/gatherer society to agriculture issued in the Neolithic (new stone) Age. Tools were still sculpted from stone in the Neolithic but the tools became increasingly more precise and more useful. Humans had learned earlier to make use of some of the softer metals such as copper and gold. In pure form these metals could be hammered, without benefit of heat, into desirable shapes and used to fabricate attractive ornaments—early status symbols. Only much later, at about 3500 BC, did they learn to convert more commonly available ores, notably copper and tin, present most often in oxidized form, into products that were more than ornamental, that could be used to fashion useful tools and weapons. Diamond (1997) suggests that the sequence that led to this advance was serendipitous rather than purposeful. People had learned earlier to use heat to convert clay into bricks and pottery. Smelting copper and tin did not require temperatures much higher than those needed to make bricks and pottery. Combining copper and tin to make bronze was an important step forward. It resulted in production of an alloy whose properties were superior to either of its components employed separately, ushering in what is referred to as the Bronze Age (from approximately 3100 BC to 1200 BC). Almost 2000 years would elapse before humans would develop the technology required to smelt iron (at about 1400 BC) signaling the development of the Iron Age.

2.6 CIVILIZATION'S RENAISSANCE

Access to wood was a critical factor in the ancient world as it emerged from the setbacks induced by the 2200 BC climatic reversal. Vast quantities of wood were consumed in the smelting of ores and a supply of metal was vital for the ability of states both to ensure adequate supplies of food (to make the tools essential for efficient agricultural production) and to project military power (to fabricate the weapons needed to ensure success on the battlefield). Wood was necessary also for the construction of buildings and to make the boats required both for trade and imperial expansion. Wood was in short supply in Mesopotamia and in Egypt. The resurgent civilizations in these regions had to reach out to obtain essential supplies elsewhere. The extensive forest resources available on the island of Crete provided the comparative advantage that led to the development of the Minoan civilization and was the stimulus later for the birth of the civilization that took hold on the Greek mainland and on the many islands of the Aegean. Access to adequate supplies of timber empowered also the development of powerful states in Asia Minor (the Hittite in particular) and in Assyria in the region of the headwaters of the Tigris River. Mesopotamia succumbed eventually to a series of invasions from the north and east. Babylon, located upriver from the site of the previously powerful city-states of Sumeria (benefiting probably from a more reliable supply of water), evolved as the center of Mesopotamian civilization surviving as a rich cultural and trading

center under a variety of different regimes. It was conquered by Persia in 539 BC but managed to maintain its privileged status even then as an important regional center of the Persian Empire. Egypt resumed its status as a major power under the New Kingdom (1550–1075 BC) when the pharaohs extended their influence deep into Africa and into the eastern Mediterranean. Eventually, though, they met a fate similar to that encountered by the Mesopotamians: they succumbed to a succession of militarily more powerful opponents—the Hittites, the Persians, the Greeks, and eventually the Romans. It is probable that the lack of natural resources, notably wood, contributed to their eventual demise just as it had earlier for the Sumerians in Mesopotamia.

2.7 MINOAN AND MYCENAEAN CIVILIZATIONS

The Bronze Age reached its peak in Crete during the era known as the Minoan (named for the legendary King Minos). As noted earlier, Crete was blessed with abundant sources of wood—oak, pine, and cedar. Indications are that by 2000 BC it had developed a lucrative trade relationship with its more developed neighbors to the east, exporting wood and importing finished goods. Soon, though, it would begin to exploit its wood resources in situ and would evolve as a major manufacturing center producing prodigious quantities of bronze both for local consumption and for export. The city of Knossos, located on the island's north shore, would flourish as a major cultural and industrial center. But humans had yet to learn the importance of conservation and the rules for sustainable forestry (perhaps we still haven't!). Within a few hundred years, local supplies of timber were depleted and the Minoans were forced to become a timber importing society. Not surprisingly, they looked to the Greek mainland for fresh supplies and turned to importing wood from Pylos in southern Greece and from even farther afield in Syria and Lebanon. They survived for some time with this strategy but the golden age of Minoan civilization was doomed to extinction. The Greeks learned that they had no need to import finished goods from Crete when they could manufacture their own with resources abundantly available in their immediate hinterland. A new civilization evolved on the Greek mainland, the Mycenaean, named for one of the cities, Mycenae, in the Greek Peloponnesus. Crete was invaded by the Mycenaeans in about 1150 BC and Crete's future course was as a sheep-rearing colony serving the needs of a more powerful, rapidly developing, mainland Greece.

The Mycenaeans were a bellicose people. They evolved an important naval capability, reaching the peak of their power between 1400 and 1200 BC, raiding and trading with regions as far-flung as Syria, Egypt, Sicily, and Italy. With increasing affluence their population grew rapidly. Forested areas was cleared both to support their ever increasing demands for wood and to satisfy needs for new land for agriculture. The environment of the region was ill suited, however, to intensive cultivation. Deprived of its protective forest cover, it was subject to enhanced erosion. Perlin (1998) suggests that the precipitous collapse of

the Mycenaean civilization during the thirteenth and early twelfth centuries BC may be attributed in no small measure to catastrophic deterioration of its environment. He quotes an ancient Greek epic to the effect that during the late stages of the civilization the region teemed with "countless tribes of men" who "oppressed the surface of the deep bosomed earth" and attributes to Zeus "in his wise heart" a decision "that the only way the earth could heal was to rid it of humans, the perpetrators of such violence on the land." The fall of the Mycenaean civilization was followed by a long period of declining population and cultural regression in Greece. Close to 400 years would elapse before civilization would return to the region.

2.8 EMERGENCE OF CLASSICAL GREECE

When civilization returned to Greece, in about 750 BC, the structure of society was very different. Gone were the hierarchical relationships and the warrior kings who had ruled earlier in Mycenaea. Society was organized now in units known as poleis (singular polis)—loosely translated as city-states—that had evolved from agricultural villages during the cultural dark ages. The citizens of individual poleis viewed themselves as members of an extended family descended from a common ancestor (Kagan et al., 2002). The polis encouraged broad participation by all of its citizens in the life of the community. First loyalty, however, was to the local polis rather than to the larger Greek community. Individual poleis competed, often by force of arms, with neighbors for advantage. At war, soldiers of the poleis fought in units known as phalanxes (singular phalanx)—tightly packed, disciplined, groups of soldiers often eight or more lines deep that could withstand attacks by less organized foes, even by cavalry (Kagan et al., 2002). The poleis established colonies, organized along the lines of the parent polis model, throughout the Mediterranean region, reaching from Spain and extending into the Black Sea. With trade came prosperity. But a challenge was to emerge to threaten the ascendancy of the Greek poleis. It came from the east in the form of a new imperial power that developed in Persia, in the region identified now with Iran, a forerunner of east–west conflicts that, as we shall see, came to dominate much of subsequent history and that continue even to today as exemplified by the current conflict in Iraq and ongoing tensions with Iran.

2.9 CHALLENGES FROM THE EAST

The first ruler of the Persian Empire was Cyrus the Great who came to power in 559 BC. The Persians quickly took control of Mesopotamia and in short order also the Greek colonies of Asia Minor. The Persians were benevolent rulers for a while but suffered defeat in 498 BC at the hands of a rebellion led by Athens that had emerged in the interim as one of most powerful of the Greek poleis, a status challenged only by Sparta, its bellicose neighbor to the south. It was a short-lived victory, however. Athens withdrew, failing to consolidate its

advantage, and the Persians resumed control in Asia Minor. They set out to punish Athens for its role in the earlier Persian reverse. The critical battle in this campaign was fought on the plains of Marathon to the north of Athens in 490 BC. Against all odds, the Athenians were victorious and Persia was forced to retreat. The battle was lost, but not the war. The Persians returned 10 years later (in 481 BC) led by Darius with an army of at least 150,000 and a fleet of more than 600 ships (Kagan et al., 2002). Greek resistance was led in this case by Sparta. The Greeks achieved a momentous victory in 479 BC and the Persians retreated in disarray.

Defeat of the Persians did not result in a period of sustained peace on the Greek mainland. Internal dissention was the rule rather than the exception as Athens and Sparta vied for power. The First Peloponnesian War lasted from 465 BC to 446 BC and was followed by an even more brutal conflict that erupted in 431 BC and continued until 404 BC when Athens was forced to surrender to the Spartan general Lander. In the final stages of the second Peloponnesian War, the Spartans received support from the Persians who, despite their earlier setbacks, maintained a significant presence in the region. Following the fall of Athens, Sparta was the dominant power in Greece. But the social structure of Sparta was geared for war not for peace. Thebes, located to the northwest of Athens, emerged as the savior of Greece, inflicting a devastating defeat on the Spartans at Lysander in 371 BC. The Theban general, Epaminondas, who engineered this victory, had a strategy to ensure the peace. He freed the Spartan slaves (the Helots), deeding to them much of the land previously controlled by Sparta, ensuring that Sparta would never again threaten the prosperity of greater Greece. But conflict did not come to an end with the defeat of Sparta. The ascendancy of Thebes proved to be short-lived as the center of Greek power shifted once again to the north, in this case to Macedonia. Perlin (1989) argues that the critical factor that led to the rise of Macedonia was the rich forest resources of the region, an abundance of wood at a time when the area to the south had been largely deforested.

2.10 ALEXANDER THE GREAT

The visionary who led the rise of Macedonia was Philip II who, as a youth, had spent several years as a hostage in Thebes. There he became familiar both with Greek politics and Greek military technology. He equipped his soldiers with long-staffed pikes, 13 ft long as compared with the conventional 9-ft spears favored by the Greeks (Kagan et al., 2002). He conquered the important city of Amphipolis, winning control of the rich gold and silver mines on nearby Mount Pangaeus. With this newfound wealth, he consolidated his power and proceeded to occupy the neighboring regions of Thessaly and Thrace (cf, Fig. 2.4). By 338 BC, Philip was the master of all of Greece and his son, Alexander, known later as Alexander the Great, who succeeded Philip as king of Macedonia at age 20 when the latter was assassinated in 336 BC, would go on to establish the greatest empire the world had ever known.

The era of the polis, in spite of its shortcomings and the interruptions imposed by frequent conflict, was a period of unprecedented cultural achievement in Greece. As Kagan et al. (2002) point out the "polis was a dynamic, secular, and remarkably free community that created the kind of cultural environment in which speculative thinking (the basis of science) could survive." Greece absorbed, refined, and advanced the accomplishments of the great civilizations that preceded it—Mesopotamia, India, Pakistan, Egypt, and Crete. It built great monuments—the temples to the gods on Mount Olympus for example. It gave birth to the essential elements of democracy and philosophical discourse and bequeathed to us a rich and varied literature including the musings of such intellectual giants as Socrates (469–399 BC), Plato (429–347 BC), and Aristotle (384–322 BC). And it spawned the cultural heritage that would later inspire forerunners of modern science and mathematics such as Euclid (330–275 BC) and Archimedes (287–212 BC)—a truly historic legacy.

Alexander the Great set out to subdue once and for all the Persian Empire. In 334 BC he crossed the Hellespont into Asia, confronting the Persians—initially in Asia Minor and subsequently in Syria. He continued to Egypt where he was greeted as a liberator and recognized as the new pharaoh. In 331 BC he moved on to Mesopotamia where again he defeated the Persians, proceeding to Persepolis, the Persian capital, which he successfully occupied in 330 BC. In 327 BC, he crossed the Kyber Pass entering what is now Pakistan, at which point, in 324 BC, under pressure from his troops, he withdrew to the Persian Gulf. A year later, he was dead, victim of a fever suffered at age 33 in Babylon. In ten short years, Alexander the Great had transformed the face of the civilized world. In the absence of an obvious successor, the new empire was divided into three distinct but interrelated dynasties: Ptolemy I assumed the position as ruler of Egypt; Seleucid I took over Mesopotamia; and Antigonus I became king of Asia Minor and the Greek mainland. For two centuries the Macedonian empire flourished but by the mid-point of the second century BC it was gone, except for Egypt, and the torch had passed to Rome.

2.11 EMERGENCE OF ROME

The origins of Roman civilization are cloaked in mystery. According to legend, the city of Rome was founded sometime between 754 BC and 748 BC by Romulus and Remus, reputably the twin sons of Mars, the god of war. Myth and legend aside, it is apparent that the Etruscans who settled initially on the coastal plain to the north of Rome, in Etruria (modern Tuscany), played a critical role in the early history of Rome and in the development not only of its physical infrastructure but also its culture, religion, and social mores. The origin of these people is obscure. Their language bears no relation to languages of the other tribes that populated Italy in the early prehistoric period. Some believe the Etruscans came from the east, possibly from Asia Minor. It is clear, though, that by the time they arrived in Etruria they had absorbed much of the culture and technology of the Greeks and learned also from the achievements of the Egyptians and Mesopotamians. They were a militaristic people, skilled both as soldiers and sailors. Through time they extended their territorial reach to include not only their original settlement area between the Arno and Tiber Rivers but also much of central Italy to the south, the fertile plain of the Po River to the north, and the off shore islands of Corsica and Elba, the latter the largest island of the Tuscan Archipelago sited 15 miles to the east of the Italian mainland. The Etruscans were accomplished civil, hydraulic, and mining engineers. They built great cities, of which Rome was the outstanding example, ringed often with protective walls. They drained regional wetlands converting swamps to fertile fields, adopting advanced techniques of irrigation to enhance agricultural productivity. They installed a system of canals to channel the waters of the Po to the Adriatic, for example, to minimize the risk of flooding. And they sunk shafts and built tunnels to exploit the mineral resources of their territory—notably tin, copper, and iron which they used to fabricate a variety of tools, weapons, ornaments, and furnishings. The Etruscans were arguably the world's first sanitary engineers: they built aqueducts to carry water to their cities and sewers both above and below ground to remove waste. Their power, however, was short-lived. The last Etruscan king of Rome, Tarquin the Proud, was perceived as a tyrant and was deposed in a revolution led by the city's aristocracy in the early years of the fifth century BC, an uprising that resulted in the formation of a more consultative form of government—the Roman Republic.

The Republic was headed by a pair of elected aristocrats, referred to as consuls, who shared executive and judicial responsibilities. They were appointed on an annual basis ensuring that their influence would be limited. Continuity in governance was provided by a Senate that controlled the finances of the state and determined foreign policy, and by a more representative body, the Centuriate, that passed on bills proposed by the Senate, made decisions on war and peace, and elected the consuls. The governance structure bears a superficial resemblance to the system that evolved much later in the United States but in its initial function it could certainly not be described as either democratic or inclusive. To put it bluntly: those with the money controlled the show and the influence of the commoners—the plebeians—was strictly limited. This would change with time, however, although slowly. What empowered the plebeians was the essential role they had to play in the army and the leverage this imparted in a society committed to military expansion. Their first success was winning the right to form a separate assembly. Their final victory, in 287 BC, was passage of a law requiring that decisions by the plebeian assembly would be binding on all Romans. Before this law would be enacted, however, Rome would have suffered its first serious reverse, a defeat in 387 BC by the Gauls, a Celtic tribe from what is now France. In face of an attack by the Gauls, who had earlier occupied the plain of the Po River in the north, the Romans abandoned their city. The city was set on fire and the Romans had to pay a heavy ransom to recover their property and rebuild their capital. Paradoxically, this defeat would provide the stimulus for Rome to commit to a major phase of expansion. The city was determined never again to be vulnerable to assault from without.

In its expansionary phase, Rome moved first to secure its immediate neighborhood. States of the Latin League, formed to defend against Roman aggression, were decisively defeated in 338 BC. By 280 BC, Rome controlled all of central Italy. Along the way it developed a masterly strategy to secure the territories it won on the battlefield. States that resisted were treated harshly. Those that accepted the Roman victory were offered generous terms and often accorded rights similar to those enjoyed by Roman citizens. As Kagan et al. (2002) put it: "this gave Rome's allies a sense of being colleagues rather than subjects. Consequently, most of them remained loyal even when put to severe tests." By 265 BC, Rome controlled all of Italy south of the Po River. Sixty-three years later, Rome was the undisputed ruler of most of the western Mediterranean following its defeat of the Carthaginian general Hannibal at Zama in North Africa in 202 BC. This was the same Hannibal who had earlier crossed the Alps with his elephants from Spain and came close to taking Rome before being forced to withdraw to defend against an attack to his rear in Spain by the Roman general Scipio. As Kagan et al. (2002) succinctly summarized the situation: "Hannibal had won every battle but lost the war. His fatal error was to underestimate the determination of Rome and the loyalty of its allies."

Having achieved dominance in the west, Rome turned its attention to the east, to the Macedonian empire ruled at this time by the successors of Alexander the Great. By 168 BC, much of this region had been added to the territory controlled by Rome. Carthage was sacked in 146 BC. The Roman world included by then all of Italy with six external provinces: Africa, Sicily, Sardinia-Corsica, Macedonia, Hither Spain, and Further Spain. Gaul was added to the empire by Julius Caesar in 50 BC. When Caesar died in 44 BC, victim of an assassination organized by enemies in the Senate whose powers he had purposely marginalized during the brief period when he reigned as absolute ruler (dictator), the Roman world extended from the English Channel in the northwest to Syria in the southeast and included also a large portion of North Africa with the exception of Egypt. It was too much for the Republic to administer and it collapsed into a period of extended civil strive from which it was rescued eventually by Gaius Octavianus, a great-nephew and original heir-designate of Caesar. Octavianus was known subsequently as Augustus (the revered), a title bestowed on him by a grateful Senate in 27 BC in recognition of his role in restoring a semblance of peace and order to Roman society. Augustus reorganized the governance of Rome and its extended territories. While he never actually claimed the title, he was in a real sense the first Emperor of Rome.

Rome prospered under Augustus. The provinces, previously largely independent, were brought under enlightened central control. Governors, answerable to Rome, were appointed to rule the far-flung reaches of the empire. Residents of the provinces were recognized eventually as full citizens of Rome and their loyalty to the central authority was cemented accordingly. From an organization of recruits and volunteers, mainly plebeians pressed into action to satisfy the needs of particular campaigns, the army was transformed to a professional body of about 300,000 soldiers recruited to serve individually for 20 years with generous remuneration and promises of pensions on retirement (Kagan et al., 2002). The primary function of the army was to defend the borders of the empire, especially the porous frontiers to the north. At the same time, great monuments were constructed—the forum for example and the temple to Apollo on the Palatine Hill. And it was a time also of historic achievements in literature and the arts with giant figures such as Virgil, Horace, and Ovid. In the latter years of the Republic and under the administration of Augustus, Rome achieved a level of affluence and cultural development the like of which the world had never seen before. The legacy of this golden age was durable: despite ups and downs its influence persists to today.

The economy of imperial Rome depended in large measure on abundant sources of slave labor and booty extracted from newly won territories, on access to adequate supplies of wood, on agriculture, and on trade. It was a time of great opulence, at least for the privileged few, but also a period of minimal technological advance. Cameron (1997) attributes this to a lack of incentive to invent anything that would save labor to a time when labor was cheap and generally poorly regarded. The Romans did pioneer in building aqueducts, roads, arches, domed buildings, monuments, and luxurious residences for their more affluent citizens. They led the way in storing and distributing water, in the application of cement for construction, and in the use of glass both for windows and ornaments (glass blowing was an early Roman art). At the peak of its power, it is estimated that the population of Rome may have exceeded a million. The city was in no sense, however, self-sufficient. It depended on trade to supply the raw materials demanded by its artisans (copper, zinc, and iron for example), the food needed to feed its burgeoning population, and the wood required to stoke its industrial furnaces, heat its homes, and warm the waters of the baths to which Romans had become addicted.

Through time a rising fraction of the economic product of the empire was siphoned off to support the excessively opulent life styles of the rich, the salaries of the military, and the bureaucracy needed to administer the empire. While the empire was expanding, the gains from conquest could support this increasingly burdensome overhead. But there was a limit to the size to which the empire could grow and to the ability of the central government to maintain control. Increasingly, resources had to be expended for defense rather than offense. Hadrian erected a wall to protect the empire's possessions in Britain from Scots to the north (although Caesar had visited Britain several times a hundred years earlier, Britain, excepting Wales and Scotland, was not finally added to the empire until 47 AD during the reign of Emperor Claudius). The army was deployed more and more to keep marauding Germanic tribes at bay. But the pressure was inexorable. The central government had difficulty in meeting its expenses and resorted to a variety of strategies to address its problems, doomed, all of them, ultimately to failure. Taxes were increased. The currency was debased. And runaway inflation ensued with consequent erosion of confidence in the coinage of the realm with serious, and inevitable, negative implications for trade. Responding to the insuperable problems in the west, the center of Roman power shifted to the east in 324 AD when Constantine established a new seat for the imperial government in Constantinople (modern Istanbul), strategically located at the border between Europe and Asia at the entrance to the Black Sea. The eastern empire, given its strategic location, was relatively immune from the attacks by the barbarian forces from the north. The previously coherent Roman Empire degenerated in the west into a series of competing local fiefdoms. Rome was sacked by the Visigoths led by their king Alaric in 410 AD. Attila the Hun led his forces into Italy in 452 AD and the last surviving western emperor, Romulus Augustulus, was deposed in 476 AD—the traditional date for the fall of the Roman Empire. The empire survived and prospered, however, in the east for close to a thousand years in its Byzantine incarnation before it too would fall, in 1453 AD, a victim in this case of the forces of the newly emerging Ottoman (Turkish) empire under the leadership of Mohammed II (known also as Mehmet II).

2.12 JUDAISM AND CHRISTIANITY

Religion played an important role in the organization of ancient civilizations and continues as a powerful force today—witness the influence of the Christian right in American politics and the sway of militant Islam over extensive regions of the Middle East, Southwest and Southeast Asia, and Africa. Religion imparted legitimacy to the ruling classes. It inspired the monumental architecture of Sumeria, the pyramids of Egypt, the temples of Greece and Rome, and even more recently the cathedrals of Western Europe. It required an impressive commitment of human labor and physical and biological resources (timber, stone, metals, and cement for construction; plants, animals, and sometimes humans for ritual sacrifice) with no obvious benefit to local economies. The religion of the ancient civilizations was intrinsically polytheistic. The gods were awesome figures, personifying, frequently, poorly understood phenomena of nature—earth, sea, sun, moon, and stars, the rhythm of the seasons, and fertility. Judaism, preaching worship of one God, posed the first major challenge to this polytheistic world. It presented a code for worship and ethical behavior encapsulated in the Ten Commandments, elaborated further in the teachings of the prophets as recounted in the Old Testament. Rome tolerated, albeit reluctantly, the aberrant teaching of the Jews: their influence was confined to a remote corner of the empire and was not perceived as a serious threat to the dominant culture and authority of the central government. Christianity, which emerged during the reign of Augustus, posed a more serious challenge. Jesus preached a gospel of charity, love for fellow man, the importance of the individual, and a future of eternal bliss in the afterlife for those who followed his precepts. In contrast to the Jews, Christians were missionaries seeking aggressively to spread their faith, to win new converts. Their message was particularly attractive to the Roman underclass, the disadvantaged.

As such, it posed a major challenge to the authority of the state. Christians were subjected to unrelenting persecution, a reaction that served only to enhance the attraction of their message. Christianity won out eventually, however, when it was adopted by Theodosius as the official religion of the Roman state in 394 AD.

The Byzantine Empire—the Empire of Rome transplanted to the east—continued to prosper following the demise of Rome, reaching the peak of its influence during the reign of Justinian (527–565 AD). The Patriarch of Constantinople emerged as a religious leader with power and influence rivaling that of the Pope in Rome. There were strong doctrinal differences between the western and eastern churches. The western (Latin) church looked to the Pope as the ultimate authority while the eastern (Greek) church sought inspiration from the bible and various church councils. Contacts between the eastern and western churches were limited as the western portion of the former Roman Empire was preoccupied with defending itself from repeated attacks by Germanic tribes in the north and later from the forces of Islam in the south (see below). The schism was complete when the leaders of the western and eastern churches, Pope Nicholas I and Patriarch Photius, excommunicated each other in the ninth century.

The Franks emerged as the regional strongmen in the west in the latter part of the fifth century. The first Frankish dynasty was founded by the warrior chief Clovis who converted to Christianity and succeeded in extending his reach over most of what is now modern France (the land of the Franks). It proved difficult however to maintain central authority in face of inevitable centrifugal forces favoring regional autonomy. The conflict was resolved only several centuries later when Charles Martel and later Pepin III and Charlemagne adopted what is referred to as the manorial system in which power was vested in regional chiefs who bequeathed land to individuals known as vassals in return for commitments on demand to military service and to various forms of tribute. The vassals in turn engaged large numbers of peasants who worked the land in return for promises of physical security.

The hierarchy in the civil domain—kings, lords, and knights—was matched by a similar hierarchical structure in the governance of the church—pope, bishops and abbots, priests, and monks. The influence of cities and towns declined in the feudal era: the agricultural surplus required to supply the militaristic ambitions of the feudal lords and to satisfy their obligations to the central government was an overarching priority. The towns survived largely as a result of their ecclesiastical function, as seats for the bishops—answerable to the Pope in Rome—who maintained a semblance of larger scale societal connectivity. The monasteries, organized according to the rule of Benedict of Nuresia as formulated in 529 AD, provided the intellectual underpinnings for the Frankish state. Through time the distinction between the civil and ecclesiastical authority became blurred. The Pope and bishops had the unique advantage that they were perceived as controlling access to happiness in the afterlife. Given this power, it was difficult for them to resist the temptation to seek influence in the here-and-now despite the edict of Jesus, recorded in the gospel of Matthew, that they should "render therefore to Caesar the things that are Caesar's and to God the things that are God's." In 753, Pope Stephen II appealed to Pepin III, successor to Charles Martel, for help in rescuing Rome from an attack by the Lombards. The Franks defeated the Lombards in 753, conferring on the Pope civil control over the immediate vicinity of Rome, over what came to be known as the Papal States. The Pope assumed a role therefore not only as head of the Church but also as the secular head of a civil state, ensuring a conflict between God and mammon that would endure for close to a thousand years (essentially unchallenged at least in the Christian world until the reformation). Pepin and his successor, Charlemagne, relied on Christian missionaries to cement their authority over the lands they conquered. As Kagan et al. (2002) put it, "the cavalry broke their bodies, and the clergy reworked their hearts and minds."

Under Charlemagne, the Frankish kingdom was extended to include all of modern France, Belgium, Holland, Switzerland, Corsica, most of western Germany and Italy, and part of Spain. Pope Leo III crowned Charlemagne Emperor on Christmas Day, 800 AD. In later years, secular leaders would claim the right to select and appoint popes, and popes would aspire likewise to anoint civil leaders, an inevitable conflict between church and state. More serious perhaps, popes would go on to claim authority subsequently to pronounce not just on matters of faith and morals but also on matters of science and technology with important consequences, negative as we shall see, for the development of the modern industrial economy: so much for the distinction between God and Caesar.

Charlemagne's reign witnessed two important advances contributing to an important increase in agricultural productivity: the introduction of an iron plow capable of turning over the heavy soils of Northern Europe; and the transition to a three-field system of crop rotation in which only one field was left fallow at any given time in contrast to the two-field system—one planted the other fallow—favored earlier. The empire, however, proved largely ungovernable following his death in 814 AD. It broke up forty or so years later ushering in the period in European history referred to as the Dark Ages.

2.13 ISLAM

At about the same time that the Franks were emerging as the dominant power in Western Europe, a new religion was taking hold in Arabia. The prophet of this new religion was Mohammad (570–632), who began life as a merchant's assistant working on caravans traveling and trading throughout the Middle East. On his travels, Mohammed would have become familiar with the tenets of the major religions prevalent in the region at the time, notably Judaism and Christianity. When he was about 40 years old, he had a deep religious experience. He was visited by the angel Gabriel who communicated to him what he received as the ultimate word of God, a series of messages complementing and extending revelations delivered to earlier prophets ranging from Abraham and Moses to Jesus. His was to be the last of God's revelations and he faithfully recorded God's words in the holy book of Islam, the Koran. Like Judaism and Christianity, Islam was a strictly monotheistic religion. It rejected, however, the Christian belief in the Trinity and the divinity of Jesus. There was no God but Allah, Mohammad was His Prophet, and there were strict rules as to diet, religious practice, and personal behavior to which His followers had to rigorously adhere. There were no priests or rabbis in the early Moslem religion. Standards were maintained by a scholarly elite of laymen, forerunners of the modern mullahs, whose authority derived from their personal piety and learning—an important difference compared with the organization of the Christian church.

Islam was successful early in attracting converts, in unifying the different Arab tribes, and in appealing to the various pagan peoples of the region who, to quote Kagan et al. (2002), "had been marginalized in a world dominated by Judaism and Christianity." It was surely a source of great pride for Arabs to learn that they had been selected singularly by God as the first recipients of His final definitive message to humanity. And they set out to quickly spread His word. The expansion of the Islamic world was facilitated by a number of fortuitous circumstances, most notably the collapse of the Persian Empire at the hands of the Byzantine emperor Heraclius. This left a convenient vacuum in Egypt, Palestine, Syria, and Asia Minor that the Arabs were quick to fill. And there was little to block their expansion to the east into the heartland of the former Persian Empire, the region defined today by Iraq and Iran. The capital of the Arab world moved first from Mecca to the more centrally located Damascus. Later, in 750, it transferred to the historically important site of Baghdad. In a little more than a hundred years following the death of Mohammad, the Muslim world had expanded to include most of Spain, all of North Africa, and much of the Middle East including the surviving portions of the Persian Empire reaching as far as India. Only the Franks in the west and the Byzantine Empire in the east blocked their way to taking over most if not all of Europe. But for these circumstances, Christianity could have been reduced, more than a thousand years ago, to a mere footnote in history!

Jihad, or holy war, was an important motivational force for early Arab expansion. Specifically targeted were the pagans or infidels who when conquered were offered a choice—to convert to Islam or die. Christians and Jews, sharing the Moslem belief in a single supreme God, were spared this fate at the outset. Jews in particular enjoyed a level of relative freedom under Arab rule, greater indeed than they would encounter later under successive Christian regimes. The relatively more favorable treatment of Jews as compared to Christians had probably less to do with doctrinal distinction as it had to do with the greater ease with which Jews were prepared to accept the ways of the Moslems. For Christians it was a matter of faith that both Moslems and Jews were on the wrong path and needed to recognize the errors of their ways and convert to the one true church if they were to have any hope of salvation.

It can come as no surprise that this led to conflict. It is interesting to note that in at least parts of the contemporary Moslem world, westerners and specifically Americans are now regarded as the infidels and there are calls again for Jihad. Infidels occupying territory in the Islamic world, notably Jews in Israel, and the Great Satan (the United States) are the specific targets of this new holy war.

As they expanded their territorial reach, the Arabs were quick to absorb and incorporate the cultures of the regions they occupied. The conquest of Egypt was particularly consequential. Despite the destruction of its great library (the major depository of Greek literature) by fire in 389 and its historic museum, (the focus of Alexandrian intellectual activity) in 415, Alexandria persisted as a major center of Greek scholarship. The Arabs absorbed this culture, translating the works of celebrated Greek scientists, mathematicians, and philosophers into Arabic, preserving thus an inheritance that might otherwise have been lost to posterity. Later, this legacy would be translated into Latin and other European languages in Spain following the demise of the Spanish Islamic state when the city of Toledo fell to Christian military forces in 1085.

The Arabs benefited also from the cultures they encountered in the east. Baghdad emerged as an important center of Islamic intellectual activity where insights from Greek scholars were combined with elements of Persian and Indian cultures to forge a unique Islamic (Arab) legacy. We owe our numerical system, for example, to the fusion of these cultures. The elements of what we refer to today as the Arabic numerical system were developed in India by Hindu mathematicians as early as 400 BC. Arab scholars learned about this innovative system and adapted it with minor modifications in 776 when they introduced the convention that the most significant digit should be located furthest to the left rather than to the right as in the original Indian formulation. Close to five centuries would elapse before the advantages of the Indo/Arabic system would be recognized in the west, when it would be publicized by the Italian mathematician Fibonacci in his treatise Liber Abaci which appeared first in 1202. Even then, the system would fail to win common acceptance and several additional centuries would elapse before it would become the standard in everyday life and commerce. In the interim, it would find application primarily in the esoteric scholarly work of professional mathematicians. Imagine the difficulty we would have today if we had to carry on commerce, or even elementary arithmetic operations, using the Roman, letter-based, numerical system, or worse, the earlier, even more cumbersome, system favored by the Greeks!

To the Arabs we owe not only our number system but also the foundations of algebra and important elements of geometry and trigonometry. They pioneered also in astronomy, chemistry, medicine, agriculture, and horticulture. They introduced, for example, rice, cotton, and sugarcane, and advanced techniques of irrigation to the territories they occupied in Europe, notably in Spain and Sicily. The period of the Arab ascendancy coincided with a low point in western European civilization and culture. As Arab navies controlled the

Mediterranean, the west progressively lost contact with its classical heritage, which continued to thrive in the Byzantine east. Trade with the richer east effectively disappeared and economies of the west depended increasingly on the relatively low, localized, outputs of the feudal/manorial system that replaced the more global organization of the earlier Roman state. Arab Spain emerged as the important point of contact between Western Europe and the more prosperous world to the east. The advance of the Arab armies into Europe was checked first by the Frankish leader Charles Martel at Poitier in France in 732. Christian reconquest of Spain picked up steam in the tenth century and by the thirteenth century most of Spain had been returned to Christian control. Moslem influence at that time was confined mainly to the most southerly region of the country. A key event, alluded to earlier, was the capture of Toledo with its important library in 1085 and the takeover later of the capital of Islamic Spain, Cordova, the largest city at the time in Europe west of Constantinople. The Christian conquest of Spain played a critical role in reconnecting Western Europe to its earlier classical tradition and in educating the west as to the critical advances in science and mathematics achieved in the interim by scholars in the Islamic world. The number of words of Arabic origin in our modern scientific and mathematical vocabularies—algebra, algorithm, almanac, azimuth, nadir, zenith, alkali, and alcohol to mention but a few—attests to their influence.

For all their success, though, the Arabs were rooted in the primitive traditions of their earlier nomadic existence on the Arabian Peninsula. Their influence declined as new peoples, mainly from the east, were introduced into the Moslem world. The Seljuk Turks, originating from north of the Aral Sea in central Asia, evolved as the dominant force in the Islamic world during the eleventh century. They controlled eventually an empire that stretched from the shores of the eastern Mediterranean (from Asia Minor) to the mountains of Central Asia. They attacked and defeated the forces of the Byzantine Christians in the battle of Manzikert in 1071 and threatened to overrun Constantinople. Alexius I Compenus, Emperor of Constantinople, appealed to the west for help, an appeal that resulted in the organization of the First Crusade, approved at the request of Pope Urban II at the Council of Clermont in 1095. Constantinople would survive the immediate Seljuk threat but would fall eventually to an even more powerful Islamic regime, the Ottoman Turks, who took over the city in 1453, changing its name from Constantinople to Istanbul signaling an effective end to the Byzantine Empire that had survived and prospered for close to a millennium.

2.14 VIKINGS

As Western Europe retreated into its manorial/feudal shell following the decline of the Frankish Empire, it became increasingly vulnerable to attacks from the outside. We alluded earlier to the menace posed by the Islamic world to the south and southeast. And there was pressure also from the Magyars

(Hungarians) to the east. But the most serious threat came from the north, from Scandinavia, from the region occupied today by modern Denmark, Norway, and Sweden.

The Scandinavian countries, on the northwestern extremity of Europe, were not part of the Roman Empire. Rather than centrally administered states, they were organized as local, fiercely independent and competitive, tribes. They were not, however, completely isolated from the larger outside world. Their lands were rich in goods of value to their neighbors to the south—notably furs of cold climate animals, seal skins, and beeswax (Diamond, 2005). Travel for trade and the contacts that ensued, limited though they were in the early stages, prompted an important transfer of technology from the more civilized south to the north. The Scandinavians discovered the benefits of applying sail to their previously, exclusively oar-driven, shallow-draft boats (a lesson learned probably as early as 600). They profited also from the advances in agriculture that had taken place in northern Europe during the Carolingian period, notably the adoption of a heavy iron plow with a moldboard that could cut deeply into the soil and could effectively turnover the heavy soils characteristic of the humid north. From trade they progressed to plunder. Given their superior military prowess, compounded by their native tendencies for aggression, it was clearly more profitable for the Scandinavians to take rather than waste time on barter. The pivotal transition occurred in 793 when a raiding party from Denmark attacked and ransacked the monastic settlement on Lindisfarne Island off the northeast coast of England. This signaled the beginning of the age of the Vikings (the name derives from the Old Norse word vikingar meaning raiders). For several centuries, the Vikings would terrorize large parts of Europe—from Ireland in the west to Russia in the east, and as far south as the countries bordering the Mediterranean (ranging as far east there as Constantinople).

The Vikings would go on to establish colonies on the Shetland, Orkney, and Faeroe Islands, in Iceland and Greenland, and for a time also in North America. Preceding Columbus by more than five hundred years, they found their way across the Atlantic to the New World, establishing settlements on Baffin Island, Labrador, and probably also on Newfoundland. Why then did they not opt to settle permanently in this rich land, a region referred to in the Icelandic sagas (Magnusson and Palsson, 1965) as Vinland or Wineland? Diamond (2005) suggests that the outpost on Vinland was too far removed from the parent colony on Greenland, that Greenland itself was scarcely self-sufficient, and that the small number of colonists who made their way to Vinland was no match for the large population of natives who resented their presence. So, they simply withdrew. Not too long after, the colony on Greenland itself disappeared, a victim, it is thought, of a change in climate—a transition from the relatively mild conditions of the Middle Ages to the much harsher environment of the subsequent period referred to by climatologists as the Little Ice Age.

While the Vikings from Norway and Denmark directed their attention mainly to the south and west exploiting their ability to navigate in the open waters of the

Atlantic, the Swedes set their sights primarily to the east. Vikings from Sweden made their way deep into Russia, penetrating as far as the Caspian and Black Seas, arriving eventually at the gates of Constantinople. There they met their match. The forces of the Byzantine state successfully defended their capital using a form of napalm (Greek fire) which they sprayed on the invaders, a challenge for which the latter had no answer. A peace treaty was arranged on terms that ensured that the Vikings could pose no further threat to Byzantine possessions. The Emperor of Constantinople was so impressed with the Norsemen, however, that he employed a group of them to serve as his personal guard and accorded rights to their associates to engage in peaceful trade, an opportunity the Vikings pursued with great success not only with the Byzantine Greeks but also with Muslim states further to the east. With this, luxury goods such as spices and silk from the east found their way north to the Baltic, greatly expanding preexisting trading arrangements. Attesting to the significance of this expanded trade is the large number of Arabian silver coins uncovered in recent archeological digs in Sweden.

The Scots, English, Irish, and Franks were ill prepared at the outset to counter the attacks of the Vikings from Norway and Denmark. Raiding parties would arrive without warning. And attacks were not simply confined to coastal communities. The Vikings in their shallow draft boats had the ability to travel large distances up rivers attacking communities far removed from the coast. A large fleet of Viking ships reached the gates of Paris in 845 AD extracting a major bounty in silver before they would withdraw. And, appreciating that they had found easy pickings, they would return frequently thereafter for more. Eventually, unable to counter these repeated assaults, the Frankish king came to an arrangement with the Viking chief Rollo (known also as Rolf the Pirate). In 911, he ceded to Rollo what developed later as the Duchy of Normandy, sealing the deal by marrying his daughter to the Viking chief. The Vikings settled in Normandy, intermarried with the natives, converted to Christianity, and within a few generations had become more French than the French adopting even their language, losing contact for all intents and purposes with their Nordic heritage.

The Vikings established also settlements in Ireland, Scotland, and England. What evolved later as the city of Dublin, the present Irish capital, began life as a Viking base in 837. The Irish cities of Cork, Limerick, Wexford, and Waterford owe their origins also to the Vikings. In England, the Vikings assumed control of large areas of land in the north and east of the country, notably in Northumberland and East Anglia, in addition to extensive portions of Northern Scotland including the offshore islands (the Hebrides, Shetlands, and Orkneys). The Vikings were forced eventually to withdraw from Ireland. They were soundly defeated in the Battle of Clontarf (a coastal community to the north of Dublin) in 1014. It is estimated that both armies in this battle, the Irish and Vikings (mainly Danes), numbered as many as 20,000. The Irish were led by Brian Boru, a legendary figure in Irish history, who regrettably (at least from the Irish point of view) lost his life in the fight. The Anglo-Saxons would eventually succeed in defeating the Vikings in England when King Harold Godwinson defeated the forces of King Harald Hardrada of Norway in the battle of Stamford Bridge in 1066. Ironically, this would prove a short-lived victory. Harold was soundly defeated in the same year by William, Duke of Normandy, successor to the Viking chief Rollo, in the Battle of Hastings. English society would evolve subsequently as a fusion of Anglo-Saxon and Norman-French cultures, a heritage attested to by the number of words of French origin in modern English. It is easy to see who won and who lost. We refer in our language today to animals in the field as pigs, calves, cows, and sheep. But when they arrive at the table, they are known as pork, veal, beef, and mutton reflecting the Norman/French origin of these words, the language of the victors.

The conquest of England by the Normans was a pivotal event in the development of the modern English nation state. The monarchy under William the Conqueror and his successors assumed an initial position of close to absolute power, a status that would not be seriously challenged until the nobles rose to assert themselves when King John, one of William's successors, was soundly defeated by the French in 1214 in an abortive attempt to retake Normandy. Led by Robert Fitzwalter, the nobles occupied London in 1214 and forced the king to negotiate. Hostilities came to an end on June 15, 1215 with a comprehensive written agreement, the Articles of the Barons, known subsequently as the Magna Carta. The document spelled out in detail (63 chapters) the rights and responsibilities of all English citizens—commoners, nobility, and churchmen—formalizing elements of what is referred to now as the Common Law, a series of legal principles based on precedent. In particular, the Magna Carta required due process for any action affecting the life or livelihood of a citizen and decreed that taxes could not be levied without the consent of "the general council of the nation." By appropriating property belonging to the barons, the king was judged to have violated an important tenet of the Common Law. The Magna Carta was destined to play a crucial role in the subsequent development of English society. Bernstein (2004) describes it as the "fuse that would detonate the later explosion of world economic growth," the critical ingredient for the early success of England and its American colonies in the formative stages of the Industrial Revolution.

The Viking Age came to an end with the Battle of Hastings. By then, Denmark, Norway, and Sweden had converted to Christianity with central governments ruled by Christian monarchs. Scandinavian countries and their colonies to the west—Iceland and for a time Greenland—were integrated fully into Western European society. Peace, though, would remain elusive as new nation states would compete for power in a struggle that would continue for close to a millennium and would end only with two world wars, the formation of the European Union and, eventually, the collapse of the Soviet Union. The legacy of the Vikings was significant nonetheless, and durable. There can be little doubt that Europe benefited from the skills of the Vikings in open sea navigation and boat building—lessons that would serve it well a few centuries later as European sailors would succeed in extending Europe's reach to the farthest corners of the globe.

2.15 EUROPE'S RENAISSANCE

The inclusion of the Scandinavian states and the reconquest of Spain with its advanced civilization incorporating not only the legacy of Greece and Rome but also the important advances achieved in the interim by the Arabs played an important role in the resurgence of Western Europe. Arguably, though, the First Crusade was the pivotal event in Europe's renaissance. In 1097, three great armies, numbering as many as 100,000 men, assembled in France, Germany, and Italy and set off for Constantinople. The soldiers were motivated by religious fervor—to recover the Holy Land from the Moslem infidels—and were inspired by the grant of a plenary indulgence by Pope Urban II to all who would take part (the plenary indulgence guaranteed immediate access to heaven to all who would die in the campaign irrespective of their prior sins). This First Crusade was spectacularly successful. The Seljuk Turks were soundly defeated in a series of battles and on July 15, 1099 the Crusaders captured Jerusalem establishing a number of strategic beachheads along the way. More important than the military accomplishments, however, was the success of the merchants who followed in the wake of the Crusaders to exploit the opportunities for trade opened up by renewed contact with the east.

The gains achieved by the Crusaders on the battlefield in the First Crusade were short-lived. The garrisons left to defend the territories acquired in the campaign were too small to prevail against the larger, and more highly motivated, Moslem forces that remained in the region. The Moslems won a major victory over the Christian forces at Edessa in 1144, prompting a second Crusade. This was spectacularly unsuccessful. Saladin, the Moslem king of Egypt and Syria, recaptured Jerusalem in 1187, triggering a third Crusade but again this proved a failure. The success of the merchants proved more durable, however. Trade flourished between the European merchants and their Arab associates with goods from as far away as India and China making their way to Europe. The merchants of the Northern Italian cities, notably Venice, Pisa, and Genoa, exploiting their sea faring skills, were the notable beneficiaries of this expanded commerce.

The affluence of the Northern Italian trading centers spread rapidly to the rich agricultural regions in their hinterland. With increased agricultural production and the new wealth flowing through the port cities, the population of the region (Tuscany and Lombardy) expanded rapidly. Important new urban centers sprung up, notably in Milan and Florence. With prosperity came a series of challenges to the authority of the Holy Roman Empire that had previously exercised feudal power over the region. The Lombard cities combined to defeat the armies of Emperor Frederick Barbarossa in 1176. Self-governance prompted further growth and expanded prosperity. By the end of the thirteenth century, it is estimated that the population of Milan (which had won its independence much earlier, in 1035) had grown to more than 200,000 with populations of Venice, Florence, and Genoa each greater than 100,000 (Cameron, 1997). By comparison, at this time, the populations of London and Cologne (the largest German city) scarcely exceeded 40,000 (Cameron, 1997).

Prosperity, though, was not confined to Italy. It spread rapidly to include much of the rest of Europe. An important textile industry took hold in the Netherlands. Merchants bought raw material, wool for example which they imported from England and the Scandinavian countries, engaging workers in the countryside (putting out) to convert the wool to cloth to be exported back to the countries from which the wool originated, and extensively to much of the rest of Europe. Technological advances, the invention of the pedal loom, and the spinning wheel for example, contributed importantly to the success of the Low Countries' textile industry. Industrialization was accompanied also by rapid growth in the market for leather and iron goods. Leather was used widely in the fabrication of saddles and harnesses for horses and as an ingredient in the manufacture of furniture and clothing (where it was required to make buttons and belts) and in a variety of industrial devices ranging from bellows to valves (Cameron, 1997). The increased demand for iron was accommodated by additional sources of ore from the German states and from Scandinavia. And the wood required to make the charcoal needed to smelt these ores was harvested by tapping the abundant resources available in the densely forested regions in the north. Trade was the essential ingredient in all of this interrelated economic activity. In the early stages, it proceeded mainly over land routes, from Northern Italy through Switzerland, to France, Germany, the Low Countries and the Baltic, with much of the goods involved in the trade changing hands in markets that sprung up in France, notably in Champagne, and in strategically located cities such as Leipzig and Frankfurt. Movement of goods by ship, from the Mediterranean through the Straights of Gibraltar along the Atlantic seaboard reaching as far as the Baltic, became important in the latter years of the thirteenth century contributing to the subsequent emergence of Portugal, Spain, France, Holland, and eventually England as important naval powers competing for influence with the northern Italians. Trade prompted also the development of essential financial instruments (letters of credit for example) and institutions, forerunners of what would evolve eventually as the critical elements of the modern banking system.

Agriculture remained, however, the backbone—the staple—of this expanded commerce. Surpluses in agriculture were required to supply the needs of the growing artisan and merchant classes in the expanding urban centers. With fewer hands to work the fields and more mouths to feed, there was need for important increases in agricultural productivity. Several factors contributed to this success. First were advances in agricultural technology—the development of the wheeled, iron plow mentioned earlier, for example, and the introduction of collars, harnesses, and shoes allowing horses capable of much greater output of work to substitute for oxen in pulling the new heavy plows. Second, the acreage available for agriculture was increased by draining swamps and clearing forests. Third, the two-field system of crop rotation favored in the relatively dry region around the Mediterranean, in which one field was left fallow for a year while the other was harvested, was replaced by a three-field system in which crops were extracted from two fields in any given year with only one left fallow, an arrangement that worked well in the organic-rich, heavy soils of the more humid regions in the north. And, fourth, farmers were able to take advantage of the expanded trade network to exploit regional comparative advantages, growing crops for which their land was best suited, relying on trade to supply goods that could be produced more efficiently elsewhere.

Agriculture, of course, is particularly vulnerable to the vagaries of weather and climate. Years of plenty lead inevitably to a sense of complacency and are accompanied often by a rise in population. When the good times end or are seriously interrupted, consequences can be serious, even catastrophic. Spring and summer weather was unusually cold and wet over a large region of Europe in 1315. The wet spring made it difficult to plow the sodden ground to plant crops and the wet summer compounded the problem by making it difficult to harvest the crops that were planted. Many were allowed to rot in the fields depleting the supply of seeds needed for planting the following year. The weather in 1316 was similarly unfavorable and by the next year the problem had evolved from serious to catastrophic with widespread famine to which even the rich were not immune. Large numbers of people died, especially the most vulnerable—the poor, the young, and the old. Compounding the problem, essential seed grains were consumed and draft animals required for plowing were slaughtered, ensuring that the difficulties would persist even with the return of better weather, The economy of the continent suffered a serious set back from which it had scarcely recovered before being faced with an even more serious challenge, the arrival of the Plague, or Black Death, in 1347.

The Plague was transmitted by a bacterium known as Y. pestis carried by rats and communicated to humans by fleas. It is thought that the disease arrived in Europe first in Sicily in October 1347 carried by rats on merchant ships returning from a trading expedition to the Black Sea. By the following summer, it had spread as far north as England killing large numbers of people along the way. Within 5 years, as many as 25 million died in Europe, reducing the population of the continent by as much as a third. Symptoms included high fever, vomiting of blood, and swollen lymph nodes. Almost invariably the impact was fatal. The swollen glands assumed a sinister black color and it was this feature that prompted the English to refer to the disease as the Black Death. It persisted for close to 5 years, returning periodically for centuries thereafter, ceasing to be a problem only in the 1600s.

Society was totally incapable of coping with the ravages of the pandemic. Contemporary medicine had no answers. Corpses were dumped in the street or buried in mass graves. Properties were abandoned as owners fled to the countryside or elsewhere seeking reprieve. But there was no escape. Commercial activity ground to a halt. Those who survived were too weak, or too few, to cultivate the fields, many of which were allowed to revert to their earlier forest cover. It was a time of unremitting social unrest with difficulties compounded by the outbreak of the Hundred Years' War (1338–1453) between England and France. But there were unexpected benefits for those who survived. With fewer people to feed, the price of grain and other agricultural commodities declined. With the rapid drop in population, the value of labor increased and the power of the aristocracy and moneyed classes went into a period of prolonged decline. Labor was in the ascendancy. As Cameron (1997) put it, "market forces resulted in the dissolution of the vestigial bonds of serfdom and the rise of wages and living standards for peasants." Dreadful as it was, the plague with its associated evils, he writes, was a "strong cathartic that prepared the way for a period of renewed growth and development beginning in the fifteenth century".

The Great Famine and the Plague aside, the High Middle Ages was a period of important technological progress in Europe. Landes (1998) refers to it as the Age for The Invention of Invention. Supporting this claim, he highlights the expansion of uses of the water wheel, the development of eyeglasses, both convex (correcting problems of farsightedness) and concave (for those who were shortsighted), the invention of the mechanical clock, the introduction of mechanized printing, and the use of gunpowder as an explosive. Power generated using the water wheel was applied to a variety of tasks including mechanized pumping of water, lifting heavy weights, milling grain, pounding cloth, shaping metal, mashing hops for beer, and pulping rags to make paper. Application of the rotary motion of the wheel to this rich variety of tasks was made possible by the development of a variety of innovative mechanical devices incorporating in many cases complex combinations of cranks and gears. Eyeglasses fitted with convex lenses extended the productive lives of artisans engaged in work requiring acute close-up vision. Access to mechanized watches and clocks allowed workers to coordinate activities substituting for earlier, less reliable, time keepers such as water clocks (subject to freezing and sedimentation) and sun dials (useless at night or on cloudy days). The printing press paved the way for the introduction of newspapers and affordable books making the written word available at modest cost to the masses. Gunpowder had been invented earlier (probably in the eleventh century) by the Chinese but had been used mainly in powder form providing a relatively weak reaction—good for spectacular firework displays but relatively useless in propelling heavy cannon balls. The Europeans learned to concentrate the powder in small pebbles enhancing its explosive power. They had access then to a powerful new military technology: an ability to kill and destroy at a distance previously unimaginable. Clearly, not all technological advances are socially constructive!

Despite the set backs of the Great Famine and the Plague, the first half of the second millennium was a period of unprecedented economic advance in Europe. The productivity of labor increased. The feudal system collapsed and the lot of the average worker improved setting the stage for further advances in the centuries that followed.

2.16 THE AGE OF EXPLORATION

As noted earlier, the age of trans-ocean exploration began with the Vikings. They made their way across the North Atlantic in the latter years of the first millennium establishing colonies in Iceland, Greenland, and as far away as Labrador and Newfoundland. The long-term benefits of this exploration were limited, however. The settlement in North America was soon abandoned: the number of colonists who found their way to this distant land was too small to sustain a permanent presence in face of committed and sustained opposition from the peoples already there. Accordingly, the newcomers elected to withdraw. The colony in Greenland survived for a time but eventually it also collapsed. A change in climate may have contributed to the demise in this case. But the Greenland settlement, as Diamond (2005) argues, was doomed most likely from the outset. The Vikings who settled there were committed to the pastoral agricultural practices of their Norwegian homeland: they knew no better. These proved ill suited to the harsher, more variable, conditions of the very different environment they encountered in Greenland. The Vikings were courageous sailors, inventive in developing the skills required to navigate across relatively large distances on an ocean subject to sudden and often violent storms. From an economic point of view, though, their North Atlantic explorations were largely unproductive. In contrast, the age of exploration that began in the fourteenth century would change the course of history, extending Europe's reach to the farthest corners of the globe, establishing along the way the basis for much of the world's modern industrial economy. It was led from the outset by Portugal, with competition a little later from Spain.

Early successes of the Portuguese, with participation by the Basques and Genoese, involved the discovery of the offshore Atlantic islands—the Azores, Madeiras, and Canaries, all but the last uninhabited. The more southerly of these islands, the Madeiras and Canaries, had climates ideally suited to the cultivation of sugarcane. By the late Middle Ages, Europeans had become addicted to the sweet taste of sugar: sugar was by then one of the most valuable commodities traded on European markets. Sugar and sugarcane were introduced to Europe first from India by the Arabs. Production of sugar from sugarcane is a labor-intensive activity, however. To capitalize on the sugar producing potential of its new territories, Portugal needed a source of low-cost labor. They chose to populate the Azores and Medeiros initially with the dregs of European society—convicts, prostitutes, and religious dissidents. The discovery of the Cape Verde Islands to the south (in the mid-fifteenth century) provided them with a more efficient and more profitable source of cheap labor: slaves from the nearby African mainland. Thus began the sorry history of European exploitation of black Africans as slaves, a practice that would spread later to the New World, consequences of which are with us still today. But for the Portuguese in the fourteenth century, traffic in African slaves represented no more than an important new opportunity for trade and profit, one they exploited with gusto.

Two factors motivated Portugal's push down the west coast of Africa. First was the drive to find a sea route to the Far East that would allow the Portuguese to undercut the monopoly in the spice trade enjoyed by Arab merchants from their base in the Middle East. The potential for profit was enormous: Landes (1998) estimates that by the time pepper arrived in Europe from India it had passed through so many hands that the price had risen by more than a factor of twenty. A second motivation was to find the source of the vast quantities of gold that made their way from Mali through Timbuktu and thence along the camel routes to North Africa (Timbuktu at that time was a thriving center of Islamic culture and learning, located on the upper Niger River; today the river is dry and Timbuktu is a ghost town depending for its existence on occasional visits by tourists interested in exploring its glorious past). Again, it was thought that the opportunity for profit was more than enough to justify the search. Landes (1998) recounts how one Mali ruler, setting out on a pilgrimage to Mecca in 1324, brought with him 80 to 100 camels each loaded with 300 lb of gold. At current prices, more than 500 dollars an ounce, this cargo would be worth more than 200 million U.S. dollars (no wonder the Portuguese licked their lips in anticipation). As a footnote to the Mali ruler's pilgrimage, Landes recounts that after 3 years when the ruler was ready to return, he had run out of money and had to borrow to finance his trip home! The Portuguese never did find the original source of the West African gold. They developed, however, an extremely profitable trade with the native African kingdom of Ghana, not only in gold but also in ivory and slaves. The Portuguese crown benefited greatly from this commerce—to the tune of more than a third of the gross extracted through a combination of license fees and taxes according to Landes.

The success of Portuguese exploration in the fifteenth century was due in no small measure to the leadership of a younger son of the Portuguese monarch, Prince Henry, better known as Henry the Navigator. Henry recruited the preeminent shipbuilders and astronomers in Europe to ensure that Portugal's sailors would have access to the best in equipment both for sailing and navigation. Ships sent out on his exploratory missions were required to maintain detailed geographic records (geometry of coast lines, water depths, potential hazards etc.) as well as data on prevailing winds and ocean currents. All of this information was incorporated in state of the art maps and logs ensuring that expeditions that ventured out later would have the benefit of the accumulated wisdom of those who went before. Henry died in 1460, triggering a brief hiatus in Portuguese exploration. The commitment was renewed, however, under King John II who came to the throne in 1481. Bartholomew Diaz rounded the Cape of Good Hope in 1488 and a few years later Vasco da Gama led a four ship expedition that, after exploring the east coast of Africa, made its way for the first time by sea from Europe to India, striking land at Calicut (a city named for the calico (cotton) cloth produced by its merchants). To his surprise, da Gama found there not a primitive society that could be induced to part with its treasures for a few Portuguese trinkets, but an advanced, savvy, community of

Muslim merchants and artisans. The Portuguese, however, had an important advantage: gunpowder and the ability to fight at a distance. Using his firepower to advantage, da Gama captured a small Muslim ship loaded with spices. Returned to Europe, these spices were more than enough to pay for the expenses of his expedition, despite the fact that da Gama lost two of his four ships along the way and close to half his original crew (Landes, 1998). Less than 6 months after da Gama's triumphant return to Portugal, a second, much larger flotilla of thirteen ships with more than twelve hundred men set out from Portugal to India under the command of Pedro Cabal: this venture was motivated solely by considerations of profit—present and future (Landes, 1998). The expedition was spectacularly successful, ensuring, at least for a time, Portugal's commanding position in the Indian Ocean and its preeminent role in the lucrative eastern spice trade.

While Portugal was proceeding on its orderly program of exploration down the west coast of Africa on its way to India, Spain realized a coup that would prove in the long run even more significant, the discovery (or, perhaps more accurately, the rediscovery since the region was already populated) of the Americas. Christopher Columbus was a Genoese sailor with a vision to find a passage to India by sailing west rather than east. He approached the Portuguese, French, and English for financial backing but failed to convince them that the venture was worth the cost. They turned him down, it is likely, for good reasons: centuries earlier, Arab astronomers had calculated the circumference of the Earth and it was clear that to go west in order to go east was to go the long way round. He found a sponsor though in Spain which had just succeeded in conquering the last Moorish outpost on the Iberian Peninsula in Grenada. Queen Isabella agreed to sponsor his expedition, in the nature of a victory celebration according to Cameron (1997). Columbus set off from the outermost Canary Island in September 1492 and in a little more than a month (33 days), propelled by the reliable easterly trade winds, he struck land on an island he named San Salvador in the Bahamas. He returned to a hero's welcome in Spain, convinced that what he had discovered was an archipelago off the coast of either China or Japan. Ferdinand and Isabella petitioned the Pope to confirm Spain's title to these new lands and the Pope agreed defining a line of longitude west of which everything would belong to Spain with Portugal's claims confined to the east. The original line would have given Spain rights to all of the Americas. Portugal petitioned for a modest westward revision of the boundary. The revision turned out to be significant, however, allowing Pedro de Cabral when he reached South America a few years later (in 1500, on his way to India taking advantage of the prevailing lower latitude easterlies and the higher latitude, southern hemispheric, westerlies) to claim title for Portugal to the land we know now as Brazil. Cameron (1997) suggests that this adjustment was not accidental—that the Portuguese were already aware of the existence of this eastward extension of South America but kept the information to themselves until they could secure their advantage.

Columbus made in all four voyages to the Caribbean, the second with seventeen ships and 1500 men carrying sufficient equipment including livestock to establish a permanent settlement. The Caribbean islands were ideally suited to the cultivation of sugarcane and, following the earlier lead of the Portuguese, the Spanish, having eliminated most of the native population through a combination of brutality and exposure to diseases for which the natives had no resistance (Diamond, 1997), imported slaves from Africa to work on the plantations they established. Later, the English and their American colonists would become important participants in this trans Atlantic slave trade. Raw materials—timber, cotton, and food stuffs—would make their way from America to England, products of English industry and agriculture would be carried to Africa where they would be exchanged for slaves to be transported west to work the sugar plantations of the Caribbean which in turn would supply the sugar needs of the Americans and English—a profitable, albeit socially reprehensible, trade network.

Spain's major gain from its territories in the New World would turn out not to be sugar from its plantations in the Caribbean but rather gold and silver from the mainland. Hernando Cortes took possession of Mexico between 1519 and 1521. Later, in the 1530s, Francisco Pizarro conquered the Incas in Peru. By the end of the century, Spain had effective control over most of the Western Hemisphere, from Florida and California in the north to Chile in the south. It would fail though to benefit from all of this newfound wealth. The Spanish crown was perpetually in debt to finance its overseas expansion, its conspicuous consumption and the unending wars in which it would become involved in Europe. The bullion from the west would be used largely to subsidize these endeavors. It would prove insufficient, however, to service even the interest on the debts owed to bankers notably in Germany and Italy. Cameron (1997) points out that on as many as eight occasions (1557, 1575, 1596, 1607, 1627, 1647, 1653, and 1680) Spain was forced to declare royal bankruptcy. Little revenue was available by way of excess for productive domestic investment and despite the riches of its overseas possessions, Spain remained an industrially undeveloped country destined to fall seriously behind industrial powerhouses that would take hold later in the north, notably in France, Holland, England, and Germany.

2.17 THE REFORMATION

The sixteenth century witnessed a number of serious challenges to the authority and governance of the Catholic Church. For close to a millennium, Roman Catholicism had been the dominant religion of Europe—since the Church was first accepted as the official religion of the Roman Empire by Theodosius in 394. The challenge to its authority came first from a German cleric, Martin Luther.

Luther was born in 1483, entered the Order of the Hermits of Saint Augustine in 1505, was ordained a priest in 1507, received a doctorate in theology from the University of Wittenberg in 1512, and was appointed soon thereafter to the faculty of that university. Luther became concerned that the Church had lost its way over the years, departing, unacceptably in his view, from the teachings of Jesus as recorded in scripture. He railed specifically (in 1517) against the practice of awarding (actually selling) indulgences, dispensations that could allow the sinner to shorten the time required to be spent in purgatory in atonement for past sins or in the case of plenary indulgences a license to proceed straight to heaven. Arguing that the bible provided the only valid authority for decisions on matters of faith and morals, Luther went on to question (in 1519) the infallibility of the Pope and the pronouncements of Church Councils. He concluded that only two of the seven sacraments (Baptism and the Eucharist) could be justified based on scripture. Luther was excommunicated for heresy in 1521—subject to sanction not only by the church but also the state. Forced to withdraw from public life to escape imprisonment or worse, he made good use of his time in seclusion, producing the first vernacular (German) translation of Erasmus' Greek text of the New Testament.

Luther's message found a sympathetic ear not only with the clergy, especially those of lower standing, but also with his German compatriots more generally. It was a time to question not only the authority of the Church but also the strictures imposed by civil society. Access to libraries and books, and expanded horizons resulting from trade-related travel with consequent exposure to new ideas and mores, led to a revolt by peasants against their civil overlords in 1524, a revolt that was brutally suppressed, resulting in the loss of as many as 100,000 lives according to Kagan et al. (2002). Luther, however, was no civil revolutionary. He declined to lend his support to this uprising.

The spirit of the Reformation spread rapidly from Germany to Switzerland and France. Growing opposition to the practice of requiring Swiss soldiers to serve involuntarily in military campaigns sponsored by the church (the forerunner of the Swiss Guards in the contemporary Vatican) provided a receptive environment for the growth of Protestantism in Switzerland. It was led there by Ulrich Zwingli, appointed in 1519 as the People's Priest in the Great Minister Church in Zurich. Zwingli's position on matters of faith and morals was similar to that of Luther: if authority could not be found in the bible, it had no standing and should be rejected. He went on to question a variety of practices and beliefs ranging from requirements for clerical celibacy, to the doctrine of transubstantiation (the presence of the real body and blood of Christ in the Eucharistic bread and wine), to the practice of the veneration of the saints. Zwingli's teachings acquired official status when they were endorsed by the city government of Zurich in 1523. Zurich's lead was followed soon thereafter by a number of the other independent Swiss cantons, notably Berne and Basel. Religious disputes and conflict resulted in a period of extended civil strife in Switzerland—eventually open war—that would come to an end only when a peace treaty was agreed in 1531 confirming the rights of individual Swiss cantons to select and institutionalize their religion of preference. If individuals did not choose to conform to the locally sanctioned religion, they were free to move. There was no place for dissenters in the new religious order of Europe. Poland was an exception to this trend, emerging as a model of religious pluralism, welcoming not only Catholics, Jews, and mainstream Protestants (Lutherans and Calvinists) but also more extreme Protestant sects such as Anabaptists and Antitrinitarians (Kagan et al., 2002).

By the third decade of the sixteenth century, Lutheranism had emerged as the dominant religion in Germany, Denmark, Norway, and Sweden and in a number of the states ringing the Baltic. Calvinism, following the teachings of John Calvin, would take hold later in the Netherlands, Scotland, and Switzerland, and in parts of France, Germany, Austria, and Poland.

Calvin was born in 1509, son of the secretary to the Bishop of Noyon in France. From a privileged French Catholic background, he converted to Protestantism at age 25. Forced to flee France to avoid religious persecution, he settled initially in Geneva, at about the time that city had officially signed on to the Reformation (in 1536). Calvin was invited to draw up rules for the governance of Geneva's new Protestant church. His rules were so austere that he was forced to leave Geneva, settling for a time in Strasbourg as pastor to a group of expatriate French Protestants. He returned to Geneva in 1540 with a new mandate to establish standards both for church and civil society. His was a take-no-prisoners version of Christianity. Either you bought and lived up to the message or you should be prepared to suffer the consequences. And punishment was meted out not just to Catholics. Michael Servetus, a leader of the Antitrinitarian sect, moved to Geneva from Spain in 1553 to escape prosecution under the Spanish inquisition. Calvin accused him of heresy and had him burned at the stake.

A different kind of Protestantism would take hold in England. Henry VIII withdrew England's allegiance to Rome when Parliament passed the Act of Supremacy in 1534 confirming the King as the "only supreme head on earth of the Church of England." Henry's dispute with Rome was based less on theological considerations as on the King's desire to obtain an annulment of his marriage to Catherine of Aragon, daughter of Ferdinand and Isabella, King and Queen of Spain, a license that would allow him to marry Anne Boleyn, one of Catherine's ladies-in-waiting. Under Henry, the Anglican Church in England was essentially Roman Catholic but without the Pope. Before he died, Henry would have six wives, two of whom, Anne Boleyn and Catherine Howard (numbers two and five), he would arrange to execute on a combination of charges of treason and adultery. Marriage to the monarch was not the safest occupation in Reformation England. Life would become even tougher for those who stepped out of line when the less forgiving Calvinist tradition took hold a century or so later under Oliver Cromwell.

Religiously inspired conflict would come to dominate the politics of Europe in the latter half of the sixteenth and for much of the seventeenth centuries. Huguenots emerged as the dominant (Calvinist) Protestant minority in France. Despite their relatively small numbers (it is estimated that at most they accounted for less than 8% of the French population), they enjoyed a position of influence in French society. They were a dominant presence in a number of important regions of France, counting among their members some of the most powerful figures of the French nobility. France was convulsed by a series of bitter religious wars in the latter half of the sixteenth century, the first of which broke out in 1562. An uneasy peace established in 1570 broke down a few years later (in 1572) when more than 8000 Huguenots were brutally murdered as they gathered in Paris to celebrate the wedding of the Huguenot Henry of Navarre to Marguerite of Valois, Catholic sister of King Charles IX. The atrocity, which came to be known as the Saint Bartholomew's Massacre, was orchestrated by Catherine de Medici, mother of the King and the power behind the throne. There ensued a period of bitter struggle between Huguenots out for revenge and conservative, Spanish-supported, Catholics pressing to secure their advantage. Henry III, who succeeded Charles IX as king in 1574, sought to find a middle ground in this dispute. He was rewarded for his pains, however, by assassination, at the hands of a Jacobin friar in 1589, an event that brought Henry of Navarre to the throne as Henry IV. The prospect of a Huguenot king and prospectively a Protestant France prompted Philip II of Spain to send troops to France to claim the French throne for his daughter, Isabella. Faced with the possibility of losing his crown, and potentially also his country, to Spain, Henry took the line of least resistance and converted to Catholicism. In 1598, he promulgated the Edict of Nantes granting limited rights to the Protestant minority. In 1610, though, he suffered the same fate as his predecessor Henry III—assassination, again at the hand of a fanatical Catholic. Persecution of Huguenots resumed in France following the death of Henry IV, led in this case by Cardinal Richelieu. The last Huguenot stronghold in France, La Rochelle, fell to (Catholic) crown forces following a prolonged siege in 1629. Louis XIV, the Sun King who ascended the throne in 1643, revoked the Edict of Nantes in 1685, forcing large-scale emigration of France's Huguenot community to Protestant countries such as England, Germany, Holland, Switzerland, and the United States. After a prolonged period of struggle, Catholicism was confirmed as the established religion of France, but at a heavy cost—loss of many of its most talented citizens.

Spain emerged in the sixteenth century as the leading power in Europe, the most committed, indeed ruthless, opponent of Protestantism. Jews who declined to convert to Catholicism were banished from Spain in 1492 and a similar fate befell the Moors of Granada in 1502. A series of strategic royal marriages contributed to the expansion of Spanish influence. Joanna, daughter of Ferdinand and Isabella, rulers of Aragon and Castile respectively, married Archduke Philip, son of Maximilian I, Emperor of the Holy Roman Empire, and Marie, duchess of Burgundy, in 1496. Charles, the son of Joanna and Philip, inherited the Low Countries and part of Northern France on the death of his father in 1506 (a claim that traced back to his paternal grandmother: the Low Countries were controlled at the time by the Duchy of Burgundy). On the death of his maternal grandfather, Ferdinand, in 1516, Charles inherited most of Spain including the Spanish colonies in the Americas, and in addition Naples, Sicily, and Sardinia. He was recognized at that point as King Charles I of Spain. When his paternal grandfather, Maximilian, died in 1519, he became the Holy Roman Emperor, serving in that role as Emperor Charles V, adding thus to his possessions the Habsburg territories in Austria and Germany. Charles was now the most powerful figure in Europe, indeed in the world. He resigned his various titles in 1556, however, turning over the western portions of his empire to his son Philip, who became King Philip II of Spain, with the eastern remnants of the Holy Roman Empire transferred to his brother Ferdinand who was installed as Emperor Ferdinand I. Charles retired to a monastery where he died 2 years later. Philip would emerge as the strongman in Europe for most of the second half of the century and as the most ardent foe of Protestantism.

Philip turned his attention first to dealing with the threat posed by the Ottoman Turks. By the 1560s, the Ottoman Turks had advanced deep into Austria having established themselves earlier as the dominant naval presence in the Mediterranean. Spain joined with Venice in 1571 to form the Holy League with a mandate from the Pope to counter the challenge to Christendom posed by the Turks. The Spanish with their Venetian allies were outstandingly successful, routing the Turkish navy definitively in the Battle of Lepanto on October 7, 1571, an engagement that took the lives of more than thirty thousand Turkish sailors with more than a third of the Turkish ships either captured or destroyed (Kagan et al., 2002). This, the greatest sea battle of the sixteenth century, established Spain as the predominant naval power in the Mediterranean. Philip was less successful, however, in efforts to impose his will on the Northern European Protestant states, notably England and the Low Countries.

In 1564, Philip took steps to require his subjects, specifically the Calvinists in the Northern Low Country provinces, to accept the decrees of the Council of Trent (1545–1563), which confirmed, among other elements of Catholic teaching, the doctrine of the infallibility of the Pope. Philip's decision provoked a minor revolt to which the King responded in 1567 by sending in troops from Milan under the leadership of the Duke of Alba. A reign of terror ensued in which many Dutch Calvinists were convicted of heresy and brutally executed. The Spanish governance, administered under what became known as the Council of Blood in the Netherlands, was ruthless, not only executing those judged guilty of heresy but imposing also punitive taxes on the Netherlanders to pay for the costs of their oppression. Not surprisingly, they resisted. The southern provinces (mainly Catholic) made peace with Spain in 1579. The Protestant northern provinces continued to resist, succeeding eventually in driving out the Spanish in 1593 taking advantage of the weakened state of Spain following the decisive defeat of its Armada by the English in 1588. The northern provinces developed into the core of the modern state of Holland. The southern provinces formed the nucleus of the modern state of Belgium.

2.18 EMERGENCE OF THE DUTCH AND ENGLISH (BRITISH) AS MAJOR POWERS

Holland emerged as a major seafaring and trading power in the sixteenth century. Landes (1998) summarized the situation thus: "by the 1560s the province of Holland alone possessed some one thousand eight hundred seagoing ships—six times those floated by Venice at the height of its prosperity a century earlier." The Dutch achieved an early position of dominance in the North Sea fishing (herring) industry, drying, salting, and smoking their catch which they distributed widely over much of Europe. They developed also an important domestic textile industry, importing rough woolen cloth from England selling more valuable finished products abroad. Dutch agriculture, specifically dairy farming, was the most advanced in Europe generating a significant surplus in the form of butter and cheese for export. Trade at the outset was limited, however, mainly to transporting goods up and down the Atlantic seaboard, carrying timber and grain south from the Baltic, bringing wine and salt north from Portugal, Spain, and France and distributing the products of Dutch domestic industry (textiles, sail cloth, and agricultural products among others). Attesting to the importance of the Dutch influence, the first cargo of spices to reach Antwerp from Portugal in 1501 arrived on a ship owned by either Dutch or Flemish merchants (Cameron, 1997). And much of the silver and gold returned to Europe in Spanish galleons from the Americas made its way to market initially in Antwerp, later in Amsterdam. But when Spain annexed Portugal in 1580 and closed off the port cities of Seville and Lisbon to Dutch shipping in 1585, it was clear that Dutch influence was at risk. The response was not to retreat but to challenge the power of Spain, not only at home but also abroad. If Holland was to be denied access to spices and other valuable goods returned from the east on Portuguese ships, it was time for the Dutch to strike out and secure an independent supply.

The Dutch East India Company (DEIC) was formed in 1602 with a mandate to develop trade in the east. The Dutch West India Company (DWIC) was incorporated in 1621 with a 24-year monopoly to operate on the west coast of Africa and in both North and South America. Within a few decades, Holland had established itself as a dominant global mercantile power. The DEIC controlled Ceylon and Taiwan and the Spice Islands in Indonesia. It built factories on the east coast of India and established a trading post as far away as Japan (on an island in Nagasaki harbor). It had a dominant position in maritime trade not only between Asia and Europe but also within Asia. The success of the DEIC was due in no small measure to the advantage it enjoyed as a result of its military might—the canons mounted on its ships for which locals had no answer—and it was seldom reluctant to exploit this advantage.

While the DEIC was establishing its preeminence in the east, the DWIC was emerging as a powerful force in Africa, Brazil, the Caribbean islands, and North America. Its American settlement, populated first with thirty, French-speaking, Walloon families in 1624, was named New Netherland and claimed authority over large parts of present day New York, Connecticut, New Jersey, and Delaware. At about the same time, English Puritans (individuals opposed not only to Roman Catholicism but also to the rituals of the established Anglican Church) had set up the Plymouth and Massachusetts Bay Colonies (in 1620 and 1630 respectively). Commercial activities in New Netherland were concentrated initially in the Hudson River Valley, between New Amsterdam (New York City) and Fort Orange (Albany). Africa provided the Company with slaves, gold, and silver, the Caribbean and Brazilian possessions supplied sugar, while New Netherland was exploited primarily for animal furs. The Director General of the Company, Peter Minuit, purchased the Island of Manhattan from the Indians for 60 guilders worth of trinkets in 1626 and proceeded to build a fortified base there which he named Fort Amsterdam. The fortunes of the Dutch in the New World would vacillate subsequently in response to the vagaries of European politics in the latter half of the sixteenth century, as Holland became involved in a series of wars with a resurgent England and also with France.

Charles II laid claim to New Netherland on behalf of the English crown in 1664, granting authority over the territory to his brother James, the Duke of York, later King James II. An English fleet was dispatched to secure this claim and in September 1664, the Director General of the Company, Pieter Stuyvesant, turned over Fort Amsterdam to the English. New Amsterdam became New York: Fort Amsterdam was renamed Fort James; and Fort Orange was reborn as Fort Albany. The Dutch would make a brief comeback in 1673 when war broke out again in Europe, this time between the Dutch and an Anglo-French force under the combined authorities of Charles II of England and Louis XIV of France. The territory was restored, however, to England in 1676, in the Treaty of Westminster that ended these hostilities. New York would remain an important English military base in the New World for more than a hundred years—until the British were forced to withdraw following the end of the Revolutionary War in 1783.

While the Portuguese, Spanish, and Dutch were extending Europe's reach to the farthest corners of the globe in the sixteenth century, the English were engaged in a program of systematic catch-up. Ultimately, though, the tortoise would overtake the hare (or hares). The beginning was modest. When Henry VIII ascended to the English throne in 1509, he inherited from his father, Henry VII, a navy with as few as ten ships. As Perlin (1989) pointed out, England depended on its continental neighbors for "both necessities and luxuries including salt, iron, dyes, glass and arms." When Henry broke with Rome, the reliability of the supply of these commodities from his largely Catholic trading partners (France and the Spanish-controlled Low Countries) was threatened. The king embarked on a program to ensure a modicum of independence. Particular priorities were to develop a source of armament that would allow England to defend itself from an anticipated continental attack and to secure access to the necessities of life.

A foundry was set up in Sussex in 1543 taking advantage of locally available high-grade iron ore and abundant sources of wood to manufacture the charcoal needed to smelt the ore. According to Perlin (1989): "fifty three forges and furnaces were operating in Sussex by 1549. They brought England to the foreground of the international arms race and created in a short space of time a national iron industry." A copper industry was established soon thereafter in Cumberland. And coal was used to evaporate seawater to produce salt in Northumberland (salt was an essential commodity used to preserve food in the pre-refrigeration era). A Venetian born glassmaker, Jacob Verselyn, was given monopoly rights to produce and sell glassware at prices that would seriously undercut potential supplies from outside. Government subsidies and incentives were used to stimulate a domestic shipbuilding industry with the understanding that the civil fleet (fishing vessels and ships carrying merchandise) could be pressed into military service in time of war. In the initial stages of its industrial development, the English had the advantage of abundant sources of wood at a time when its primary rivals on the Continent had largely depleted their native supplies (Dutch ships were built for the most part with timber imported from the Baltic). By the end of the century (1600), though, the easily accessible sources of wood were largely depleted (accessible in this case meant that the trees were growing within easy reach of a river; overland transport was essentially impossible until improved roads and specifically the railroad were introduced more than a century later). England soon joined its continental neighbors as a net importer of timber. In the interim, though, English industry had developed to the point where it was competitive internationally. And the country had emerged as an increasingly important naval power. Sir Francis Drake successfully circumnavigated the globe between 1577 and 1580 and a few years later (in 1588) the English navy inflicted a devastating defeat on the celebrated Spanish Armada.

England continued to develop militarily and industrially under the Stuarts (who succeeded the Tudors) and later under the House of Hanover. Queen Elizabeth, the last of the Tudors, died in 1603 and was succeeded by James VI of Scotland—James I of England—instituting the Stuart line of succession to the English throne. The Stuart line would endure for close to a century, except for a brief hiatus (1649–1660) when Cromwell's Rump Parliament abolished the monarchy. Its rule would be interrupted, however, when William of Orange invaded England, taking up the throne in 1688 in response to an invitation from Parliament, forcing James II into exile in France. Parliament took the step of replacing James with William, fearing that under James, a devout Catholic, England might revert to Roman Catholicism. Roman Catholicism was outlawed in 1689. The Protestant Elector of Hanover assumed the throne as George I in 1714, succeeding the last of the Stuarts, Queen Anne. The Anglican Church, headed by the British monarch, remains the established Church of England to the present day. And English law (contested, however, briefly during the parliamentary elections of 2005) continues to hold that a Roman Catholic may not succeed to the English throne, nor is it permissible for a present or potential future monarch to marry a Roman Catholic.

Two factors contributed to the emergence of England as a globally dominant naval and industrial power in the late seventeenth and early eighteenth
centuries. First was a series of protectionist measures enacted by its newly assertive Parliament, exemplified most notably by the Navigation Acts. Second was
the access the country enjoyed to an almost inexhaustible supply of wood and
raw materials from its American colonies, most importantly the tall, straight,
tree trunks required to fashion the masts of the great new ships emerging as the
standard ships-of-the-line of the Royal Navy. These vessels ranged in length
from 100 to 140 ft, were 35–44 ft wide and 17–19 ft deep and carried a complement of 60–100 heavy cannon (Perlin, 1989). Masts to fit these ships required
trees with diameters of between twenty-nine and thirty-seven inches. By the
end of the seventeenth century such trees were essentially unavailable in
Europe—the King of Denmark had earlier prohibited the export of trees with
diameters greater than twenty-two inches. Perlin (1989) summarized the situation thus; "by the end of the eighteenth century, all [English] ships of the line
were masted with timber from New England."

The first of the Navigation Acts, passed in 1651, mandated that goods arriving in England from Asia, Africa, or America must be transported in English
ships. Goods from Europe had to be carried either in English ships or in ships
of the country of origin of the material transported. The second Act, passed in
1660, required that specified goods emanating from the English colonies—
notably tobacco, sugar, cotton, and wool—must be transported on either
English or colonial ships and must be brought first to England where they
would be subject to a significant tax before they could be transshipped to other
destinations. The Navigation Acts posed a serious threat to the dominance of
the Netherlands in intra-European trade and led to the first of the Anglo-Dutch
wars in 1670, resulting indirectly, as indicated earlier, in the surrender to the
English of most of the Dutch possessions in North America. They contributed
however in important respects to the early development of New England as an
independent naval power and as an important industrial state.

Taking advantage of the Navigation Acts, ships built in New England (the
Yankee Clippers) came to dominate trade across the Atlantic, in addition to
trade between New England and the English colonies in the Caribbean. The
New England whaling fleet provided the world with as much as 400, 000 lbs of
whale oil candles per year while fish from the rich fishing grounds off the New
England coast were marketed as far away as the Caribbean and southern Europe
(Perlin, 1989). Eventually, though, New Englanders would rise up, objecting to
protectionist regulations imposed by the homeland, measures they saw as
exploitive, inhibiting economic development in the colonies while subsidizing
growth in England. The Continental Congress adopted the Declaration of
Independence on July 4, 1776. Hostilities continued, however, for more than
seven years, involving not only the English and the colonists but also the French
and Spanish fighting on the side of the colonists. Peace was restored only in 1783
when the Treaty of Paris acknowledged the independence of the thirteen former
English colonies reconstituted by that time as the United States of America.

By the middle of the seventeenth century, Holland had emerged as the richest country in Europe, indeed in the world. According to data presented by Maddison (2001) as discussed by Bernstein (2004), the gross domestic product (GDP) per capita of Holland—a useful if imperfect measure of national wealth—increased at an average annual rate of 0.52% per annum between 1500 and 1700 (from $754 in 1500 to $2100 in 1700, adjusted for inflation and expressed in 1900 dollar terms). Italy was richer than Holland in 1500 ($1100 GDP per capita) reflecting the glory days of the rich city-states, but failed to advance significantly over the subsequent several centuries. GDP per capita rose at a compound annual rate of 0.15% in France between 1500 and 1700 (from $727 to $986). Progress in England was more impressive (0.28% per year, from $714 to $1250 between 1500 and 1700) but still lagged behind Holland.

What were the secrets of the Dutch success? First and foremost, as discussed above, was the success of the Dutch traders—the wealth returned to Holland from its overseas operations. Trade provided Holland with a rich source of capital available at relatively low interest rates for investment both domestically and internationally. According to Bernstein (2004), quoting a private communication from Richard Sylla, between 10 and 20% of the debt associated with the American Revolutionary War was held by Dutch bankers. Second was the absence of a predatory central government: Dutch taxation rates were modest and provincial and municipal authorities shared power with significant influence vested in the merchant class. Third was an efficient canal-based system of inland transportation, a beneficiary of geography and the ready availability of domestic capital. And fourth, Bernstein suggests, was the liberating effect of the Reformation: the Dutch were empowered to a greater extent than before to think and act for themselves. Paradoxically, economic growth began to slow in Holland soon after the Dutch finally won uncontested independence from Spain (in 1648 with the signing of the Peace of Westphalia ending the eighty-year war). Historians disagree as to the precise cause of the slow-down. There can be little doubt though that increased competition from England, empowered initially by the Navigation Acts, and instability engendered by the various wars Holland became involved in during the latter half of the seventeenth century and the early part of the eighteenth (not only with the English but also with the French) must have played an influential role.

England gradually supplanted Holland as the richest country in Europe over the course of the eighteenth century. A number of developments contributed to England's ascendancy. First was an important increase in the efficiency of English agriculture. Enclosure of previously open land and private ownership encouraged advances in animal husbandry (experiments in selective breeding for example), while improved strategies for crop rotation, in which land was used alternatively for crops and pasture planted often in the interim with nitrogen fixing legumes, contributed significantly to increases in soil productivity. By the end of the century, as much as a half of England's arable land was enclosed. By 1851, more than a third of the cultivated acreage consisted of farms measuring more than 300 acres (Cameron, 1997) and while the absolute number of workers engaged in agriculture continued to increase up to 1850, the fraction of the population so employed declined steadily, from about 60% in 1700, to about 36% in 1800, dropping to as low as 22% by 1850, and further, to less than 10%, by the beginning of the twentieth century (Cameron, 1997; Bernstein, 2004).

Improvements in agriculture alone could not account for England's economic resurgence. According to Cameron (1989), trade contributed 12% of the country's national income in 1688. By 1801 it had risen to 17.5%, an increase of close to 50%. The comparable percentages for agriculture in 1688 and 1801 were 40 and 32.5% respectively, reflecting a net decline in the relative importance of agriculture. Access to credit at favorable rates encouraged both private and public investment. The surplus from international trade was a factor here but more important was the domestic political environment.

The power of the monarchy was severely limited following the so-called "glorious revolution" when William of Orange was invited to replace James II in 1688. Parliament assumed direct control of the government's finances in 1689 and, in one of its earliest initiatives, authorized establishment of a national debt, interest and principal payments which would be backed by excise taxes that could be levied only by Parliament. Members of Parliament were drawn largely from the more affluent classes of English society and thus had a vested interest in encouraging responsible management of the nation's finances. The Bank of England was set up as a private joint-stock company in 1694, authorized to issue shares to the public (the Bank was converted to a public institution only in 1946 when it was nationalized by the then Labor government). This and other initiatives facilitated the availability of capital.

Between 1750 and 1820, 3000 miles of canals were added to the already existing 1000, funded largely by private sources, greatly facilitating internal distribution of goods, notably coal, demand for which rose markedly after 1750 as wood became scarce and as coke increasingly substituted for charcoal in the smelting of iron ore, exploiting an innovation achieved earlier (in 1709) by Abraham Darby, an ironmaster from Shropshire (Cameron, 1997). Parliament authorized also a series of turnpike trusts that markedly improved the nation's road system. Expenses for construction and maintenance of these facilities were met by a series of tolls paid by users. The system was administered by locally constituted trustees, individuals with most to gain from its success. From 3400 miles in 1750, the road network grew to 15,000 in 1770, reaching a peak of 22,000 in 1836 by which time railroads were beginning to supplant roads as the least-cost method for overland transport of goods (Cameron, 1997).

All of these factors contributed to England's economic resurgence, as did the emergence of a large class of people with funds sufficient to allow them to purchase at least a portion of the goods produced by the country's growing economy. As we shall discuss later, however, it was developments in the textile industry that would eventually spark the transition from what economists refer to as the proto-industrial period in England to the modern age of mechanized industrialization.

A summary of the key dates and key events in the history recounted here is presented in Table 2.1.

Table 2.1 Chronology: 14,000 BP–1750 AD

14,000 BP	Hunter-gatherers migrate from Africa to Middle Eastern region known as the Fertile Crescent.
12,000 BP	Tool use evolves during the beginning of the Neolithic Age.
11,800 BP	Climatic event known as the Younger Dryas produces cold, adverse, climatic conditions in the Fertile Crescent; triggers development of agriculture and animal husbandry replacing hunter/gatherer lifestyle.
11,400 BP	Younger Dryas cold period ends.
8400 BP	Another cold climatic episode prompts migration toward the delta of Euphrates River.
3500 BC	Conversion of copper and tin to bronze ushers in the Bronze Age.
3100 BC	Menes unites Upper and Lower Egypt.
2900–2371 BC	The Early Dynastic Period of Sumeria; development of an impressive, complex civilization.
2371 BC	Akkadians conquer Sumeria; first Mesopotamian empire established.
~2200 BC	Global aridification event results in famine and the collapse of Akkadian and Egyptian empires.
~2000–1150 BC	Minoan civilization flourishes in Crete as an industrial and cultural Center.
1550–1075 BC	Egypt recovers regional dominance under the New Kingdom.
1400–1200 BC	Mycenaean civilization flourishes as a major naval power.
750 BC	Greece returns to dominance as a collection of *poleis*.
~750 BC	Rome founded; Etruscans develop the basis for future Roman infrastructure.
559 BC	Expansion of the Persian Empire results in a series of Greek wars.
~500–200 BC	Period of great intellectual achievement in Greece.
510 BC	Last Etruscan king of Rome is expelled; Roman Republic forms.
~500 BC	Hindu mathematicians develop forerunner of Arabic numerical system.
336 BC	Alexander the Great succeeds Philip as king of Macedonia.
334–324 BC	Alexander the Great conquers Egypt and Persia, proceeds into Pakistan.
324 BC	Alexander the Great dies; Macedonian empire splinters.
287 BC	Formation of plebeian assembly transfers some political power to Commoners.
338–202 BC	Rome subdues most of the Mediterranean region.
44 BC	Julius Caesar is assassinated.
27 BC	Augustus rises to power; Rome prospers during Golden Age.
324 AD	Roman power shifts to Constantinople; Western Rome declines.
394 AD	Theodosius embraces Christianity as the official religion of Rome.
410 AD	Rome is sacked by Visigoths.
486 AD	Clovis consolidates the Frankish tribes.
527–565 AD	Byzantine Empire reaches peak of prosperity under Justinian.
570–632 AD	Life and travels of Mohammed inspire the birth of Islam.
~600 AD	Vikings begin to implement sail technologies in their rise to military power.
732 AD	Arab advance into France is stopped by Charles Martel at Poitiers.
753 AD	Frankish king grants Pope civil control over Papal States.

Table 2.1 continued

800 AD	Charlemagne is crowned Emperor of the Holy Roman Empire; Manorial system creates hierarchical political structure
845 AD	Viking raids escalate; Paris is sacked.
863 AD	Leaders of western and eastern churches excommunicate each other, resulting in a long-lasting religious schism.
911 AD	Vikings settle in Normandy.
1066 AD	Anglo-Saxons defeat Vikings in England; English society evolves hereafter with mixed cultural influences.
1085 AD	Christian military forces recover classical texts preserved by Arabs.
1099 AD	First Crusade—Christians capture Jerusalem.
1187 AD	Jerusalem returns to Muslim control.
1215 AD	Magna Carta restricts previously unchallenged monarchial power in England.
1453 AD	Ottoman Turks capture Constantinople, signaling fall of Byzantine Empire.
~1300 AD	Urban centers flourish in Italy.
1347 AD	Black Death appears in Europe and begins to devastate populations.
1338–1453 AD	Hundred Years' War between England and France complicates existing social challenges of famine and disease.
1460 AD	Cape Verde Islands are discovered; slave labor is introduced into Portuguese society.
1410–1460 AD	Henry the Navigator pioneers Portugal into the Age of Exploration.
1492 AD	Christopher Columbus sets sail for Americas.
1492 AD	Spanish Inquisition under Ferdinand and Isabella.
1497 AD	Vasco da Gama discovers sea route to India; results in major trade profits for Portugal.
1517 AD	Martin Luther publishes 95 Theses in protest of corrupt practices within the Catholic Church; initiates Protestant Reformation.
1519 AD	Hernando Cortes invades Mexico; Spain begins to amass wealth from the New World.
1523 AD	Ulrich Zwingli receives political endorsement for Protestantism in Zurich, Switzerland.
1540 AD	John Calvin leads Reformation in Geneva.
1534 AD	Henry VIII declares independence of Anglican Church.
1543 AD	Iron smelting is developed; England gains technological advantage in European arms race.
1562 AD	First of several religious wars break out between Catholic and Protestant groups in France.
1571 AD	Spanish navy routs Turks at Lepanto; Spain achieves naval superiority in Mediterranean.
1588 AD	English navy defeats Spanish Armada; England supplants Spain as dominant naval power.
1593 AD	Protestant provinces in Netherlands succeed in driving out Spanish influence.
1501 AD	Dutch emerge as prominent European traders.
1602 AD	Dutch East India Company is formed.
1624 AD	Dutch West India Company establishes its first American settlement.
1648 AD	Peace of Westphalia; Dutch achieve total independence from Spain.
1651 AD	England enacts protectionist Navigation Acts; mercantilism thrives.
1664 AD	Pieter Stuyvesant seizes Dutch settlement, establishes New York.
1688 AD	Glorious Revolution in England; monarchy is limited further by Parliamentary authority.
1694 AD	Bank of England formed; bolsters English capital structure.
1750 AD	England consolidates road and canal networks.

2.19 RETROSPECTIVE

We began this chapter with a discussion of the origin of agriculture and con-cluded with an account of the role agriculture played in the run up to the modern industrial age in England. For most of human history, production of food was the most important human activity, requiring the greatest single com-mitment of human labor (indeed it remains so today in many of the poorer parts of the world, notably in sub-Saharan Africa). Domestication of animals, in particular oxen and later horses, provided humans with a useful early ancil-lary source of labor (and fertilizer) but of course these animals had to be fed in order to be able to work. If 100% of the population of a region was engaged in producing food, there was no scope for investment in other activities (assum-ing that the region was not a net exporter of excess food production) and little opportunity for a society to advance either socially or economically. As noted above, the fraction of the English population employed in agriculture dropped from about 60% in 1700 to less than 10% in 1900. In the United States today, only 2% of the workforce is engaged in agriculture and despite this the country is a net exporter of agricultural produce.

The success of the Mesopotamian, Egyptian, Indus, and early Chinese civi-lizations was due in no small measure to the improved efficiency of their irri-gation-based agriculture as compared to the less productive rain-fed farming favored by earlier societies. But to realize the advantages of the irrigation system, these societies had to develop complex social structures to support the wide range of essential related activities—those involved for example in the construction, management, and maintenance of their irrigation works. Along the way, the Sumerians developed a system of standards for measurement, introduced the world's first known written language, and recorded a series of important advances in the manufacture of textiles and the production of a vari-ety of metallurgical goods. Similar successes were enjoyed by the other great river basin-based civilizations. Agricultural surpluses—the ability of an agri-cultural workforce to produce food sufficient to feed not only the immediate families of those directly engaged in the farm work but also a significant number of others—were what allowed all of this to function. These surpluses could not have been realized, however, absent a supportive infrastructure, and vice versa.

The productivity of an agricultural system depends not only on the skills of its workers, on the richness of the natural environment and on a cooperative human and technical infrastructure (draft animals, iron-forged farm imple-ments, or, today, access to fossil sources of energy to drive tractors and harvest-ers or to pump water from underground aquifers), it depends also on a compatible, supportive, climate regime. Too much, or too little, water can pose a problem. As we saw, a change in climate, evidently global in scale, led, 4200 years ago, to an apparently simultaneous demise of the irrigation-based civili-zations in Mesopotamia, Egypt, and the Indus Valley. There is a natural ten-dency for human populations to expand in good times. When the good times

end, especially if they end abruptly, the impact is likely to be all the more serious. And this is not just a story from antiquity. A few bad years for crops in
Europe in the early part of the fourteenth century resulted in the widespread
starvation of the Great Famine. Closer to present, a blight on the potato crop in
Ireland in the late 1840s took the lives of close to a million people forcing a
comparable number to emigrate in order to survive: the population of Ireland
fell from a little more than eight million in 1841 to less than six million
in 1861.

The term "industrial revolution" is used commonly to refer to the series of
developments responsible for the increase in the efficiency with which consumer goods were produced in England in the nineteenth century. These
included specifically: the introduction of mechanically driven machines (powered initially by running water) as substitutes for human labor in the textile
industry; James Watt's improvements in the efficiency of the steam engine and
the variety of applications of steam power developed by Watt and others subsequently; and the introduction of the factory-based method of production.
Landes (1998) defines the factory as "a unified unit of production (workers
brought together under supervision), using a central, typically inanimate
source of power," contrasting the factory with what he termed the "manufac-
tory" which operates in the absence of central power. Cameron (1997) considers the term "industrial revolution" a misnomer, arguing that the changes that
occurred in the nineteenth century were not "merely industrial, but social and
intellectual as well. Indeed, they were also commercial, financial, agricultural,
and even political." He concludes that the most fundamental changes were
intrinsically intellectual. The advances that revolutionized production of goods
in England in the nineteenth century, and almost simultaneously elsewhere
(notably in the United States), should be viewed in their larger historical context, the culmination of forces set in motion many centuries earlier. The pace
of change certainly picked up, but there was no singular event, in Cameron's
opinion, to justify the accolade "revolution." The point may be to some extent
semantic. There is no doubt that the changes that took place in the nineteenth
century, and those that ensued later (a story to be told in subsequent chapters
of this book), markedly and permanently altered the course of human history.
In an earlier age, humans had to rely on personal muscle power, on the muscle
power of animals, or on power harnessed from running water or wind, to do
work. These power sources have been supplanted to a large extent in our
modern world, at least in the so-called developed parts of it, by energy extracted
from coal, oil, natural gas, and the atom. These new sources of power and
energy, while conferring important benefits, have not been, as we shall see,
entirely cost-free.

The story of human history as recounted in the preceding has been dominated by a number of powerful recurrent forces. Societies that were successful
invariably had access to adequate supplies of energy and essential resources.
Those that won out did so often by pillaging, by stealing from their neighbors
either their goods or by exploiting their human capital as slave labor. This was

true in Mesopotamia, and later in Egypt, Greece, Persia, and Rome, and closer to modern times it was the rule also on the sugar and cotton plantations of the Caribbean and the American South. In a sense, we have progressed little from the days of our hunter/gatherer ancestors. If we see it and we need it and we can't trade for it, we move all too often to simply take it. Invariably, military might has been essential for success. Societies that were successful historically had to invest large fractions of their wealth to preserve what they had. This was the story of Rome. The Empire grew by accretion. Wealth acquired by conquest was essential to Rome's success. Eventually, though, the Empire grew so large that it could no longer afford the resources necessary to pay the fighters needed to defend its borders. It collapsed ultimately in a state of anarchy. Outsiders became insiders and the rule of Rome was replaced by the ascendancy of the Germanic Franks. The eastern empire survived longer but ultimately it too would succumb, in this case to the Ottoman Turks. Nation states fought nation states in Europe even after the larger society succeeded in repelling the forces of Islam. Throughout this long history, religion has been an important organizing force.

Religion provided legitimacy for leaders, considered in many early societies as critical intermediaries to the gods, in others as gods themselves, as was the case for example with the pharaohs of Egypt. Great monuments were erected to honor or appease the deity, requiring, often, inordinate expenditures of scarce economic resources. This was true not only in antiquity but also more recently—consider the great Gothic cathedrals of the Late Middle Ages. The influence of religion has not always been positive, even under circumstances where the underlying message may have been constructive. On too many occasions, religious belief has been used as a sword to punish those who would dissent. It was invoked to justify the atrocities of the Crusades and later the barbarities of the Inquisition, despite the injunction of the Second Great Christian Commandment that "you shall love your neighbor as yourself." And it is used today by zealots in the Islamic world to justify killing innocents in the name of Jihad and a quick assured path to heaven.

The problem, of course, is probably not with religion itself but rather with those who would subvert it, knowingly or unknowingly, for invidious purpose (although at times one has to wonder). Human societies have enjoyed greatest success when individuals have been free to think for themselves without fear of retribution, at least so long as their thoughts or actions did not adversely affect the lives of their compatriots. The intrusion of Church authorities on the world of science in the seventeenth century, notably the condemnation of Galileo in 1633 for his perceived heretical views as to the place of the earth relative to the sun in the solar system, set back the course of intellectual enquiry and technological progress in southern regions of Europe for several hundred years. More than three hundred and fifty years would elapse before the Church would admit (in 1992) that it had erred in its treatment of Galileo. The Reformation provided increased license for unfettered intellectual enquiry, so long as it did not infringe on matters or faith or morals. It was not an accident, perhaps, that modern science took root first in northern regions of Europe where the influence of the Catholic Church was least pervasive—in France, Holland, Germany, and most notably in England. It may be more of a stretch to argue, though, as did Weber (1958), that the Protestant ethic with its emphasis on self-discipline and hard work was the critical factor that sparked the modern age of capital-intensive industry.

Bernstein (2004) suggests that four conditions are required for a society to get rich in the modern sense of the word. He attributes it all to technological innovation. For societies to innovate, innovators must be assured first of an ability to profit from the fruits of their innovation. Second, the society must be committed to scientific rationalism. By this he meant a commitment to the scientific method where new ideas are subjected routinely to empirical test, where "honest intellectual inquiry [does not] place life and property at grave risk from the forces of state and religious tyranny." Third, he argues, innovators must have access to capital to allow their inventions to move from concept to practice. And fourth, he suggests, is the need for an efficient communication and transportation infrastructure to allow the products of innovation to be transferred to the greatest possible number of potential consumers able and willing to pay to enjoy their benefits. All of these factors were uniquely present in nineteenth century England. To Bernstein's list we might add the importance of a secure energy source to power the emergent technology and a supportive climate. England had coal.

Today we are hooked for better or worse, not only on coal but also on oil and gas and vulnerable to twin threats—uncertain supplies and the danger of potentially irreversible changes not just in regional but also in global climate induced by emission of the products of these fuels, most notably the greenhouse agent carbon dioxide. And we must confront these challenges in a world where the human population has never been greater, where distances have contracted, and where an ever-greater number of people aspire legitimately (and often forcefully) to share the benefits of the good life. Fluctuations in climate that interrupted the progress of civilizations in the past were for the most part natural in origin. The changes in climate we confront today, however, are for the most part manmade, a product of our modern fossil fuel intensive industrial economy. The choices we make for our energy resources in the future will be linked inextricably to the future of the global climate system and thus inevitably to the future prospects for the human enterprise. We have attempted here and in the pages that follow to provide some perspective on the origins of our current dilemma. As George Santayana famously remarked, those that forget their history are fated to repeat the mistakes of the past. But we should heed also the advice of Samuel Taylor Coleridge: If men could learn from history, what lessons it might teach us! But passion and party blind our eyes, and the light which experience gives is a lantern on the stern, which shines only on the waves behind us!

REFERENCES

Bar-Yosef, O. and Belfer-Cohen, A. (1992). From foraging to farming in the Mediterranean Levant. In: *Transitions to Agriculture in Prehistory* (eds. A.B. Gebauer and T.D. Price), pp. 21–48. Madison, WI: Prehistory Press.

Bar-Yosef, O. (1998). On the Nature of Transitions: The Middle to Upper Paleolithic and the Neolithic Revolution. *Cambridge Archaeological Journal*, 8(2), 141–163.

Bentaleb, I., Caratini, C., Fontugne, M., Marie-Therese, Morzadec-Kerfourn, Pascal, J.-P., and Tissot, C. (1997). Monsoon regime variations during the Late Holocene in Southwestern India. In: *Third Millennium B.C Climate Change and Old World Collapse* (eds. H. N. Dalfes, G. Kukla, and H. Weiss), pp. 193–244. NATO ASI Series no.49, Springer, Berlin.

Bernstein, W.J. (2004). *The Birth of Plenty: How the Prosperity of the Modern World Was Created.* New York: McGraw Hill.

Bond, G. et al. (1997). A Pervasive Millennial-Scale Cycle in North Atlantic Holocene and Glacial Climates. *Science*, **278**, 1257.

Cameron, R. (1997). *A Concise Economic History of the World: From Paleolithic Times to the Present.* Third Edition. New York: Oxford University Press.

Cullen, H.M., P.B. deMenocal, S. Hemming, G. Hemming, F.H. Brown, T. Guilderson, and Sirocko, F. (2000) Climate change and the collapse of the Akkadian Empire: Evidence from the deep-sea, *Geology*, *28* (4), 379–382.

deMenocal, P.B. (2001). Cultural responses to climate change during the Late Holocene. *Science*, **292**, 667–673.

deMenocal, P., Ortiz, J., Guilderson, T., and Sarnthein, M. (2000). Coherent High- and Low-Latitude Climate Variability During the Holocene Warm Period, *Science*, **288**, 2198.

Diamond, J. (1997). *Guns, Germs and Steel: The Fates of Human Societies.* New York: W.W. Norton and Company.

Diamond, J. (2005). *Collapse: How Societies Choose to Fail or Succeed.* New York: Viking Press.

Frumkin, A., Carmi, I., Zak, I., and Margaritz, M. (1994). Middle Holocene environmental change determined from salt caves of Mount Sedom, Israel. In: *Late Quaternary Chronology and Paleoclimates of the Eastern Mediterranean* (eds. O. Bar-Yosef and R. Kra), pp. 315–332. Tucson: Radiocarbon, University of Arizona.

Gasse, F. (2000). Hydrological changes in the African Tropics since the Last Glacial Maximum. *Quaternary Science Reviews*, **19**, 189–212.

Gasse, F. and van Campo, E. (1994). Abrupt post-Glacial events in West Asia and north Africa monsoon domains. *Earth and Planetary Science Letters*, **126**, 435–456.

Hassan, Fekri, (2001) http://www.bbc.co.uk/history/ancient/egyptians/apocalypse_egypt_03.shtml

Johnson, T.C. and Odada, E.O. (1996). *The Limnology, Climatology and Paleoclimatology of the East African Lakes.* New York: Gordon and Breach.

Kagan, D., Ozment, S., and Turner, F.M. (2002). *The Western Heritage*, Brief edition, Combined volume, Third Edition. Englewood Cliffs, NJ: Prentice Hall.

Lamberg-Karlovsky, C.C. and Sabloff, J.A. (1995). *Ancient Civilizations. The Near East and Mesopotamia.* Second edition. Long Grove, Illinois: Waveland Press, Inc.

Landes, D.S. (1998). *The Wealth and Poverty of Nations: Why Some are so Rich and Some are so Poor.* New York: W.W. Norton.

Lemcke, G. and Sturm, M. (1997). δ^{18} O and trace element measurements as proxy for the reconstruction of climate changes at Lake Van (Turkey): preliminary results. In: *Third Millennium B.C. Climate Change and Old World Collapse* (eds. H.N. Dalfes, G. Kukla, and H. Weiss), pp. 193–244. NATO ASI Series no.49, Berlin: Springer.

Magnusson, M. and Palsson, H. (1965). *The Vinland Sagas: The Norse Discovery of America*. New York: Penguin Books.

Maddison, A. (2001). *The World Economy: A Millennial Perspective*. Paris: OECD.

McElroy, M.B. (2002). *The Atmospheric Environment: Effects of Human Activity*. Princeton: Princeton University Press.

Moore, A. and Hillman, G. (1992). The Pleistocene to Holocene transition and human economy in Southwest Asia: the impact of the Younger Dryas. *American Antiquity*, 57, 482–494.

Perlin, J. (1989). *A Forest Journey: The Role of Wood in the Development of Civilization*. Cambridge: Harvard University Press.

Weber, M. (1958). *The Protestant Ethic and the Spirit of Capitalism*. New York: Charles Scribner's Sons.

Weiss, H. (2001). Beyond the Younger Dryas: collapse as adaptation to abrupt climate change in ancient West Asia and the Eastern Mediterranean. In: *Environmental Disaster and the Archaeology of Human Response*, Maxwell Museum of Anthropology, Anthropological Papers No. 7, 75–98.

3

Energy: What Is It and How Do We Measure It?

3.1 BASIC CONSIDERATIONS

Think of the number of times you have occasion in everyday life to refer to the word energy. I have no energy: I'm tired. The United States has an energy problem. The world faces a looming energy crisis. We need an energy program to protect us from uncertain supplies of foreign oil. We need to reduce our use of fossil sources of energy to protect us from the perils of global warming. The solution is to switch from fossil to renewable forms of energy. Maybe solar energy is the answer. Or perhaps it is nuclear power, or energy from the wind. Is nuclear power really a form of energy? Is there a distinction between power and energy or are they really two sides of the same coin? What exactly do we mean by energy and power?

Since the concepts are clearly important it would be useful to have a convention with which to measure them. How much energy do we need? How much power? What are the units we should use to express these quantities?

We might look to the dictionary for answers. Webster's New World Dictionary of the American Language offers five options for energy. Only one, however, comes close to the mark—option 5. Prefaced with an italicized "in physics," it defines energy as the capacity for doing work and overcoming resistance. But what do we mean by work, or resistance? For power, we find no fewer than fifteen choices. Closest to the mark in this case is option 6: the capacity to exert physical force or energy, usually in terms of the rate or results of its use. We need to look elsewhere for more precise definitions. Clearly we can do better.

A useful point of departure is to define what we mean by force. Force (write it as F) is a quantity that causes an object of mass m to undergo an acceleration of magnitude a in the direction of the force. Force and acceleration are vector quantities (we denote vector quantities here by bold face symbols): that is to say, to completely specify them one needs to define both their magnitude and their direction. The relationship is expressed in terms of Newton's Law of Motion (named in honor of its discoverer, one of the world's all time great scientists, England's Sir Isaac Newton, 1642–1727):

$$F = ma \tag{3.1}$$

Acceleration identifies the rate of change of the velocity of the object—the change in velocity per unit of time (velocity is also a vector). Velocity in turn is a measure of the distance moved by the object in a particular direction in unit time. To associate a number with velocity, we need a system of units with which to measure both distance and time. The standard in general scientific use today is Le Systeme Internationale, SI for short, which came into common practice first officially in October 1960. In this system, distance is expressed in meters (abbreviated as m), time in seconds (s).

As originally introduced, the meter was identified as a certain fraction (one-tenth of a millionth or 10^{-7}) of the distance around one-quarter of the Earth's circumference. This was good enough a hundred years ago but it is wholly inadequate for modern scientific or engineering analyses today (not to mention the problem that the Earth is not a perfect sphere; its equatorial radius is greater than its polar radius by more than 22 km, 22,000 m). The second was defined originally as a specified fraction of the length of a day. We know now that the length of the day is not constant; it can vary measurably in response to changes in the motion of the atmosphere and ocean (winds and currents). The standard for the meter is defined today with respect to the distance traversed by light in a vacuum over a well-defined (very small) unit of time. The second is referenced to the frequency of radiation emitted by an isotope of cesium, cesium-133.

Mass is a measure of the amount of material contained in an object. The unit of mass in the SI system is the kilogram, defined ultimately with respect to a specific physical object, a platinum–iridium cylinder maintained in a laboratory in Sevres, a suburb of Paris. We have a tendency in everyday language to confuse the concepts of mass and weight. Suppose you were to order a kg of potatoes at the supermarket. To fill the order, the supermarket typically might place the potatoes on a scale and observe the displacement. The scale, however, is measuring not mass but the response of the scale to the pull of gravity exerted on the mass by the Earth. If the potatoes were weighed with the same scale on the moon (where the strength of gravity is 6 times less than on the Earth) the scale would indicate 167 g rather than a kg. For most purposes, we will have little reason to refer directly to these basic standards. We may be content to rely on a good ruler to measure length, a good watch to measure time, and a high quality calibrated scale to record mass.

An acceleration of 1 unit in the SI system implies a situation in which the velocity of an object increases by 1 meter per second every second. Using mathematical shorthand, we write this as $(1\ ms^{-1})\ s^{-1}$ or $1\ ms^{-2}$. Equation (3.1) can now be used to define the unit of force in the SI system. A mass of 1 kg undergoing an acceleration of $1\ ms^{-2}$ is consistent with a force of 1 SI unit. Expressed in terms of the basic units of the SI system, m, kg, and s (mks), force in the SI system has dimensions of $kg\ ms^{-2}$. Rather than simply repeating these units, it is convenient to give the unit of force a name. We refer to the unit of force in the SI system as the Newton (N), honoring again the aforementioned Isaac Newton. A force of 1 N is sufficient to impart an acceleration of magnitude $1\ ms^{-2}$ to a mass of 1 kg.

Energy is expressed in a variety of different forms. A moving mass possesses what we refer to as kinetic energy. The kinetic energy (K) of a mass m moving at speed v (speed defines the magnitude of the velocity v) is given by:

$$K = 1/2\ mv^2. \tag{3.2}$$

Equation (3.2) can be used to define the basic unit for kinetic energy, indeed for energy more broadly in the SI system. A mass of 1 kg moving at a speed of 1 meter per second $(1\ ms^{-1})$ is defined (in the SI system) to possess an amount of kinetic energy equal to 1 joule (J), a unit named in honor of the English physicist James Joule (1818–1889).

If you throw a ball up in the air, or bat a baseball, you use muscle energy to impart initial kinetic energy to the ball. As the ball rises, its speed and consequently its kinetic energy decrease. Its kinetic energy is converted to what we refer to as gravitational potential energy. Conversion of kinetic to gravitational potential energy is complete when the ball reaches the apex or high point of its trajectory. As it returns to the ground, its kinetic energy increases and its potential energy decreases accordingly. When the ball reaches the ground, the kinetic energy it possesses is about the same as when it started up, but not quite. As a consequence of the resistance of the air to its motion, a small part of the initial energy is converted to heat in the atmosphere. Heat is also a form of energy. Energy, in toto, is conserved: the sum of the kinetic energy of the ball when it returns to the ground combined with the energy converted to heat in the atmosphere is the same as the kinetic energy the ball had when it left the hand or bat (ignoring for the moment the small fraction of the energy that may have been used additionally to heat or compress the ball).

As the ball rises, its upward motion is opposed by the force of gravity. We say that the ball must do work against the force of gravity in this case. Kinetic energy is expended in the process. As the ball returns to the ground, the force of gravity does work on the ball leading to an increase in its kinetic energy. Work, in the scientific context, is defined as the product of the distance an object moves, D, in the direction of the applied force times the magnitude of the force responsible for its displacement:

$$W = DF \tag{3.3}$$

If the force is applied in the same direction as the displacement, the value of F is positive: W is positive in this case. The work supplied is used to increase the kinetic energy of the mass. If the force is applied in a direction opposing the motion of the mass, F is negative and the mass must do work to achieve a given displacement: W is negative and kinetic energy is depleted accordingly.

The increase in (gravitational) potential energy as the ball rises is a measure of kinetic energy lost temporarily due to work done against the force of gravity, energy that can be recovered as the ball returns to the ground. Energy and work are exchangeable commodities. They have equivalent dimensions: in the SI system, both are measured in units of joules. We began this chapter looking for a definition of energy. The concept of work provides an answer. The energy of a system can be defined as the capacity of the system to do work. The kinetic energy of running water can supply the force needed to turn the paddles of a water wheel. Or, the potential energy of water at the top of a dam, converted to kinetic energy as the water falls, can supply pressure to turn the blades of a turbine and generate electricity (another important form of energy). Energy is expended to supply oxygen to our lungs and pump blood through our veins. It is used to heat our houses in winter and cool them in summer and to drive the machines that make our every day lives easier. For the most part this energy is derived ultimately, primarily, from the sun, degraded through a variety of work-supplying transformations, returned eventually to space, as we shall see, in the form of long wave radiation. The solar energy we use can be harnessed, either immediately—as biomass, wind, or running water—or recovered from the past stored as chemical potential energy in the form of coal, oil, and gas.

To increase the temperature of air or water or the temperature of any material system, we need to supply energy. Conversely, if the temperature of a system is observed to decrease, energy must be removed (transferred elsewhere). To raise the temperature of a gram of water (one thousandth of a liter or 1 milliliter) by one degree centigrade (at a temperature of $15°C$) requires an input of energy equal to 4.184 J. In everyday language, this unit of energy is referred to as a calorie (cal). A second unit in common use is the British Thermal Unit or Btu. A Btu defines the energy needed to raise the temperature of a pound of water (1 pound defines a mass of 0.454 kg) by one degree Fahrenheit (Americans continue to quote masses in pounds and temperature in degrees $°F$: the British use pounds for mass but have switched to the centigrade system when referring to temperature). To convert from one system of energy units to another, we need to apply appropriate scaling factors: 1 cal = 4.184 J = 3.968×10^{-3} Btu; 1 Btu = 1055 J = 252 cal.

Power defines the quantity of energy delivered in unit time (to your house for example). The unit of power in the SI system corresponds to delivery of energy at a rate of 1 J per second (1 Js^{-1}). This unit is known as the watt (W), named in honor of James Watt (1736–1819), the Scottish inventor/scientist credited with the perfection of the steam engine. To confuse matters further, if you check your electricity bill you will find that the energy for which you are charged is quoted in units of kilowatt-hours (kWh). A kilowatt-hour denotes the quantity of energy consumed in an hour at a power level of 1 kilowatt. The energy corresponding to a kilowatt hour is given thus by multiplying the energy delivered per second (the power in kW) by the number of seconds in an hour, $1 \text{ kWh} = 9.48 \times 10^{-1} \times 3.6 \times 10^3 \text{ Btu} = 3.41 \times 10^3 \text{ Btu}$.

Watt, in marketing his steam engine, found it convenient to introduce a unit that could give his potential customers a measure of the rate at which his machines could put out work. Prior to the invention of the steam engine, much of the work humans relied on was provided by animals, either horses or oxen. Watt introduced horsepower as a unit of power designed to approximate the rate at which a healthy horse could do work over a typical 8-h working day: 1 horsepower = 0.746 kW. The 4.8 l V-8 engine on a BMW 6 series car is rated at 360 horsepower. Just imagine: 360 horses working full out at your service for 8 h a day! And you can harness equivalent power simply by feeding your BMW a steady diet of gasoline!

If you check your (natural) gas bill (at least in the United States) you will find that you are charged in this case in units of therms. One therm corresponds to a supply of gas with an energy content of 10^5 Btu, delivered in a volume of 100 cft at a pressure of 14.73 lb per sq. in. at 60°F, just a little higher than the pressure of the atmosphere at sea level, 14.70 lb per sq. in. Pressure measures the force applied to a unit area of surface (in a direction normal to the orientation of the surface). In the SI system, it has dimensions of Nm^{-2}. The unit of pressure in the SI system is referred to as the Pascal (Pa) named in honor of the French mathematician and philosopher Blaise Pascal (1623–1662). Expressed in the fundamental units of the SI system, pressure has dimensions of $\text{kg m}^{-1}\text{s}^{-2}$. A pressure of 1 Pa would impart a force of 1 N to a surface area of 1 m^2. Engineers prefer to express pressure in units of pounds per square inch (a distinctly British tradition!): 1 psi corresponds to the force applied by a mass of one pound weighing on a surface of area one square inch. The fact that the pressure of the gas supplied to our homes is slightly higher than the pressure of the atmosphere ensures that when you turn on the gas on your stove the gas will emerge, but not so fast as to cause an unacceptable build-up of combustible methane in your kitchen. A summary of conversion factors for different energy units is presented in Table 3.1.

Given our newfound skills in converting from one set of energy units to another, it may be instructive to apply this expertise to an analysis of the bills we receive from the utility companies that supply our home needs for gas, electricity, and oil. The bill for gas delivered to my house in Cambridge, Massachusetts, in January 2006 reflected a total charge per therm of $1.70261, of which $1.40165 was attributed to the cost of supply with the additional $0.30096 assessed as the cost of distribution. By way of comparison, the spot price for gas in early February 2006 was $0.7735 per therm, having declined from a high of close to $1.5 per therm in mid-December of 2005. The difference between the cost of supply indicated on my January bill and the price on

Table 3.1 Conversion factors for different energy units. To convert from one of the units listed in the left vertical column to one of the units listed in the horizontal, multiply by the appropriate entry in the table. For example: 1 J = 9.48 × 10⁻⁴ Btu.

Unit	Joule	Calorie	Btu	Kilowatt hour	Therm	Quad
Joule	1	2.39×10^{-1}	9.48×10^{-4}	2.88×10^{-7}	9.48×10^{-9}	9.48×10^{-19}
Calorie	4.18	1	3.97×10^{-3}	6.89×10^{-8}	3.97×10^{-8}	3.97×10^{-18}
Btu	1.06×10^{3}	2.52×10^{2}	1	2.93×10^{-4}	10^{-5}	10^{-15}
Kilowatt hour	3.60×10^{6}	8.60×10^{5}	3.41×10^{3}	1	3.41×10^{-2}	3.41×10^{-12}
Therm	1.06×10^{8}	2.52×10^{7}	10^{5}	2.93×10^{1}	1	10^{-10}
Quad	1.06×10^{18}	2.52×10^{17}	10^{15}	2.93×10^{11}	10^{10}	1

the spot market reflects presumably the fact that the supplier, NSTAR in this case, was obliged to commit to the higher price that prevailed earlier in order to ensure continuity of supply in the face of an extremely volatile market for gas in the wake of the damage caused by the hurricanes that impacted supplies from the Gulf of Mexico in late August and September of 2005. If there is any justice, my bill for February should be significantly lower than what I had to pay for January! It was!

The charge on my January 2006 bill for electricity came to $0.19 per kWh of which $0.12 per kWh reflected the cost for the original source of the electricity with the balance attributed to a variety of charges for transmission and distribution, subsidies for renewable energy, and costs imposed on the supplier to encourage conservation by consumers. Expressed in monetary units and converting from kWh to equivalent numbers of therms, my electricity cost $5.57 per therm, more than three times what I paid on an equal energy basis for my gas. Why the discrepancy? The difference reflects to a large extent the relatively low efficiency with which the energy of a fuel such as gas, coal, or oil is converted to electricity, typically less than 40%, and the high cost of the capital invested in the power plant in which my electricity was generated.

My home is heated with a gas-fired hot water system. I am obviously much better off heating my house with gas than with electricity. Would I have been better off if my house was heated with oil? The spot price for home heating oil in the United States in mid-January 2006 averaged about $1.50 per gallon. For oil delivered to my house I would have paid about $2.00 per gallon (the difference between wholesale and retail). This issue is addressed in Box 3.1. The rapid decrease in spot gas prices for gas in the U.S., by almost a factor of two from mid-December 2005 to mid-February 2006, reflected the influence of unusually mild temperatures in the U.S. over this period and the impact energy market participants judged that this would have on the future demand for gas.

Box 3.1

As indicated above, I paid $1.70 per therm for gas delivered to my house in Cambridge in January 2006.

Solution:

The spot market for home heating oil in the United States in January 2006 averaged about $1.50 a gallon. The (retail) price for oil delivered to private homes averaged about $2.00 a gallon. To answer this question we need to estimate the costs per therm for gas as compared to the cost per therm for oil.

The energy content of a gallon of heating oil is equal to 1.39×10^5 Btu. The cost per therm may be calculated as follows:

$$\text{Energy content of 1 gallon} = 1.39 \times 10^5 \text{ Btu} = 1.39 \text{ therm}$$
$$\text{Price per therm} = \$2.00/1.39 = \$1.44$$

I would have been better off using oil in January but would not be advised to shift given the expectation that gas prices should drop significantly in February (gas prices have been more volatile recently than prices for home heating oil which tend to track international prices for crude oil: gas prices in contrast tend to reflect to a greater extent local issues of supply and demand in the United States). As indicated above, they did.

3.2 ADDITIONAL PERSPECTIVES

The conventional unit used to measure commercial energy is the Btu (except of course for the units favored by my electricity supplier). National energy statistics are quoted usually in units of Quads: a Quad corresponds to 10^{15} or a quadrillion Btu (see Table 3.1). Annual consumption of energy in the U.S. in 2002 amounted to 98 Quad. Global consumption for the same year was at a level of 412 Quad. The U.S., with approximately 5% of the world's population, accounts for close to 24% of the world's total commercial energy consumption. On a per capita basis, with a population of 296 million, our annual consumption amounts to 330 million Btu as compared to 64 million Btu per capita for the world as a whole. If we assumed that all of our U.S. energy was supplied in the form of home heating oil with an energy content of 1.39×10^5 Btu per gallon (of course we also use coal and natural gas and oil to generate electricity and for a variety of other purposes such as heating, cooking, and driving), this would imply that we would consume every year in the U.S.—every man, woman, and child of us—close to 2300 gallons of oil. At $2 a gallon, the bill for a typical family of four would come to $18,400. Since home heating oil is significantly cheaper on an energy basis than electricity, and most other sources of energy we consume (gas and gasoline for example), the expense for a typical U.S. family would actually be much higher than this.

The daily nutritional requirement for a healthy human, expressed in energy terms, corresponds to about 40 cal per gram of body mass (Odum and

Barrett, 2005)[1]. A representative 70 kg (154 lb) adult over the course of a year would require therefore close to 4×10^6 Btu in the form of ingested food energy. Allowing for the nutritional requirements of animal intermediaries, this number should be multiplied by about a factor of 4 to account for the overall dietary requirements of a representative American adult (the animals we rely on also have to eat!). The energy consumed annually per adult in the form of food corresponds in this case to 1.6×10^7 Btu, equivalent to approximately 5% of the energy employed commercially to fuel all other functions of our society. The former is supplied of course mainly by photosynthesis (see Chapter 4). The latter depends largely on fossil sources—coal, oil, and natural gas—with modest additional contributions from biomass, hydro, wind, and nuclear.

How much food (how many kilograms?) do we need per year to meet our nutritional requirements? The energy content of the food we ingest is equal to about 5 kilocalories (5×10^3 cal) per gram of dry matter (slightly less than this for carbohydrates, a little more for protein). The quantity of food consumed over the course of a year by a 70 kg (155 lb) adult amounts therefore to about 200 kg, almost three times the individual's biomass. The total quantity of dry food mass required to satisfy the annual nutritional needs of a 70 kg adult American (allowing for implicit animal intermediaries) adds up therefore to about 800 kg. To put this in context, the net primary productivity (NPP), of cultivated land (what it produces that we can harvest) is equal to about 0.65 kgm^{-2}year^{-1} (cf. Chapter 4). Assuming that half of this NPP is available as food for either humans or animals (an optimistic assumption: the actual number is probably much less), it follows that we need at least 2.5×10^3 m^2 or 0.6 acres of farmland (cultivated plus pasture), to supply the nutritional needs of an average adult American (requirements for individuals in other developed countries are not much different). This amounts to approximately 2 acres for a typical family of four allowing for the slightly lower food demands of the children. The computations underlying these considerations are summarized in Box 3.2.

Box 3.2

How much land is required to supply the nutritional needs of a typical American adult?

Solution:
Assume that the quantity of food consumed by an individual over a year is equal to 200 kg.

Allowing for animal intermediaries, the annual requirement for dry food mass to feed an individual is equal to 800 kg.

Assume annual production of useful biomass equal to 0.32 kg m^{-2}.

Land required to feed the typical adult is equal thus to 2.5×10^3 m^2 = 0.6 acres.

Feeding a typical family of four would require a little less than 2.4 acres, perhaps 2 acres given the lower nutritional demand for children.

3.3 HOW MUCH ENERGY DO WE RECEIVE FROM THE SUN?

The energy from the sun reaches the Earth in the form of light, or more technically speaking electromagnetic radiation. What is the origin of this light (radiation) from the sun?

I remember, as a child, my parents took me to visit a blacksmith's forge. The blacksmith was busy forging a piece of iron into a shoe for a horse. He put the piece of iron into his forge and used a bellows to fan the flames of his fire. As the temperature of the fire increased (a consequence of the enhanced rate of combustion facilitated by the additional oxygen supplied by the bellows), the iron began to glow. Initially it was red hot and eventually its color changed to white. Only then was the iron sufficiently malleable for the blacksmith to be able to fashion the horseshoe. I learned later when I took my first serious course in physics that what I was witnessing was the fact that an object (the piece of iron in this case) emits energy in the form of light such that as the temperature increases the wavelength corresponding to the peak emission of this light shifts steadily to shorter wavelengths. Initially the light radiated by the iron was invisible to my eye (our eyes are sensitive mainly to light in the so-called visible portion of the electromagnetic spectrum). The wavelength of light in the red portion of the spectrum is longer than the wavelength in the green or blue. As the temperature of the iron increased, the wavelength of peak emission shifted gradually from (invisible) infrared, to red, to green, incorporating eventually some blue. By this time, I was viewing a mix of colors my eyes interpreted as white (a mixture of the primary colors, red, green, and blue). I know now that not only did the wavelength of the light shift to shorter wavelengths as the temperature of the iron increased but I understand that this increase was accompanied by a significant increase in the energy content of the light reaching my eyeball. In fact the radiant energy emitted by an object varies as the fourth power of its temperature.

The light that arrives at the Earth from the sun in the visible portion of the electromagnetic spectrum originates from a region of the solar atmosphere where the temperature is significantly higher than the temperature of the fire in the blacksmith's forge, close to 5600°C as it happens. What is the source of the energy in the light emitted by the sun?

The energy the sun emits as light is produced by nuclear reactions in the hot, high pressure, central region of the sun known as the core. Albert Einstein established that there is a basic equivalence between mass and energy. The essence of this equivalence is captured in his famous $E = mc^2$ equation where c denotes the velocity of light (most people have heard of this equation but few are aware of its implications). Temperatures near the center of the sun reach values as high as 15 million °C with pressures as much as 300 billion times greater than that of our atmosphere at sea level. Under these conditions, protons (the nuclei of hydrogen atoms) are squeezed so close together that they can combine to form heavier nuclei, initially deuterium (^2H), with associated release of vast quantities of energy (physicists refer to this process as nuclear fusion). Subsequent nuclear reactions result in net conversion of four protons to an alpha particle, the nucleus of the helium atom with atomic mass number 4 (^{4}He). The mass of a helium nucleus (known also as an alpha particle) is slightly less than the combined mass represented by the four protons from which it is formed, by about 0.7%. This mass deficit is converted to energy, accounting for the dominant source of the fuel that keeps our sun shining.

The energy emitted by the sun, largely in the form of visible light, emanates from a region of the solar atmosphere known as the photosphere. The temperature of the sun decreases steadily from the core to the photosphere, a necessary condition to ensure that the energy liberated in the core can make its way eventually to the outside (energy flows from regions that are hot to those that are relatively cool). The sun is an omni-directional light beacon—there is no preferred direction to its luminosity. Little of the energy it emits is lost, however, in the comparatively empty expanse of interplanetary space. If you imagine surrounding the sun with a series of imaginary spheres (with the sun at the center), the energy crossing the surface of any one of these spheres in unit time will be the same as the energy crossing any other (irrespective of the radius of the sphere). The energy crossing unit area of the surface of a particular sphere will decrease, however, in proportion to the increase in the area of its surface, i.e. as the square of the distance from the sun. By the time the light reaches the orbit of the Earth, the energy crossing unit area in unit time, referred to as the energy flux, has dropped to a level of about 1.37×10^3 Wm^{-2} (1370 Wm^{-2}) as compared with the flux of 6.31×10^7 Wm^{-2} emitted at the photosphere. The flux of solar energy crossing unit area in unit time at the Earth's mean distance from the sun is known as the solar constant.

The Earth offers a large (circular) target to intercept energy from the sun. Think of aiming a bright lamp from one side of a large (dark) room at a small spherical object on the other side of the room (a globe for example). A shadow will form on the wall behind the sphere. It will appear as a dark circle with radius equal to the radius of the sphere. The area of the shadow reflects the size of the target presented by the sphere and consequently the fraction of the incident light intercepted by the sphere. The target the Earth offers to incident sunlight is given by πR^2 where R is the radius of the Earth (actually a little larger than the radius of the solid Earth: think of the atmosphere as a thin skin surrounding the Earth, not thick enough though to add more than a fraction of a percent to its radius). Approximately 30% of the sunlight that reaches the Earth is returned to space, reflected from a combination of the atmosphere (6%), clouds (20%), and bright colored regions of the surface (4%). The balance is absorbed by the atmosphere/ocean/surface system where it provides the fuel for photosynthesis, where it is used to evaporate water, to cause the winds to blow, and contribute generally to the hospitable conditions we take for granted for life on our planet.

The energy absorbed by the Earth is distributed over the entire area of the planetary surface, $4\pi R^2$. The energy available per unit surface area per unit time is equal thus to one-quarter of the energy intercepted by the Earth per unit time per unit area normal (perpendicular) to the incident solar beam (the factor of 4 reflecting the difference between the total surface area of the Earth and the area of the circular target offered by the Earth to the incident sunlight). Accounting for the fact that 30% of the incident light is reflected back to space, the solar energy absorbed per unit area per unit time averaged over the entire Earth's surface and over the entire year is equal to about 240 Wm^{-2}. The total quantity of solar energy absorbed by the Earth over the course of a year is equivalent to 3.66 million Quad. By way of comparison, as noted earlier, society on a global scale consumed 460.14 Quad of commercial energy in 2005 (Energy Information Administration, 2007), a little more than 1 part in 10,000 of the total quantity of energy absorbed from the sun. Details of the computations underlying these considerations are presented in Box 3.3.

Box 3.3 Absorption of solar energy by the Earth

Solar constant (energy from the sun crossing unit area in unit time at the Earth's average distance from the sun):

$$S = 1370 \text{ Wm}^{-2}$$

Total rate at which solar energy is intercepted by the Earth:

$$
\begin{aligned}
E &= (\pi R^2)\, S \\
&= \pi\,(6.38 \times 10^6 \text{ m})^2\,(1.37 \times 10^3 \text{ Wm}^{-2}) \\
&= (3.14)(4.07 \times 10^{13} \text{ m}^2)(1.37 \times 10^3 \text{ Wm}^{-2}) \\
&= 1.75 \times 10^{17} \text{ W}
\end{aligned}
$$

Total rate at which solar energy is absorbed by the Earth:

$$A = (0.7)E = 1.23 \times 10^{17} \text{ W}$$

Distributing this energy over the entire surface area of the Earth implies a rate of energy supply per unit area equal to

$$
\begin{aligned}
R &= \frac{A}{4\pi R^2} \\
&= \frac{(1.23 \times 10^{17} \text{ W})}{4\pi\,(4.07 \times 10^{13} \text{ m}^2)} \\
&= \frac{(1.23 \times 10^{17} \text{ W})}{(5.11 \times 10^{14} \text{ m}^2)} \\
&= 2.41 \times 10^2 \text{ Wm}^{-2}
\end{aligned}
$$

continued

Box 3.3 Absorption of solar energy by the Earth (*continued*)

The quantity of solar energy absorbed by the Earth per second is given by:

$$A = 1.23 \times 10^{17} \text{ W} = 1.23 \times 10^{17} \text{ Js}^{-1}$$

It follows that the energy absorbed by the Earth per year, Y, is given by:

$$
\begin{aligned}
Y &= (0.7)(1.75 \times 10^{17} \text{ W})(3.15 \times 10^{7} \text{ s year}^{-1}) \\
&= 1.23 \times 10^{17} \text{ W} (3.15 \times 10^{7} \text{ sec}) \\
&= 3.87 \times 10^{24} \text{ J} \\
&= 3.67 \times 10^{21} \text{ Btu}
\end{aligned}
$$

Energy, Z, absorbed per unit area per year is given by:

$$
\begin{aligned}
Z &= \frac{3.67 \times 10^{21} \text{ Btu}}{5.11 \times 10^{14} \text{ m}^2} \\
&= 7.2 \times 10^{6} \text{ Btu m}^{-2}
\end{aligned}
$$

As indicated above, 70% of the energy incident on the Earth from the sun is absorbed by the Earth. The balance is reflected back to space. Of the energy that is incident, 20% is captured directly by the atmosphere, by a combination of H_2O (water vapor), O_3 (ozone), dust, and clouds, with the balance (50%) absorbed at the surface. Approximately 48% of the energy absorbed at the surface (24% of the incident solar energy) is used to evaporate water, mainly from the ocean, while the rest, 52% of the total energy absorbed at the surface, is converted to heat (12%) or emitted in the form of radiation at infrared wavelengths (40%). A diagrammatic illustration of the overall energy balance of the Earth in response to incident solar radiation is illustrated in Fig. 3.1.

Approximately 1.26×10^{21} Btu of incident energy per year is used to evaporate water, mainly from the ocean (24% of the energy contained in the incident solar radiation or 34% of the total energy absorbed by the Earth, see Fig. 3.1). Significant quantities of energy are required to change the phase of water, to transform the substance from its liquid form to vapor or gas. The energy required to evaporate 1 kg of liquid water is equal to 2.5×10^{6} J or 2.4×10^{3} Btu. It follows that the total quantity of water evaporated per year as a consequence of the absorption of solar radiation, roughly equivalent to the quantity that falls on the surface in the form of either rain or snow, is equal to about 6×10^{17} kg or a little more than 1170 kg of liquid water per m^{-2} averaged over the entire surface area of the Earth, corresponding to about 117 cm, or 46 in. of rain per year. Obviously, there are large areas of the Earth where it does not rain, the deserts for example, or high latitudes where precipitation is minimal and occurs

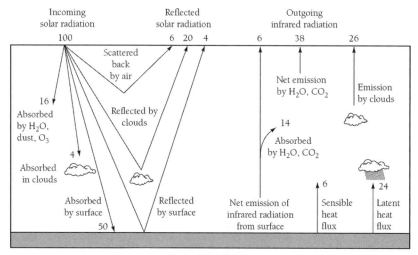

Figure 3.1 Earth's radiation balance (from McElroy, 2002).

largely in the form of snow. The average annual precipitation inferred here may be compared with an average annual precipitation in the Northeastern United States of about 45 in. or about 10 in. in the desert Southwest. Precipitation rates for a number of selected locations around the globe are presented in Table 3.2.

The flux of solar energy incident at any given position on the Earth's surface depends on its latitude and on the time of year (the combination of which determine the elevation of the sun in the sky at any given time of the day, a determining factor influencing the flux of sunlight illuminating unit area of the surface). It is sensitive also to cloud cover (more clouds mean that a larger fraction of the incident sunlight is reflected, thus less is transmitted to the surface). The average flux of solar radiation illuminating unit area of surface in the Northeastern United States varies from a minimum of 0.5×10^4 Btu m^{-2} per day in November–January to a maximum of 2.1×10^4 Btu m^{-2} per day in June and July. By way of comparison, the flux reaching the surface in the Southwest ranges from a minimum of 1.0×10^4 Btu m^{-2} per day in December to peaks of close to 2.8×10^4 Btu m^{-2} per day in June and July (Reifsnyder and Lull, 1965; Odum and Barrett, 2005). Averaged over an entire year, the flux of solar energy reaching unit area of surface in the Northeast is about 45% lower than that reaching the surface in the Southwest: 4.6×10^6 Btu m^{-2} as compared to 7.1×10^6 Btu m^{-2} (Reifsnyder and Lull, 1965; Odum and Barrett, 2005). To put this in perspective: as noted earlier, annual demand for energy in the United States amounts to 98 Quad (9.8×10^{16} Btu) serving a population of 296 million,

Table 3.2 Annual average temperatures and precipitation for selected
           locations (Pidwirny, 2006)

Location	Annual average temperature (°C)	Annual average precipitation (cm)
Barrow USA (72°N)	−13	11
Berbera, Somalia (10.5°N)	30	5
Buenos Aires, Argentina (34.5°S)	16	103
Calcutta, India (22.5°N)	27	158
Darwin, Australia (12.5°S)	28	156
Denver, USA (40°N)	10	39
London, England (51.5°N)	10	54
Los Angeles, USA (34°N)	18	37
Monterrey, Mexico (26°N)	22	61

corresponding to average annual per capita energy consumption of 3.3×10^8 Btu. Suppose we were to try to satisfy per capita energy demand by capturing sunlight with an efficiency of 1%. How much land would we require to meet this demand? The answer for the Northeastern United States is 7.2×10^3 m^2 or 1.8 acres per person (1 acre is equal to 4.047×10^3 m^2). For the Southwest the requirement would be a more modest 4.6×10^3 m^2 or 1.1 acres per person.

3.4 SUMMARY

We introduced here a variety of physical quantities. Force, velocity, and acceleration are examples of what we refer to as vectors. To specify these quantities we need to define not only their magnitude but also their direction. Mass, energy, work, and power are examples of scalars. A scalar is specified completely by a number, its magnitude in appropriate units.

The SI system identifies a set of units in which length is expressed in meters, mass in kilograms, time in seconds. In this convention, force is measured in newtons, energy and work in joules, power in watts. Most scientific studies are conducted using the SI system. Engineers, however, favor a more motley selection of units and we need to be careful doing calculations with these units to ensure consistency.

Our primary interest here is in energy. The unit in most common use for commercial energy is the British thermal unit or Btu. Energy consumption on a national or international level involves large numbers of Btu. The unit in common use in this case is the Quad (10^{15} Btu). Energy consumption in the U.S. in 2001 amounted to 98 Quad corresponding to per capita use of 3.3×10^8 Btu. Consumption of natural gas is quoted usually in units of therms (10^5 Btu). Electrical energy is quoted in units of kilowatt hours (kWh) defining the energy supplied over an interval of an hour at a power level of 1 kilowatt. Power is a measure of the rate at which energy is supplied (joules per second in the SI system). The power of most engines, notably automobile engines, is quoted in units of horsepower: 1 horsepower = 0.746 kW.

Crude oil is traded in barrels (1 barrel = 42 gallons), heating oil No.2 and unleaded gasoline in gallons, with natural gas quoted in MMBtu (1 MMBtu = 10^6 Btu = 10 therm). Spot prices of these commodities, as quoted from Bloomberg, the Energy Information Agency, the Intercontinental Exchange, and the Official Nebraska State Website as of October 28, 2008, are presented in Table 3.3. For ease of comparison, the table includes also prices quoted in units of dollars per MMBtu.

NOTE

1 Calories as quoted here refer to conventional physical units (1 cal defines the energy required to raise the temperature of a gram of water by 1°C at 15°C). The energy content of food is often quoted misleadingly in units referred to as calories. We need to remember that 1 food calorie is actually equal to 1 kilocalorie in real energy units.

Table 3.3 Energy commodity spot prices

Commodity (Location)	Price in Dollars per trading unit			
	WTI crude oil (Cushing)	Heating oil No. 2 (New York Harbor)*	Natural gas (Henry hub)**	Electricity (Mid-Columbia, day ahead on-peak)***
Unit	$/Barrel	$/Gallon	$/MMBtu	$/MWh
July 14, 2008	145.16	4.0331	11.58	85.67
October 28, 2008	62.73	2.148	6.36	51.27

Commodity (Location)	Price in Dollars per MMBtu			
	WTI crude oil (Cushing)	Heating oil No. 2 (New York Harbor)*	Natural gas (Henry hub)**	Electricity (Mid-Columbia, day ahead on-peak)***
Unit	$/MMBtu	$/MMBtu	$/MMBtu	$/MMBtu
July 14, 2008	25.03	29.12	11.58	25.11
October 28, 2008	10.82	15.51	6.36	15.03

Source: EIA, Bloomberg, Official Nebraska Government Website, Intercontinental Exchange
  *Peaked at $4.0833 on July 3, 2008
   **Peaked at $13.31 on July 2, 2008
  ***Peaked at $92.21 on July 16, 2008

REFERENCES

Energy information Administration (2007). Annual Energy Review http://www.eia.doe.gov/emeu/aer/inter.html (read May 09, 2009).

McElroy, M.B. (2002). *The Atmospheric Environment: Effects of Human Activity*. Princeton: Princeton University Press.

Odum, E.P. and Barrett, G.W. (2005), *Fundamentals of Ecology*. Fifth Edition. Thomson Brooks/Cole.

Pidwirny, Michael. 2006. "Temperature". Encyclopedia of Earth, Editor Cutler J. Cleveland. http://www.eoearth.org/article/Temperature. Topic Editor, Budikova, Dagmar. Environmental Information Coalition (EIC) of the National Council for Science and the Environment (NCSE), Washington D.C., USA.

Reifsnyder, W.E. and Lull, H.W. (1965). Radiant energy in relation to forests. Technical Bulletin Number 1344. United States Department of Agriculture and Forest Service, Washington, D.C.

4

Wood, Photosynthesis, and the Carbon Cycle

4.1 INTRODUCTION

As discussed earlier, access to wood played a critical role in the success of civilizations over much of human history. It served not only as a critical energy resource but also as source for essential raw materials employed in a myriad of applications: in the construction of buildings and bridges; in the fabrication of boats and wheeled vehicles (chariots, carts); as masts for sailing ships; even as track for some of the early railroads. Arguably, though, the most important use of wood in the preindustrial modern world, as we noted, was as a source of charcoal consumed in vast quantities in the smelting of ores such as copper, iron, and tin. Charcoal continued to enjoy this status for much of the world until 1709 when Abraham Darby succeeded in England in developing an alternative—coke produced from coal. Darby's innovation was the development of a procedure to remove impurities such as sulfur from coal that otherwise impeded the smelting process. By that time, however, England and much of continental Europe had largely depleted their forest resources and wood, as discussed in Chapter 2, was in critical short supply.

The energy content of wood is derived ultimately from the sun. As discussed in Section 4.2, sunlight absorbed by green plants through photosynthesis serves to convert carbon from the low energy form of CO_2 in the atmosphere to the higher energy form of carbon incorporated into the tissue of what we refer to as biomass. Carbon is present in nature in a variety of chemical forms, from the highly reduced condition of carbon in methane (CH_4) to the higher oxidized condition in CO_2. A simple recipe serves to define the oxidation state of carbon in a particular chemical compound. We assign a number of $+1$ to the hydrogen atoms in the compound; to the oxygen atoms a number of -2. If the compound

---

Box 4.1

Calculate the oxidation states of carbon in CH_4, biomass (CH_2O), and CO_2 and HCO_3^- (the dominant form of C in the ocean).

Solution:

Let x denote the oxidation state of C.

$$CH_4: \quad x + 4(+1) = 0 \Rightarrow x = -4$$
$$CH_2O: x + 2\,(+1) + (-2) = 0 \Rightarrow x = 0$$
$$CO_2: \quad x + 2(-2) = 0 \Rightarrow x = +4$$
$$HCO_3^-\ x + (+1) + 3\,(-2) = -1 \Rightarrow x = +4$$

This accounts for the negative charge in HCO_3^-.

---

is electrically neutral (as is the case both for CH_4 and CO_2), the number assigned to the carbon atoms (or atoms) in the compound should be such that the oxidation number associated with the totality of the atoms in the compound should add up to zero (Box 4.1). The oxidation state (number) for the carbon atom in CH_4 by this recipe is –4; the oxidation state of the carbon atom in CO_2 is +4.

Carbon in compounds characterized by negative oxidation numbers are said to be reduced; elements with positive oxidation numbers are said to be oxidized. The significance of this convention has to do with the nature of the electrostatic forces that serve to bind the atoms together in a particular compound. The carbon atoms in CH_4 are stabilized by borrowing electrons from each of the neighboring hydrogen atoms; we can think of the carbon atom in CH_4 as behaving thus as though it had a net electric charge of –4 (hence the convention for definition of its oxidation state) bound as a consequence strongly to the (temporary) positive charges on the neighboring hydrogen atoms. The carbon atom in CO_2 on the other hand delivers two electrons to each of the component O atoms acquiring in this case an apparent net electric charge of +4 (hence an oxidation state of +4). The CO_2 molecule is stabilized by the strong electrostatic (attractive) forces resulting from the positive charge on the carbon atom interacting with the negative (–2) charges on the neighboring O atoms.

Section 4.2 presents an overview of the global carbon cycle, the complex suite of processes that combine ultimately to determine the abundance of CO_2 in the atmosphere. The bulk of the Earth's reservoir of reduced carbon is present below ground. It was formed by photosynthesis either on land or in the ocean many millions of years ago, buried subsequently in sediments but destined to return to the atmosphere as a consequence of the geological processes responsible for the slow but inexorable recycling of surface materials associated with plate tectonics (Chapter 1). The average carbon atom has cycled between sediments on the one hand and the biosphere/soil/atmosphere/ocean system on the other as many as ten times over the course of Earth history. Carbon in sediments is returned under natural circumstances to the atmosphere either as sediments are uplifted (raised back to the surface) or as they are subducted (withdrawn to the mantle). In the latter case, they are cooked and the carbon they include is transferred to the atmosphere as a component of either hot springs or volcanoes. Mining the carbon contained in subsurface reservoirs of coal, oil, and natural gas serves to accelerate the rate at which carbon is returned to the atmosphere from sediments. As a consequence, humans have evolved now as a dominant influence on the global carbon cycle. It is not surprising in this case that the abundance of CO_2 in the atmosphere is observed to be rising rapidly with concurrent changes in the physical and chemical state of the ocean and the global climate system. We conclude with summary remarks in Section 4.4.

4.2 CAPTURING SOLAR ENERGY BY PHOTOSYNTHESIS

The energy stored in wood is derived ultimately from the sun. Photosynthesis refers to the process by which sunlight is absorbed and used to convert atmospheric carbon dioxide (CO_2) and water (H_2O) into plant material with associated release of molecular oxygen (O_2). The overall reaction may be written as:

$$\text{Sunlight} + CO_2 + H_2O \rightarrow (CH_2O) + O_2 \tag{4.1}$$

where (CH_2O) denotes a carbohydrate compound incorporated in the plant.

The photosynthetic process is not a simple one-step reaction. It proceeds by a complex biochemical pathway. As many as twenty photons (the basic elements of light) are absorbed before a single carbon dioxide molecule can be converted to carbohydrate. Transformation of solar to chemical energy by way of photosynthesis may be considered analogous to the conversion of kinetic to potential energy when, as discussed earlier, a batted or thrown ball rises up in the air. When the ball reaches the apex of its trajectory, the kinetic energy it possessed initially is converted to potential energy. When it returns to the ground, most of this potential energy is transformed back to the kinetic form. So also with photosynthesis: a fraction of the solar energy absorbed in reaction (4.1) is converted to chemical potential energy. This chemical potential energy is stored primarily in the form of the carbohydrate molecules that compose the plant and is released when these molecules react with oxygen, effectively reversing the initial photosynthetic process. The reverse reaction may be summarized as:

$$(CH_2O) + O_2 \rightarrow CO_2 + H_2O + (\text{energy, mainly heat}) \tag{4.2}$$

The energy of the food we eat is derived either directly or indirectly from photosynthesis. If our dietary preference is vegetarian, photosynthesis can satisfy our needs entirely. If on the other hand we like to eat meat, drink milk, or eat cheese we must rely on animals as intermediaries. These animals also have to eat: demands for primary photosynthesis increase accordingly. The chemical potential energy of the food we ingest is released through the oxidation process summarized by reaction (4.2). The energy so released is what keeps us alive and empowers us to carry out the variety of different forms of work we discharge in everyday life. To oxidize the carbon in the food we eat, we breathe in oxygen, react the oxygen with the carbohydrate of the food (left hand side of reaction (4.2)), and respire carbon dioxide with associated production of water (right hand side of reaction (4.2)).

Plants also have to breathe (they inhale oxygen and exhale carbon dioxide) to satisfy their requirements for energy. To survive they must transform the initial products of photosynthesis to other chemical compounds (the chemicals that compose the woody structures of a tree for example). They need to maintain their temperature at an acceptable level (they cool typically by evaporating water). And they must be able to draw nutrients and water from the soil through their root systems to satisfy their requirements for essential elements such as nitrogen and phosphorus. At most a few percent of the solar energy incident on a healthy ecosystem is converted through photosynthesis to the chemical potential energy of biomass. Approximately half of the energy captured by plants growing under cold climate conditions is consumed by respiration. Even more, up to ten times as much, is required to satisfy the respiratory demands of plants growing in warm climates (the thought is that hot-climate plants need to expend more energy in respiration than plants in a warm environment in order to survive and prosper).

It is important to distinguish between gross and net productivity. Gross primary productivity (GPP) refers to the initial utilization of solar energy to convert CO_2 and H_2O to biomass (the quantity of biomass produced as a result of the initial absorption of light). Net primary productivity (NPP) refers to the quantity of organic material that remains after satisfying the plant's requirements for respiration. Estimates of mean NPP for a number of globally important ecosystems are summarized in Table 4.1 (data adapted from Whittaker, 1975). Natural ecosystems include not only plants but also populations of bacteria, animals, fungi, and other organisms (heterotrophs). They also must be fed and the plants, the primary producers, provide the nurture they need to prosper. Accounting for the requirements of the heterotrophs, allowing also for the respiratory demands of the plant and its roots, the net quantity of carbon incorporated in an ecosystem is referred to as net ecosystem productivity (NEP) (this identifies the total quantity of biomass available potentially for sustainable harvest).

As discussed earlier, for much of human history wood was the dominant source of energy people used to cook and to stay warm in winter. It was the primary source of energy (other than human and animal muscle power) employed in the United States until recently. Only over the past 100 years or so has it been supplanted by coal, oil, and gas.

Table 4.1 Estimates of average net primary productivity (NPP) for selected ecosystem types (after Odum and Barrett, 2005)

Ecosystem type	Area (10^6 km^2)	NPP (g m^{-2}year^{-1})
Tropical rainforest	17.0	2200
Tropical seasonal forest	7.5	1600
Temperate evergreen forest	5.0	1300
Temperate deciduous forest	7.0	1200
Boreal forest	12.0	800
Savanna	15.0	900
Temperate grassland	9.0	600
Tundra	8.0	140
Desert and semi-desert	18.0	90
Cultivated land	14.0	650
Total global land	149	773

When first harvested, wood contains a considerable amount of moisture. Even when the wood is well seasoned, the moisture content is still significant, accounting for as much as 20% of the total mass of the wood. When the wood is burnt, energy must be used to evaporate the residual moisture. This requires an expenditure of energy equal to 1050 Btu per pound of water. A realistic estimate of the energy available from well-seasoned wood is about 6050 Btu per pound or 13,367 Btu per kilogram (slightly more for hardwood, less for softwood).

Firewood is sold today in the U.S. in units known as cords. A standard cord consists of a stack of wood measuring 8 ft by 4 ft by 4 ft corresponding to a volume of 128 cft (note again the pervasive use of English units). The volume occupied by the wood ranges from about 60 to 110 cft (the rest is air space). The wood content of a typical cord weighs about 4000 lb or about 1.8 metric tons with the energy content for a well-seasoned standard cord of wood ranging from about 18 to 24 million Btu. For present purposes we shall assume that the energy content of a representative cord of wood is equal to about 20 million Btu. Historical data suggest that consumption of wood in New England in 1800 amounted to about 5.5 cords per person per year. Converting this figure to energy units implies a value of 120 million BTU per capita per year for wood-based energy consumed in New England in 1800 (roughly equivalent to total energy use). This may be compared with per capita annual energy use in the contemporary United States (as indicated above) of about 330 million Btu.

It may come as a surprise that per capita energy use today is only about three times greater than the rate at which energy was used New England in 1800. Think of the impressive quantities of energy we consume today to generate electricity and drive our cars, trucks, boats, and planes and fuel all of the other components of our economy. We are richer today but we use much less energy to produce a specific unit of economic product. How can we account for this circumstance?

Most of the wood consumed in 1800 New England was used to heat New Englanders' homes during the cold New England winter. Winters are comparably cold today but our houses are much better constructed and much better insulated. Also, 1800 New Englanders did not have the benefit of the efficient heating systems we rely on today—gas or oil fired hot water or hot air systems with which up to 80 or 90% of the energy released by burning the fuel is communicated as heat to the interior of our houses. In 1800, New Englanders had to rely on open fires where most of the energy liberated by burning the wood fuel was wasted, exhausted as heat up the chimneys. We need much less energy to heat our homes today (at least on a per unit area basis) than was the case in 1800. We use much more, however, for transportation for example, and we have found a myriad of additional applications for energy that contribute to the much greater affluence of our twenty-first century lifestyles.

Approximately half of the dry mass of wood consists of carbon. Gross annual uptake of carbon by photosynthesis at Harvard Forest (located about 40 miles west of Cambridge in rural Massachusetts) amounts to about 4.5 tons C year^{-1} acre^{-1} (Wofsy, 2006). Only about 25% of the carbon fixed by photosynthesis is retained by the forest (NPP is equal to about 25% of GPP) and of this approximately half is stored in biomass above ground with the balance converted to biomass (mostly root material) below ground (Wofsy, 2006). The carbon available for sustainable harvest amounts thus to $(0.5)(0.25)(4.5)$ tons C year^{-1} acre^{-1} = 0.56 tons C year^{-1} acre^{-1}. To meet 1800 New Englanders' per capita requirements for wood by sustainable forestry (assuming that the NPP of 1800 New England forests was comparable to the NPP observed in Harvard Forest today) would require harvesting annually the product of more than 7 acres to satisfy requirements of a single individual. If we were able to substitute wood for fossil fuel on a one for one basis to meet contemporary per capita requirements for commercial energy (assuming unrealistically that the efficiency with which the energy of wood could be employed to produce a given outcome was comparable to that that could be achieved with currently available mix of fossil sources), we would need the product of close to 24 acres per person. Given the current population of the United States, 296 million, we would need to harvest annually a forested area of more than 7 billion acres (assuming NPP comparable to Harvard Forest). To put this in context, the total area of the continental United States amounts to only about 2.4 billion acres (a little more than 8 acres per person). And of course there are functions of our modern society that could not be supplied with energy furnished by wood—driving our trucks and cars or flying our airplanes for example. The numbers underscore the extent to which we are dependent now on other sources of energy, notably coal, oil, gas, and nuclear.

4.3 THE CARBON CYCLE

As we have seen, carbon plays a critical role in production of wood and other forms of biomass. As CO_2, it provides the fuel for photosynthesis. In chemically reduced form, denoted symbolically above by (CH_2O), it represents a major component of the structural material of all living (and dead) organisms. Carbon is a vital component not only of the atmosphere and living organisms, it is an important constituent also of soils, the ocean, and global sediments (the material deposited at the bottom of the ocean). The quantity of carbon in the atmosphere is comparable to that included in the structural material of all living land-based plants and animals (most of this incorporated in the cellulose and lignin that compose the trunks and branches of trees). The carbon content of soils is almost twice as much as that contained in the totality of all living land-based organisms (we refer to the global aggregate of living organisms as the biosphere).

While the rate of photosynthesis in the ocean is comparable globally to that on land, the biomass maintained by photosynthesis in the ocean is much less than that on land (the lifetime of a typical marine phytoplankton—the base of the marine food chain—is measured in days as compared to the lifetime of terrestrial plants that is measured in months or even centuries—think of the stately, thousand-year-old California Redwoods). Despite the low abundance of organic matter in the ocean, the overall abundance of carbon in the ocean is much greater than that on land—more than 14 times greater than in the composite atmosphere–biosphere–soil system. Carbon is present in the ocean mainly in inorganic form, dissolved in seawater primarily as the bicarbonate ion HCO_3^-. The content of carbon in sediments exceeds that in the entire atmosphere–biosphere–soil–ocean system by more than a factor of 2000. The abundance of carbon in these different compartments of the Earth is illustrated in Fig. 4.1.

The data in Fig. 4.1 are intended to provide a representation of the disposition of carbon under normal, unperturbed, conditions. The quantity of carbon indicated here for the atmosphere reflects the concentration present a few hundred years ago. The abundance of atmospheric carbon, present mainly as CO_2, has increased, however, over the past several centuries, from about 280 parts per million (ppm) to more than 380 ppm today (concentrations are quoted conventionally in terms of the number of CO_2 molecules expressed as a fraction of the total number of molecules—mostly oxygen and nitrogen—included in the atmosphere excluding the variable contribution from H_2O). The increase reflects primarily the extra carbon added to the atmosphere as a consequence of our mining and burning of fossil fuels—coal, oil, and natural gas.

The arrows connecting the boxes in Fig. 4.1 indicate rates at which carbon (absent the recent human-induced disturbance) is moving from one reservoir to another. We can estimate the time carbon resides on average in any particular reservoir by dividing the content of the reservoir by the rate at which the element is either entering or leaving the reservoir. Consider for example how

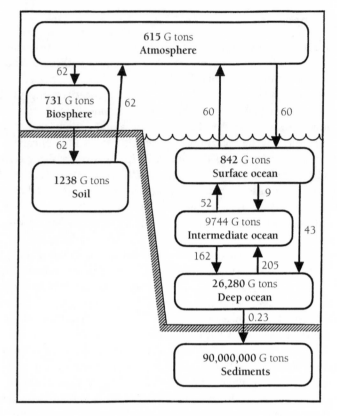

Figure 4.1 Composite model for the global carbon cycle. Reservoir contents are in units of 10^9 tons C; transfer rates are in 10^9 tons C year^{-1}. Carbon is deposited in sediment both as $CaCO_3$ and as organic matter. There is a small release of CO_2 in steady state from the ocean; this source is employed in weathering of crustal rocks (from McElroy, 2002).

we would go about calculating the residence time (referred to commonly as the lifetime) for carbon included in the land-based biosphere. According to the data in Fig. 4.1, the land-based biospheric reservoir contains 731×10^9 tons of carbon supplied by net photosynthesis at a rate of 62×10^9 tons carbon per year. How long would it take photosynthesis to fill the reservoir, ignoring for the moment the fact that carbon is not only being added to the reservoir by photosynthesis but that it is also at the same time being lost to the soil? The answer is 11.8 years. We would obtain the same answer had we asked how long it would take the biospheric reservoir to lose its carbon to the soil in the absence of the photosynthetic source.

Times calculated in this fashion (dividing reservoir contents by either input or output rates) provide an informative estimate of how rapidly reservoirs turn

---

Box 4.2 Lifetimes for carbon in different terrestrial reservoirs

Land-based biosphere:

$$T_1 = \frac{\text{content(tons)}}{\text{turnover(tons year}^{-1})} = \frac{731 \times 10^9}{62 \times 10^9} \text{year} = 11.8$$

Ocean:

$$T_2 = \frac{\text{content(tons)}}{\text{turnover(tons year}^{-1})} = \frac{(842+9744+26{,}280) \times 10^9}{(60+0.23) \times 10^9} \text{year}$$
$$= 612 \text{ year}$$

Entire atmosphere–biosphere–soil–ocean system:

$$T_3 = \frac{(615+731+1238+842+9744+26{,}280) \times 10^9}{(0.23) \times 10^9} \text{year}$$
$$= 171{,}522 \text{ year}$$

Sediments:

$$T_4 = \frac{(90{,}000{,}000) \times 10^9}{(0.23) \times 10^9} \text{year} = 391 \text{ million year}$$

---

over their stock. The lifetime calculated here for carbon in the land-based biosphere gives a reasonable estimate for the average life of the average tree on the Earth—11.8 years according to the present analysis (remember, trees account for the bulk of the carbon included in the land-based above-ground biosphere). Lifetimes for carbon in the ocean, for the entire atmosphere–biosphere–soil–ocean system and for sediments are developed in Box 4.2.

As indicated in Box 4.2, the lifetime of carbon in sediments is measured in hundreds of millions of years. Organic compounds account for about half of the carbon deposited in the sediments (residues of calcareous shells, $CaCO_3$ account for the balance), about 0.12 billion tons per year. The sediments provide, however, only a temporary reservoir for this carbon. Were it not so, all, or almost all, of the world's carbon would end up in sediments, a circumstance that would result in a serious shortfall in meeting the carbon requirements of the global life support system: life has been present on Earth for close to 4 billion years. Sedimentary carbon is cycled back to the atmosphere, thanks to the influence of plate tectonics. The sediments are eventually either uplifted and weathered or drawn down into the mantle. In the latter case the sedimentary material is cooked. The carbon is vented back to the atmosphere as a component of volcanoes and hot springs. Assuming that the carbon cycle has operated more or less as today for the entire 4.6 billion year history of the Earth, we must conclude that the average carbon atom has cycled through sediments and back to the more mobile compartments of the Earth (atmosphere, biosphere, soils, and ocean) more than 10 times over the history of the Earth.

Mining coal, or drilling for oil and gas, may be considered as a human-assisted acceleration of the rate at which carbon is returned to the atmosphere from the sedimentary reservoir. The preceding discussion attests to the significance of this influence. Combustion of fossil fuel on a global basis accounts at the present time for an additional source of CO_2 to the atmosphere–ocean–biosphere–soil system amounting to more than 7 billion tons C per year. The rate at which carbon is transferred from sediments to the atmosphere as a consequence of human activity today (our use of fossil fuel) exceeds thus by more than a factor of 50 the rate at which the same task is accomplished by nature. It should come as no surprise therefore that the abundance of atmospheric CO_2 is increasing rapidly at the present time. The concentration of atmospheric CO_2 is greater now than at any time at least over the past several million years. The increase in CO_2 over the past 50 years is presented in Fig. 4.2. Changes over the past 650,000 years, inferred from analyses of gases trapped in ancient ice in Antarctica, are illustrated in Fig. 4.3.

Given our reliance on fossil sources of organic carbon (energy captured from the sun by photosynthesis hundreds of millions of years ago) it is unlikely

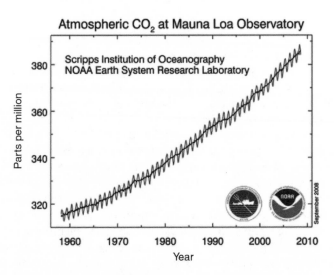

Figure 4.2   Monthly mean atmospheric carbon dioxide (red curve) at Mauna Loa Observatory, Hawaii. The black curve represents the seasonally corrected data. (http://www.esrl.noaa.gov/gmd/ccgg/trends/co2_data_mlo.html, read September 28, 2008).

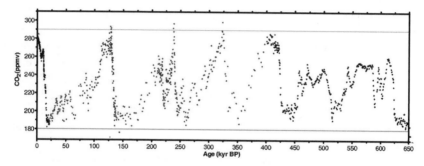

Figure 4.3 A composite CO_2 record back to 650,000 years BP. The record results from the combination of CO_2 data from three Antarctic ice cores: Dome C (black), Vostok (blue), and Taylor Dome (light green). [Siegenthaler et al., 2005].

that the current upward trend in atmospheric CO_2 will be reversed at any time in the near future. The concern about human-induced climate change is directly related to the significance of the contemporary human impact on the global carbon cycle.

4.4 SUMMARY AND CONCLUDING REMARKS

Access to adequate supplies of wood, as we have seen, played a critical role in scripting the success and failure of specific societies over much of human history. It enjoyed a status roughly equivalent to that of oil in the modern world.

Biomass, specifically wood, may be considered as a convenient means to capture and store solar energy for future use. Only a small fraction of the incident solar energy, a few percent at most, is absorbed, however, and converted to biomass in specific ecosystems. Plants, like animals, need to breathe. They have a requirement for energy to supply the work they need to accommodate essential systemic functions. This energy is supplied by oxidation of the organic carbon formed by conversion of CO_2 to the higher potential energy state of organic matter during photosynthesis (gross primary productivity or GPP). The organic carbon that remains in an ecosystem after accounting for respiration is referred to as net primary productivity (NPP). Average values of NPP for specific ecosystems range from a high of 2200 gm^{-2} $year^{-1}$ for tropical rainforests to a low of 90 gm^{-2} $year^{-1}$ for desert and semi-desert environments as summarized in Table 4.1.

Wood was the dominant source of energy in the U.S. up to the end of the nineteenth century. The energy yield from wood harvested sustainably on an annual basis from a New England forest was estimated at 1.4×10^7 Btu per acre (1.1 tons seasoned wood: 1.3×10^7 Btu per ton). To satisfy the energy needs of a typical New Englander in 1800 required harvesting wood produced by approximately 8 acres of forest assuming no net depletion of forest resources (this is what we mean by sustainable). To supply the energy needs of the current population of the U.S. (296 million) at current rates of energy consumption, we would need 7 billion acres of forest with NPP comparable to that of Harvard Forest, close to three times the total available land in the country (2.4 billion acres). And of course not all of this land could accommodate the magnitude of NPP realized at Harvard Forest (much of the western and southwestern parts of the country are desert or semi-desert in the absence of irrigation).

In principle, we could burn wood, or biomass more generally, to satisfy our requirements for electricity. And wood could be used to produce fuels, ethanol for example, which could substitute in part at least for current consumption of oil in the transportation sector. This would be accomplished, however, at the expense of a significant reduction in the efficiency of energy use. Demands for production of biomass would increase accordingly.

The economy of the U.S. is fueled today largely by fossil sources of energy—coal, oil, and natural gas—with additional contributions, specifically in the electricity sector, from nuclear, hydro, and wind. In mining fossil fuels, we tap solar energy captured by the biosphere hundreds of millions of years ago. In doing so, however, we accelerate significantly the rate at which organic carbon would be returned under natural conditions (absent the human influence) from sediments to the atmosphere (by a factor of more than 50 at the current time). Not surprisingly, the concentration of atmospheric CO_2 is increasing rapidly at the present time as illustrated in Fig. 4.2. It is greater now than at any time over the past 650,000 years (Fig. 4.3) and the trend is unlikely to reverse in the foreseeable future. Unless we act promptly to reduce our reliance on fossil fuels, or sequester the additional carbon (capture it before it is released to the atmosphere and deposit it in some long-lived reservoir, the deep ocean, or the sediments for example), we will be forced inevitably to confront the challenge of a very different climate—global warming, with uncertain implications for future global prosperity.

REFERENCES

Odum, E.P. and Barrett, G.W. (2005). *Fundamentals of Ecology*, Fifth Edition. Thomson Brooks/Cole.

Siegenthaler, U. et al. (2005). Stable carbon cycle–climate relationship during the late Pleistocene. *Science*, **310** (5752), 1313–1317. DOI: 10.1126/science.1120130.

Whittaker, R.H. (1975). *Communities and Ecosystems*, Second Edition. New York: Macmillan Publishing Company.

Wofsy, S.C. (2006). Private communication.

5
Coal: Origin, History, and Problems

5.1 INTRODUCTION

Global (commercial) consumption of energy increased by 82% between 1973 and 2004 (IEA, 2006). The proportion of total world energy derived from coal remained relatively constant over this period, at a little less than 25%. The fraction of global energy consumption contributed by oil in 2004 was actually less than in 1973, 34% as compared to 45% (IEA, 2006). Greatest gains between 1973 and 2004 were registered by natural gas and nuclear power, the fractional contributions of which increased from 16% to 21% and from 0.9% to 6.5% respectively (IEA, 2006).

China is responsible for most of the recent growth in global coal combustion. It accounted for 18.7% of total world production in 1973. By 2005, its fractional contribution had risen to 44.8% (IEA, 2006). Over the same period, production of coal in OECD countries (the political grouping including 30 of the world's more developed economies, excluding however, countries in the former Soviet Union, Africa, Latin America, the Middle East, China, and less developed regions of Europe and Asia), expressed again as a fraction of the global total, declined from 50% to 29.6%.

Coal accounted for 34.9% of world industrial production of CO_2 in 1973, rising to 40% in 2004 responsible for a net increase in emissions over this period of 95% reflecting the overall increase in coal consumption. The contribution of oil to CO_2 emissions as a fraction of total fossil fuel use declined from 50.7% in 1973 to 39.9% in 2004. Given the increase in global oil consumption over this period, however, related emissions of CO_2 rose by 34% (IEA, 2006).

Trends in coal production for different regions of world from 1972 to 2005 are illustrated in Fig. 5.1. Pie charts summarizing the breakdown of the

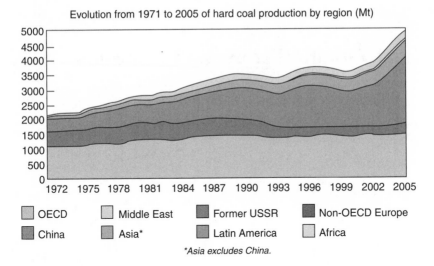

Evolution from 1971 to 2005 of hard coal production by region (Mt)

☐ OECD	☐ Middle East	■ Former USSR	■ Non-OECD Europe
■ China	■ Asia*	■ Latin America	☐ Africa

*Asia excludes China.

Figure 5.1 From IEA (2006).

different components for 1973 and 2005 are presented in Fig. 5.2. Note the rapid recent growth in production (and consumption) of coal in China. Rates for production of coal in the U.S. and China were comparable in the late 1990s. Production in China now exceeds that in the U.S. by more than a factor of 2. China and the United States together accounted for 64% of the total quantity of coal mined world wide in 2005: 2200 million metric tons (Mt) in the case of the former, 951 Mt for the latter (IEA, 2006).

Recoverable coal reserves in the U.S.—by this we mean coal that can be mined using current technology—are estimated to be large enough to maintain present levels of consumption for several hundred years. China's recoverable reserves are thought to be a little less than those of the U.S. There is little danger, however, that either country will soon run out of coal. The available coal resources in both cases exceed by more than a factor of 10 quantities currently defined as recoverable. With advances in mining technology, there is no doubt that current levels of output can be maintained in both countries for the foreseeable future. And additional, abundant, reserves are available elsewhere—comparable to those of China—notably in the combined coal-rich regions of Russia, Ukraine, and Kazakhstan. If potential world supplies of coal were to be exploited to their maximum extent in the future, associated release of carbon would be sufficient to more than double, possibly triple, current levels of CO_2 in the atmosphere. The concentration of atmospheric CO_2 would increase in this case to levels not seen since dinosaurs roamed the surface of the Earth 50 million years ago!

Worldwide (recoverable) coal reserves are estimated at about 10^6 Mt. Assuming an energy content per metric ton of 23 million Btu, a little less than the average energy content of a metric ton of U.S. coal (about 25 million Btu) but more than the average energy content of a metric ton of lower grade

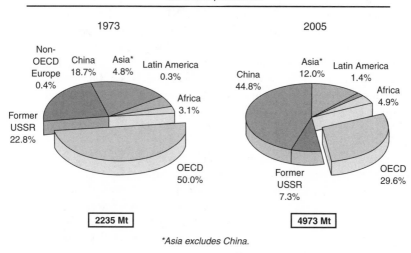

1973 and 2005 regional shares of
hard coal production

Figure 5.2 From IEA (2006).

Chinese coal (about 21 million Btu), we conclude that coal over time could provide an energy source as large as 2.3×10^{19} Btu or 2.3×10^5 Quad. Given present global annual energy consumption, 460 Quad in 2005 as discussed earlier, it follows that coal could accommodate current levels of worldwide total energy demand for at least 500 years.

We begin in Section 5.2 with an account of the origin of coal, a description of the different types of coal and their different energy contents. Section 5.3 presents a brief history of coal use—what Freese (2003) describes in the title to one of the chapters of her book as the Rise and Fall of King Coal. Environmental consequences of coal use are discussed in Section 5.4. Summary remarks are presented in Section 5.5.

5.2 ORIGIN AND ENERGY CONTENTS OF DIFFERENT TYPES OF COAL

Most of the world's coal was formed during the geological era known as the Carboniferous Period, approximately 300 million years ago. The climate of the Carboniferous was not very different from that of the most recent geological epoch, the Pleistocene/Holocene (the past few million years). That is to say, it was characterized by intermittent ice ages punctuated by periods of comparative warmth. The landmasses of the Earth were arranged very differently, however, during the Carboniferous Period as compared to the present. South America, Africa, India, Australia, and Antarctica and a few sundry extra pieces were combined to form a super continent, Gondwanaland, located near the South Pole. North America, Europe, much of Asia, and Indonesia were located closer to the equator (Fig. 1.1), a circumstance that was important, as we shall see, in setting the stage for where the major coal deposits are to be found today.

The plants that captured the solar energy we mine today as coal grew for the most part on coastal margins of the tropical and subtropical landmasses of the Carboniferous Period, most probably in tidal, deltaic, environments. Freese paints an imaginative picture of the landscape. Think of peculiar looking trees growing to heights of as much as 175 ft with trunks up to 6 ft in diameter at their base covered in what she describes as "a beautiful lizard-skin bark," a feature that accounts for their name (lepid means scale in Greek). Think of a few short branches at the top with "narrow leaves up to a yard long." Think of other trees known as sigillaria with "long trunks forking near the very top like a two-headed monster, each head crowned with a large spray of strap-like leaves."

Imagine these plants growing in swampy wetlands. As the plants died, the material of which they were composed fell into the waters of the swamps where insufficient supplies of oxygen inhibited decomposition of their residues. Changes in climate—transitions between Carboniferous warm periods and ice ages—would have radically altered these environments (sea level during the most recent ice age was approximately 110 m lower than today). Plant material would have been buried subsequently, covered by sediments derived from increased erosion of continental rocks. Rivers which followed a meandering path to the sea during periods when topographic relief was relatively low (when climate was warm and sea level high) would have been transformed to fast moving torrents carrying higher loads of land-based sediment to the ocean burying the plant material that accumulated in the coastal swamps that existed earlier. With onset of a new interglacial warm period, sea level would have risen again, coastal swamps would have been reestablished, trees would have regrown, and organic sediments would have again accumulated, a sequence would have repeated many times over millions of years. Reflecting this history, coal is found today often in discrete beds with vertical thicknesses ranging from as little as a few feet to as great as 25 ft or more sandwiched between layers of mineral-rich, organic-poor sediment.

Tectonic processes played an important role subsequently in reshaping the landscape, pushing material from the surface deep into the Earth in some regions, thrusting it to the skies in others forming chains of high mountains. The crustal plate supporting the European landmass converged on the plate supporting North America resulting in the appearance of the mountain chain we know now as the Appalachians. When first formed, the Appalachian Mountains were probably as high as the Himalayan Mountains today. Over time they weathered and were eroded. But they include still some of the world's richest reserves of coal, legacies of the plant materials that grew earlier on the coastal soils and rocks of the Carboniferous North American landmass. Similar geologic (tectonic) processes were involved in shaping the conditions for coal formation in Europe and Asia and elsewhere, the environments where we find the world's richest reserves of coal today.

The first step in coal production involved deposition of a relatively thick layer of plant residue forming peat. Bacterial activity in the anaerobic environment of the swamp resulted in production and release of methane. Since the methane molecule contains four hydrogen atoms for every carbon atom, the relative abundance of carbon in the peat was increased accordingly. Peat is a relatively low-density, water-rich, material. To form a 1 m thick seam of coal requires approximately 20 or 30 m of peat. The transition between layers of coal and overlying sediment is often very sharp suggesting that changes in the swamp environment played an important role in terminating the peat forming process. These changes could be triggered either by climatically induced changes in sea level as discussed above (associated with formation and collapse of continental ice sheets on landmasses at high latitude) or by tectonically induced changes in local and regional topography, or by a combination of both.

Subjected to an increasing burden of overlying sediment, peat was compacted. Its water content was reduced accordingly (water was squeezed out of the interstices of the peat). Gradually it was transformed into the slightly higher energy content material we identify as low-grade brown coal or lignite. Exposed to increasingly higher temperatures and pressures, lignite turned into sub-bituminous coal, and with subsequent processing to bituminous coal and eventually to the highest grade of coal, anthracite. The fractional carbon and energy contents of the coal increased moving along this sequence.

Lignite is brownish black in color with a relatively high component of water, as much as 45%. Its energy content, computed on a moisture/mineral free basis ranges from about 9 to 17 million Btu per metric ton. Sub-bituminous coal is dull dark brown to black in color with an energy content of between 17 and 24 million Btu per metric ton (computed again on a moisture/mineral-free basis). The Powder River Basin in Wyoming is a major contemporary source of sub-bituminous coal in the U.S. The most abundant source of U.S. coal is bituminous with a moisture content of less than 20% with an energy content of between 21 and 30 Btu per metric ton. Anthracite is high in carbon, low in volatile gases, with a moisture content of less than 15% and an energy content of 22 to 28 Btu per metric ton. Anthracite is hard to ignite but once lit it burns; it gives off impressive heat with relatively low emission of smoke. When England moved to ban use of coal in households a few decades ago (see below) it permitted at least in some regions use of what was referred to as smokeless fuel—anthracite. A summary of the properties of different coal types is presented in Table 5.1.

Table 5.1 Properties of different coal types

Type	Energy content (Btu per metric ton) (mineral/moisture-free basis)	Moisture	Color
Lignite	9–17	up to 45%	Brownish black
Sub-bituminous	17–24	20–30%	Dark brown to black
Bituminous	21–30	<20%	Black
Anthracite	22–28	<15%	Black, lustrous

5.3 HISTORY OF COAL USE

We are familiar with the role coal played in energizing the modern industrial state in England, a process that began with Thomas Newcomen's delivery to the Earl of Dudley's colliery of the first full-scale, coal-fired, steam engine in 1712 AD. Less well known is the role coal played in fueling an industrial revolution in China almost eight hundred years earlier.

The Northern Song Dynasty was established in China in 960 AD when Zhao Kuangyin was acclaimed by his troops as the New Emperor. The capital of the Northern Song was located in Kaifeng near the junction of the Grand Canal and the Yellow River with access to resources (food and materials) that could be drawn from an extensive region of Northern China taking advantage of the comparative ease of water-borne transportation. Fairbank (1992) estimates that the population of Kaifeng living within the walls of Kaifeng numbered close to 500,000 in 1021 AD. By 1100 AD it had increased to 1.4 million (including the army). To put this in context, Milan, the most populous city in Europe at the end of the thirteenth century, had a population of about 200,000: the population of London numbered no more than 40,000 a hundred years later in 1377 AD (Cameron, 1989).

As discussed earlier, for much of history charcoal produced from wood was the critical ingredient determining the success or failure of human societies. The early Song had access to abundant sources of iron and wood and exploited these resources to the full to develop a flourishing metallurgical industry. By 1000 AD, the wood resources of the region were largely depleted. Chinese artisans developed, however, the means to produce coke from coal with purity sufficient to allow coal-based coke to replace charcoal as the fuel that could be used to smelt iron ore. Almost seven hundred years would elapse before Abraham Darby would accomplish a similar feat in Shropshire in England. By 1078 AD, Fairbank (1992) reports, Song ironworkers were producing more than 114,000 tons of pig iron annually—twice what England would realize 700 years later.

The Romans were reputably the first to use coal as a fuel in England. Freese (2003), quoting a history of England written by Saint Bede the Venerable in 731 AD, recounts that use of coal as a fuel in England ended effectively with the collapse of the (western) Roman Empire in the fifth century AD. It would revive, however, in the twelfth century when coal had a renaissance as the critical fuel in London as the city outstripped its regionally available sources of wood. The coal that energized London at that time came from the northeast of England, from Newcastle on the banks of the Tyne River. Coal was readily accessible there. Rich seams of coal extended in many instances to the surface. Equally important, there was a ready means to transport Newcastle's coal by boat to meet demands of the growing market for the commodity in London—not just for household heating but also to fuel the fires of the city's legion of blacksmiths and brewers. Freese (2003) summarized the situation thus: "once you got the coal to the water, the world opened up; shipping it three hundred miles from Newcastle to London cost about the same as carrying it three or four miles overland." More than 500 years would elapse (until the late 1700s) before England would develop the system of canals, and eventually the network of roads and railroads, that would facilitate inland (onshore) transport of goods and commodities.

Londoners were quick to realize the downside of the switch from wood to coal to fuel the city's forges and brewers. Burning low-grade coal (the coal produced from surface resources in Newcastle) caused a rapid deterioration in the city's air quality. When winds were light, the city was enveloped in a layer of black, sulfurous, smoke (more on this later) that led to an inevitable response. Freese (2003) tells the story that during the reign of King Edward I (Edward the Confessor) "commissions were set up in London to address the problem of coal smoke which complainants said had infected and corrupted the air." Laws were enacted instituting "steep fines and the destruction of furnaces" to mitigate the situation, arguably the first regulations instituted anywhere in the world to address the challenge of unacceptable air quality. But the problem persisted. As it happened, the issue was resolved, not by legislation, but by the onset of an even more serious set back, the Black Plague that struck Europe in late 1347 and England in the summer of 1348 (see Chapter 2). Economic progress came to an abrupt halt as the Plague took the lives of close to a third of the European population, permitting forests to regrow, and demand for coal to decline as a consequence. But the die was cast. As population recovered from the ravages of the plague, so also did demand for coal. Industrialization increased to fever pitch in the late eighteenth and early nineteenth centuries in England and so also did demand for coal.

Steam produced by burning coal would provide the motive force for England's industrial renaissance (Chapter 10), most notably for the mechanization of the textile industry that took hold in cities such as Manchester in the north of England in the nineteenth century. Coal would supplant wood as the dominant fuel much later in the U.S. New England's textile industry, for example, was powered not by coal but by running water. Only in the last decade of the nineteenth century did coal replace wood as the dominant source of energy in the U.S. (as forest resources went into decline as land was cleared for agriculture among other purposes).

When European settlers first arrived in the Americas they discovered what Freese (2003) describes as "one of the largest stretches of woodland ever to grow on the planet." What they had yet to discover was that beneath this forest lay one of the world's greatest reserves of coal, described by the Christian Review of April 1856 (again quoting Freese) as a bounty "scattered by the hand of the Creator with very judicious care, as precious seed, which, though buried long, was destined to spring up at last, and bring forth a glorious harvest." Fifty-seven years later, Katherine Lee Bates would complete the final version of her hymn America the Beautiful that would include the refrain: "America! America! God shed his light on thee." It is good to believe that God is on your side. But affluence, should it beget arrogance, can lead to indolence and potentially decadence. Remember the exhortation in the Gospel of Saint Luke: every one to whom much is given, of him much will be required.

The distribution of the major U.S. coal reserves is illustrated in Fig. 5.3. The eastern coalfield extends more or less continuously from western Pennsylvania to Alabama. The coal reserves in this belt are mainly of the bituminous type except for a pocket of anthracite in eastern Pennsylvania (more on this later). There are extensive deposits of bituminous coal also in the Mid West—in Illinois, Iowa, Kansas, Missouri, and Oklahoma. If that were not enough, the west-central states of North Dakota, Montana, Wyoming, Utah, and New Mexico are also richly endowed although the coal there consists generally of either the sub-bituminous or lignite varieties (with consequently lower energy content). Western coal has an important advantage, however, with respect to eastern coal (as we shall see) in that it contains relatively low concentrations of polluting sulfur.

Pittsburgh would emerge as the first significant center of coal-based industry in the U.S. Located at the confluence of three major rivers, the Ohio, the Monongahela, and the Allegany, the city was founded in 1758 when the British defeated the French and established a fort they named after their prime minister, William Pitt. The early industries in Pittsburgh included glass making and smelting of iron to forge the tools needed by the hordes of settlers traveling down the rivers that flowed through the city to populate the virgin territories of the west and south. By 1817, Pittsburgh would boast a population of 6,000 with more than 250 factories (Freese, 2003) and would soon earn the unenviable reputation as the most polluted city in North America. It would develop later as the most important steel-manufacturing center in the Western Hemisphere, source of the fortune of one of America's first industrial multi-millionaires and one of its first great philanthropists, the Scottish immigrant Andrew Carnegie (1835–1919). Carnegie in later life moved to New York. Freese (2003) quotes him warning then that "if New York allows bituminous coal to get a foothold, the city will lose one of her most important claims to pre-eminence among the world's great cities, her pure atmosphere." Environmental quality matters—especially when you are rich!

The anthracite resources of the U.S. are concentrated in eastern Pennsylvania where the peat deposits of the coastal swamps were thrust deepest into the Earth by the convergence of the European and American crustal plates, subjected to the highest range of temperatures and pressures. The richest deposit was located in a mountainous region known as Summit Hill near the Lehigh River and extended to a depth of more than 35 ft (Freese, 2003). This was the energy source Carnegie favored to supply the needs of his pristine New York. The energy content of anthracite was particularly high and, as noted earlier, it burned with minimal emission of smoke. Getting this resource to market, however, posed a formidable challenge. What made this possible eventually was the construction of the Schuylkill Canal, completed in 1825. The canal provided a means to transport the anthracite to market, initially to Philadelphia. A few years later, the Delaware and Hudson Canal Company would succeed in bringing

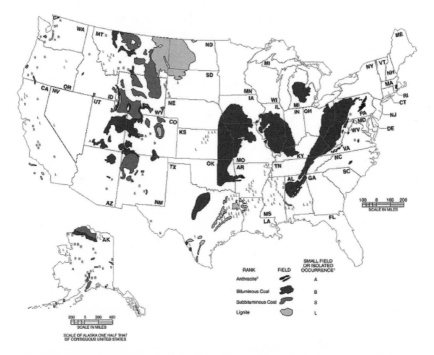

Figure 5.3 Coal-bearing areas of the United States (from EIA). http://www.eia.doe.
gov/cneaf/coal/reserves/chapter1.html, read May 6 2009.

anthracite from Pennsylvania to New York but only after a major cost overrun
in construction costs that would precipitate what Freese (2003) describes as
"one of the [nation's] largest public bailouts." Eventually, rail transport would
win out over water transport in bringing Pennsylvania's anthracite to the east and
Carnegie would realize his dream to ensure continuity of the "pure" atmosphere
of New York. As an interesting footnote, the nation's first trains were fueled,
not by coal, but by wood.

5.4 ENVIRONMENTAL PROBLEMS POSED BY THE USE OF COAL

Environmental impacts of coal consumption are both obvious and subtle.
Obvious is the smoke emitted in burning coal, especially when the coal consumed is of lower grade. This smoke consists of a complex mixture of soot
(carbon compounds formed as a result of incomplete combustion) and a
variety of mineral elements. In addition to the particulate matter, emissions
from coal burning include also a variety of toxic gases, notably sulfur dioxide,
nitrogen oxides, and carbon monoxide. Breathing this complex mixture poses
a problem for public health. Inhaling the air in even a moderately polluted
coal-fired city in the late nineteenth and early twentieth centuries had health
effects equivalent to smoking several packs of cigarettes a day. The impact

was even greater for more polluted cites such as Manchester and Pittsburgh—three packs a day or even worse. Not surprisingly, lung-related diseases were a major cause of sickness and premature death for populations living in these environments.

I experienced these effects firsthand (but survived) growing up in Belfast, Northern Ireland, in the 1940s and 1950s. Belfast was an industrial city boasting one of the world's largest shipbuilding facilities (the Titanic was built there between 1909 and 1912) with scores of factories producing textiles, machine goods, and even aircraft. Homes were heated almost exclusively by coal. Every week or so, people would come around to clean the windows of houses in Belfast: otherwise it would have been impossible to see to the outside. Thick layers of soot would accumulate in chimneys posing an ever-present danger of house fires. Chimney sweeps were engaged several times a year to take care of this potential problem. As a boy, I would go off to school in the morning equipped with a clean handkerchief. When I returned in the evening the handkerchief was jet black, just from blowing my nose during the day. Gentlemen wore shirts with detachable collars: easier to wash the collar everyday rather than have to wash the entire shirt (these were the days before washing machines when all washing had to be done by hand). Colds were common and everyone coughed. And facades of city buildings were covered invariably in a thick layer of black grime. Surprisingly, there were few objections. People accepted this state of affairs as simply the price of progress.

Public consciousness was raised, however, in England in 1952 when an unusual meteorological inversion over London (temperature increasing with altitude causing air near the surface to be trapped in place) caused smoke and noxious fumes from the city's homes and factories to accumulate in the atmosphere over a period of 4 days between December 4 and December 8. Visibility dropped to the point where people had trouble seeing more than a few feet ahead. More than 4000 people died during this episode, many of them simply falling down dead in the streets. Parliament reacted by banning the use of soft coal in the city. The legislation was expanded subsequently to prohibit burning of all coal in all homes in all British cities with exceptions for people living in rural areas to burn so-called smokeless fuels—the more expensive anthracite introduced earlier. Air quality improved markedly as a result of these initiatives.

The wakeup call to the dangers of coal-related air pollution was sounded a few years earlier in the U.S. Donora was a small town located south of Pittsburgh in the Monongahela Valley. It was the site of a number of steel mills and zinc smelters burning cheap, locally available, bituminous coal. An atmospheric inversion set in on October 26, 1948, much like the inversion that occurred a few years later in London, and persisted in this case for 5 days. By the end of this period more than half of the residents of this small town (population about 14,000) were seriously ill and a sufficient number died to draw national attention. The problem in Donora was caused not by emissions from private homes as in the case of London but by the smoke stacks of local industry. The solution in this case was not to ban use of coal but to insist that factories build higher smoke stacks. Local residents would be protected then. The problem was not resolved, however, but simply exported elsewhere.

Steps were taken soon after the problems of Donora and London to mitigate the problem of black, soot-rich, smoke emanating from industrial sources in the U.S. and Europe. Industrial furnaces were made more efficient to ensure that a larger fraction of the carbon in the coal would be burnt ensuring thus a reduction in the emission of the carbon compounds responsible for the production of black soot. Devices were installed—electrostatic precipitators—to remove particulate matter before it was released to the air. If you look at the smoke stacks of power plants or other industrial facilities burning coal in the U.S. or Europe today the smoke you see is white in color, consisting mainly of water vapor condensing as it cools leaving the stack. But further problems, even more recalcitrant, were soon to raise their ugly heads. These had to do with the fate of the nitrogen and sulfur oxides released as the coal was consumed—the problems of acid rain and so-called fine (extremely small) particulates—and the consequences of emission of mercury, and more recently the problem of global climate change induced by conversion of the energy yielding carbon of the coal to a higher burden of the atmospheric greenhouse agent CO_2.

Sulfur and fixed nitrogen are important ingredients of coal. As best we can tell, the sulfur content of a particular coal deposit is a function of the extent to which the waters of the swamp that regulated conditions for the initial decay of the plant material (that would be transformed eventually to coal) were invaded by water from the neighboring ocean. Sulfate, after chloride, is the second most abundant anion (negatively charged component) of seawater. The greater the salt content of the decay medium (the greater the ocean water contribution to the decay medium), the greater the sea-derived sulfur content of the coal. Western U.S. coal, as noted earlier, is relatively low in sulfur: eastern coal is comparatively high. Coal also contains fixed nitrogen, a residue of the original plant material. Subjecting the coal to high temperatures during the combustion process results in additional production of fixed nitrogen, formed in this case by breaking the tight triple bond that binds the nitrogen atoms together in atmospheric N_2 during the high temperature combustion process. As the coal is consumed, the sulfur is released and emitted to the atmosphere as a component of the exhaust gases of the combustion process, largely in the form of sulfur dioxide, SO_2. The fixed nitrogen content of the coal, enhanced by the contribution from the high temperature decomposition of the atmospheric N_2 (admitted to the combustion site in conjunction with the supply of oxygen from the atmosphere used to oxidize the carbon), is emitted primarily in the form of NO.

Once in the atmosphere, the sulfur is converted, on a time scale of a few days, to sulfuric acid. This conversion process is thought to proceed largely in small droplets of liquid water that are ubiquitously present in the atmosphere. Nitric oxide (NO) is converted to nitric acid (HNO_3) through reactions that take place in the gas phase. Eventually, both of these acids are removed from the air, either by contact with the surface or by incorporation in precipitation (either rain or snow). This is what gives rise to the phenomenon of acid rain or, more accurately, acid precipitation.

As the acidified precipitation falls on the surface, it percolates into the soil triggering decomposition of some of the naturally occurring mineral components of the soil. Included among the elements mobilized in this manner are calcium and magnesium essential for healthy plant growth and toxic elements such as aluminum which when transported to rivers and lakes can result in serious damage to fish (the fish are deformed first, then they die) and mercury. Problems of fish kill, attributed to acid precipitation, were observed first in rivers and lakes of southern Scandinavia and New England. Acid precipitation has been implicated also in the die back of forests, specifically in regions where the chemical composition of soils is such that it is unable to neutralize the effect of the added acid. Lime, or alkaline, rich soils are comparatively immune to damage from acid precipitation. Regions underlain by thin soils and granitic bedrock are particularly vulnerable. Areas of North America that are unusually sensitive to effects of acid precipitation are illustrated in Fig. 5.4.

Consumption of coal in the U.S. is concentrated to a large extent in the Midwestern region of the country. Sulfur and nitrogen compounds emitted by coal-fired power plants in the Ohio River Valley region are transported by prevailing winds to the northeast which, as indicated in Fig. 5.4, is ill equipped to accommodate the associated inputs of acids. Rain in the northeast is characterized by pH values as low as 4.0 or even lower (the pH of pristine rain in equilibrium with atmospheric CO_2 has a value of 5.6: for every unit decrease of pH, the acid content increases by a factor of 10). Not surprisingly, with increasing

Figure 5.4 Regions of North America that are particularly sensitive to acid rain due to low buffering capacity of soils. Source: Jacob (1999).

public awareness of the acid precipitation phenomenon, there was a call for action to reduce the offending emissions. Why should the Midwest benefit from cheap electricity produced from dirty coal when New England would have to pay the price? Similar inter-regional disparities existed in Europe (fish dying in Scandinavian rivers as a consequence of emissions from Germany and Poland) and provoked there also calls for a coordinated (multinational in this case) policy response.

This response focused on reducing emissions of sulfur (automobiles in both the U.S. and Europe account for a source of NO greater than that from coal). A number of strategies are available to decrease emissions of sulfur. Most obvious is to switch from high to low sulfur coal. Legislation passed by the U.S. Congress in 1990 mandated a 40% reduction in sulfur emissions by 2010 and proved a boon not only for western coal producers but also for U.S. railroads. Today, more than half of the coal consumed in the U.S. is mined in the west and consumed elsewhere, shipped by rail to feed the coal burning electrical utilities of the east and south. Wyoming is now the country's largest coal-producing state with production of 396.4 million short tons in 2004: West Virginia was number two in 2004 with 147.7 million short tons followed by Kentucky with 114.2 million short tons (1 short ton is equal to 0.907 metric tons). Living with high sulfur coal while reducing emissions of sulfur, requires (at present) the installation of systems to spray chemicals (calcium oxides for example) into the exhausts of coal burning furnaces converting the sulfur in the exhaust to chemical forms that can be recovered prior to emission to the atmosphere (as calcium sulfate, or gypsum, for example in the case where calcium oxides are added to the exhaust), products that can be either used productively (as would be the case for gypsum) or deposited in some acceptable geological reservoir. More futuristic schemes involve converting the energy content of the coal to gaseous products such as CO and H_2, disposing of the residues subsequently, or washing the coal prior to combustion to reduce the content of undesirable impurities. All of these options require additional expenditures of energy with potentially adverse effects, either economic or environmental or both (increased emissions of CO_2, increased demands for water, and requirements either to treat contaminated water or to dispose of it in an environmentally acceptable fashion).

The steps taken in the U.S. and Europe to reduce emissions of sulfur have been notably successful: the sulfur content of precipitation has decreased steadily since 1970. Paradoxically, though, the pH of precipitation has remained uncomfortably low. Hedin and Likens (1996) attribute this bothersome result to the fact that the reduction in emissions of sulfur has been accompanied by a simultaneous decrease in the release of acid-neutralizing elements of the coal such as calcium, magnesium, sodium, and potassium (recall that coal consists not simply of pure carbon, that it includes also a variety of additional trace minerals, components of the rocks that contained the coal). The decrease in release of these elements occurred as a consequence of the steps taken to reduce the particulate component of coal-related emissions. Further, even more aggressive, reductions in emissions of sulfur (and nitrogen) will be required in the future if we are to succeed eventually in achieving an acceptable resolution to the problem of persistent acid precipitation.

There are other good reasons to reduce emissions of sulfur. The sulfate particles produced when sulfur dioxide is transformed in the atmosphere to sulfuric acid are responsible for a significant reduction in atmospheric visibility. Further, these particles are typically very small in size. When respired, they penetrate readily into our lungs bypassing the cilia of our noses and bronchial tubes that can exclude particles of larger size. Once in the lungs, these particles can persist for a very long time and are responsible for a variety of pulmonary problems including, all too frequently, premature death (Wilson and Spengler, 1996). As discussed later (Chapter 13), sulfate aerosols (the term used to describe particles suspended in the air) have implications also for climate change.

Combustion of coal is associated further with emission of hazardous levels of mercury. The mercury content of coal is greatest in the highest grades of coal (reflecting presumably the fact that these grades were exposed to higher temperatures and higher concentrations of mercury in subcrustal rocks where they formed; the mercury content of granite—an igneous rock—is approximately twice that of average crustal material). Data summarizing measurements of mercury for a variety of different coal samples are displayed in Fig. 5.5.

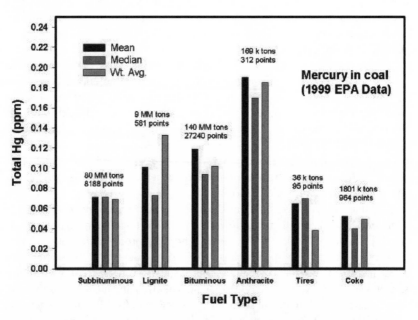

Figure 5.5 From U.S. Environmental Protection Agency, Unified Air Toxics Program: Electric utility steam generating units hazardous air pollutant emission study (2000), http://www.epa.gov/ttn/atw/combust/utiltox/utoxpg.html#DA3).

Concentrations of sulfur are generally more variable than those of mercury in coal, as indicated in Fig. 5.6 for bituminous coal, the type of coal most commonly consumed in the U.S. The greater range of concentrations of sulfur observed as compared to mercury reflects most likely the different origins of these species in coal—seawater incorporated in the swamps where the original plant material grew in the case of sulfur, crustal rocks to which the plant material was exposed during subsequent metamorphism in the case of mercury.

Mercury is emitted to the atmosphere in a variety of different forms in conjunction with coal combustion: as a component of particulate matter, as mercury oxide, or as elemental (gaseous) mercury. Particulate and oxidized forms of mercury are removed from the atmosphere relatively efficiently by precipitation and by contact with surface materials (dry deposition): the problems that ensue as a consequence are either local or at most regional in scope. Elemental mercury is more persistent, however. Its lifetime in the atmosphere can be as long as a year or more, ensuring that the impact of emissions from any particular region can have consequences that extend to global scale (air is

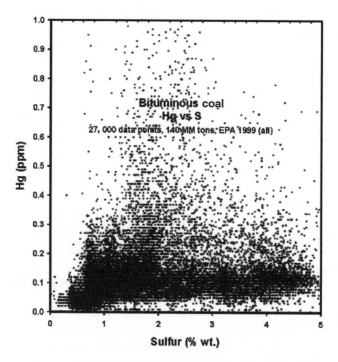

Figure 5.6 Hg and sulfur in bituminous coal (from U.S. Environmental Protection Agency, Unified Air Toxics Program: Electric utility steam generating units hazardous air pollutant emission study (2000), http://www.epa.gov/ttn/atw/combust/utiltox/utoxpg.html#DA3).

well mixed within a hemisphere over a period of a few months: a source of pollution from one hemisphere—north for example—can penetrate to the other—south—over a period of a year or so if not previously removed). Elemental mercury from China can contribute therefore to problems in the U.S. At the same time, elemental mercury from the U.S. can be carried by the prevailing winds to Europe, reaching potentially all the way back to China. The problem is clearly global and its resolution will require a global response.

A significant fraction of the mercury emitted by burning mercury-rich coal ends up in aquatic systems, in rivers, lakes, and the ocean. Once there, it can be converted to methyl mercury, the form most harmful to organisms. Ingested by fish, methyl mercury accumulates in their tissue. As small fish are eaten by larger fish, the concentration of mercury in the tissue of the fish increases. The problem is most serious for humans when fish constitutes a major component of their diet, notably for native peoples living at high northern latitudes (the Inuit for example). The developing fetus is particularly sensitive to the toxic effects of methyl mercury. Methyl mercury is a neurotoxin (it can cause damage to the body's nervous system). Babies born to mothers who have ingested excessive quantities of methyl mercury in their diet when pregnant, can suffer from a range of disabling deformities[1] including difficulties in learning to speak, in processing information, and in coordinating visual and motor functions. EPA suggests that ingesting methyl mercury at a level less than 0.1 microgram per day per kilogram of body weight of the person should be safe. Problems arise (particularly for unborn children) if quantities consumed are much greater than this limit.[2]

Amendments to the Clean Air Act enacted in the U.S. in 1990 included mercury as one of 189 substances designated as "hazardous air pollutants". In December 2000, EPA ruled that it was "appropriate and necessary" to use "maximum achievable control technology" (MACT) to reduce emissions. This ruling was reversed under the Bush Administration on March 15, 2005, with a decision to set a limit on annual emissions at a level of 26 tons by 2010, decreasing to 15 tons by 2018. In addition, EPA proposed a cap-and-trade mechanism that could be adopted by emitters to meet these standards in a fashion judged to be most cost-effective from an economic perspective on a national scale. This would permit power plants, or other sources, that chose not to expend funds to install systems to limit their emissions to purchase rights to emit from facilities elsewhere that chose to make such investments. The strategy was similar to that introduced earlier (judged to have been successful in that case) to reduce emissions of sulfur.

Environmental groups objected to what they perceived as a sop to industry in the 2005 ruling arguing that since technologies are available (although expensive) to achieve much greater immediate reductions in emissions that they should be applied promptly to protect public health (consistent with the earlier 2000 ruling). The debate continues, likely to be resolved in the future either by court rulings as to the validity of EPA's 2005 revision to the 2000 ruling, or by additional legislation in Congress.

5.5 CONCLUDING REMARKS

We presented an account of how coal was produced several hundred million years ago, tracing its origin from plants growing in tropical coastal swamps to the subsequent production of the increasingly energetic forms of lignite, sub-bituminous, bituminous, and anthracite. We noted that the energy content of anthracite is greater than that of lignite by as much as a factor of 2 or 3. We traced the use of coal from China to England to the U.S. describing the problems that ensued and the responses. We outlined the issues underlying the more subtle environmental consequences of coal use—acid precipitation, sulfate aerosols, and the persistent problems of mercury. We opted to skip problems associated with production of coal: the deaths of miners occasioned by collapse of mine walls; poisoning of miners from inhalation of toxic carbon monoxide; death through exposure to excessive concentrations of carbon dioxide; dangers occasioned by the build-up of explosive concentrations of methane; premature death due to black lung disease induced by long-term accumulation of small carbon particles in miners' lungs; and the loss of life and property occasioned by the collapse of unstable heaps of slag built up at the surface from the accumulation of the inevitable uneconomic by-products of mining. These issues are discussed in some detail by Freese (2003). We postpone until later (Chapter 14) addressing the question of whether there could be an environmentally benign future for coal (clean coal technology) including environmentally acceptable options for disposal (other than to the atmosphere) of the inevitably large related source of carbon dioxide.

NOTES

1 (http://www.epa.gov/mercury/exposure.htm)
2 (http://www.epa.gov/iris/subst/0073.htm)

REFERENCES

Cameron, R. (1997). *A Concise Economic History of the World: From Paleolithic Times to the Present*, Third Edition. New York: Oxford University Press.

Fairbank, J.K,(1992). *China: A New History*. Cambridge, MA: Belknap Press of Harvard University Press.

Freese, B. (2003). *Coal: A Human History*. New York: Perseus.

Hedin, L.O. and Likens, G.E. (1996). Atmospheric dust and acid rain, Scientific American, December 1996.

IEA, International Energy Agency (2006). Key world energy statistics.

Jacob, D.J. (1999). *Introduction to Atmospheric Chemistry*. Princeton: Princeton University Press.

Wilson, R. and Spengler, J. (1996). *Particles in our Air: Concentrations and Health Effects*. Cambridge: Harvard University Press.

6

Oil: Properties, Origin, History, Problems, and Prospects

6.1 INTRODUCTION

Oil is arguably the single most essential ingredient of modern industrial society. Turn off the oil spigot for more than a few weeks and our transportation system would grind to a halt. Cars, trucks, trains, ships, and airplanes would be grounded. Cities would run out of food (there would be no way to get food from the countryside to the city and no way to get people from the city to the countryside). Factories, deprived of raw materials, would be forced to shut their doors. Large numbers of people would die of starvation. It could prompt a disaster comparable to the climatic reversal in 2200 BC that triggered the simultaneous collapses of the Old Kingdom in Egypt, the Akkadian Dynasty in Mesopotamia, and the advanced Harappan Civilizations in India and Pakistan (Chapter 2). It could wreak havoc comparable to what ensued following the onset of the Plague in Europe in 1347 AD (Chapter 2). And the damage would most likely not be confined locally. Given the interconnectivity of global economies, we might expect it to extend quickly to global scale.

This, to be sure, paints an excessively apocalyptic picture of the consequences of a short-term loss of oil. After all, there are multiple sources of oil in the world today and it is unlikely that all would be turned off simultaneously. Deprived of foreign sources of oil, the U.S. could tighten its belt and fall back on domestic resources including supplies maintained in the Strategic Petroleum Reserve (SPR). Legislation authorizing the SPR was enacted in the U.S. on December 22, 1975 in response to the energy crisis prompted by the 1973 Arab–Israeli war and the subsequent decision by Arab countries to impose a boycott on exports of oil to the U.S. (also to Canada and the Netherlands). As of October 13, 2006, the SPR contained 688.3 million barrels of crude oil, sufficient to satisfy current

U.S. demand (about 20 million barrels a day) for a period of approximately 34 days or to replace current imports (now close to 70% of total consumption) for about 52 days. There can be no doubt though that an unanticipated loss of access to foreign oil would result in a major immediate economic crisis for the U.S. (and for Japan and for many of the scores of countries depending on oil imports). The U.S. would adjust presumably in the short-term by prioritizing functions for which oil (and oil products) could be deployed. In the longer-term it could respond by developing alternative sources of liquid fuel, from coal or plant materials (Chapter 15) and oil shale for example (more on this below), but decades would elapse before the infrastructure to support these activities could be in place. In the meantime, the economy would suffer severe damage and would take a long time to recover. It goes without saying that we are dangerously dependent on a potentially unreliable—critical—resource, access to which could be denied at any time for reasons beyond our control in the case of an international emergency.

Table 6.1 presents a summary of the major producers, exporters, and importers of crude oil (data for 2005 from EIA, 2006). Data are quoted here in units of millions of metric tons (Mt) (in units of mass). Quantities of crude oil are quoted more commonly in units of volume—barrels where a barrel is equivalent to 42 U.S. gallons. The number of barrels that can be filled from a ton of crude oil depends on the density of the crude—more for light crude, less for heavy—varying from a high of about 8 barrels per ton to a low of 6.6 barrels per ton, averaging about 7.4 barrels per ton (EIA, 2007a). Note that four Middle

Table 6.1 Major worldwide producers, exporters, and importers of crude oil, 2005 (Million barrels per day) (EIA, 2007a)

Producer	Total oil production*	Exporter	Net oil exports	Importer	Net oil imports
Saudi Arabia	11.1	Saudi Arabia	9.1	United States	12.4
Russia	9.5	Russia	6.7	Japan	5.2
United States	8.2	Norway	2.7	China	3.1
Iran	4.2	Iran	2.6	Germany	2.4
Mexico	3.8	United Arab Emirates	2.4	South Korea	2.2
China	3.8	Nigeria	2.3	France	1.9
Canada	3.1	Kuwait	2.3	India	1.7
Norway	3	Venezuela	2.2	Italy	1.6
United Arab Emirates	2.8	Algeria	1.8	Spain	1.6
Venezuela	2.8	Mexico	1.7	Taiwan	1
		Libya	1.5		
Kuwait	2.7	Iraq	1.3		
Nigeria	2.6	Angola	1.2		
Algeria	2.1	Kazakhstan	1.1		
Brazil	2	Qatar	1		

* Total oil production includes crude oil, natural gas liquids, condensate, refinery gain, and other liquids.

Table 6.2 Sources of crude oil imports to the U.S. (top 15 countries) (Thousand barrels per day)(EIA, 2007b)

Country	Nov-06	Oct-06	YTD 2006	Nov-05	Jan–Nov 2005
Canada	2065	1704	1778	1756	1609
Mexico	1462	1481	1606	1658	1542
Saudi Arabia	1444	1322	1417	1267	1445
Venezuela	1069	1125	1146	1009	1246
Nigeria	919	1049	1046	1163	1068
Iraq	589	505	567	572	540
Angola	505	506	504	658	458
Algeria	253	449	352	265	230
Kuwait	253	234	180	273	223
Ecuador	243	315	274	264	270
Brazil	156	171	134	65	88
United Kingdom	119	74	131	229	241
Chad	118	109	93	33	78
Norway	81	120	97	103	124
Azerbaijan	77	88	23	0	0

Table 6.3 Conversion factors for different measures of oil amount. To convert from one of the quantities included in the left vertical column to one on the horizontal list, multiply by the appropriate entry in the Table. For example: 1 barrel = 42 gallons = 159 liters = 0.136 tons.

Unit	Gallon	Liter	Barrel	Ton
Gallon	1	3.785	2.38×10^{-2}	3.25×10^{-3}
Liter	0.264	1	6.29×10^{-3}	8.58×10^{-4}
Barrel	42	159	1	0.136
Ton	308	1166	7.33	1

Eastern states (Saudi Arabia, Iran, United Arab Emirates, and Iraq) accounted for close to 30% of total global exports in 2004. The U.S. is the world's largest importer (exceeding Japan, number two, by more than a factor of 2), ranking third in production but by far and away number one in consumption (reflecting the sum of entries for production and import minus export in Table 6.1). Sources of U.S. oil imports in 2005 and 2006 (through August) are summarized in Table 6.2 (EIA, 2007b). Units used to quantify various amounts of oil with appropriate conversion factors are summarized in Table 6.3.

We begin in Section 6.2 with an account of the chemical composition of crude oil and the processes by which it can be treated to yield a variety of useful products. The geologic processes responsible for production of oil are discussed in Section 6.3. The historical background to its use is treated in Section 6.4. Environmental consequences of the use of oil are discussed in Section 6.5. Prospects for future supplies are treated in Section 6.6 with concluding remarks presented in Section 6.7.

6.2 CHEMICAL COMPOSITION AND PRODUCTS OF CRUDE OIL

The composition of crude oil varies from geological reservoir to reservoir (variations in composition are responsible for the differences in density noted earlier). In general, crude oil consists of a complex mix of hydrocarbons (molecules composed of a combination of carbon and hydrogen atoms with the carbon atoms linked to form chains of varying length) with variable quantities of sulfur, oxygen, and trace metals. Crude oil containing high concentrations of sulfur is described as "sour"; oil with low sulfur is classified as "sweet." An important fraction of the molecules in the hydrocarbon component fall into the chemical class referred to as the alkanes (or paraffins). Molecules in the alkane family are identified by the chemical formula C_nH_{2n+2} where n is an integral number[1]. A list of names, chemical formulae, and boiling points for some of the lower molecular weight alkanes is presented in Table 6.4.

Approximately a third of the hydrocarbons in crude oil are present in the low mass range, numbers of carbon atoms (n) less than 10. A third fall in the intermediate range with n between 10 and about 18 with the balance in the high n range ($n>18$). The individual components of crude oil can be separated by distillation, exploiting the fact that different species evaporate at different temperatures. The separate components can be concentrated then by condensation as they are exposed subsequently to lower temperatures. A modern oil refinery includes not only a facility for distillation but also a means to convert the products of the distillation process to a variety of desirable end products.

Gasoline is made up of hydrocarbons with numbers of carbon atoms in the approximate range 6 to 12 including notably heptane, octane, and nonane (see Table 6.4). Kerosene is composed of molecules in the C_{10-15} range while diesel and heating oils are formed by molecules in the range C_{10-20}. Heavier species are combined to produce lubricating fluids (engine oil for example) while the heaviest of all are employed to make tar and asphalt. Boiling points increase with increasing carbon number, from gasoline (40–200°C), to kerosene

Table 6.4 Low molecular weight alkanes (paraffins): common names, chemical formulae, and boiling points (°C)

Name	Formula	Boiling point (°C)
Methane	CH_4	−162
Ethane	C_2H_6	−89
Propane	C_3H_8	−42
Butane	C_4H_{10}	−0.5
Pentane	C_5H_{12}	36
Hexane	C_6H_{14}	69
Heptane	C_7H_{16}	98
Octane	C_8H_{18}	126
Nonane	C_9H_{20}	151
Decane	$C_{10}H_{22}$	17

(180–320°C), to diesel (250–350°C). Lubricating fluids, tar, and asphalt top the list with boiling temperatures higher than 300°C.

We can think of the products of the refinery distillation process as providing a series of Lego blocks. The blocks can be combined to form a variety of new compounds. Larger structures can be produced from a combination of smaller units, a process known as polymerization or alkylation. Better than that, blocks can be broken apart, forming new blocks, a process known as cracking, and combined with other units to form totally new compounds. The protocol of oil refineries is adjusted under normal circumstances in the U.S. and elsewhere to optimize for production of heating oil in winter, switching to output of gasoline in summer when demand for this product is typically greatest. Even the composition of gasoline can be fine-tuned to minimize its role in production of photochemical smog: the formation of ozone, the most recalcitrant component of air pollution over large regions of the industrial world (peaking specifically in local summer, more on this later). The modern oil refinery represents a triumph of chemical engineering skill. The operator of the plant has a much greater variety of options to play with than the kid with even the most elaborate Lego kit!

6.3 FORMATION OF CRUDE OIL

Formation of economically recoverable concentrations of crude oil depends on a number of (to a large extent) unrelated geological circumstances. In contrast to the case of coal, the organisms responsible for production of crude oil lived for the most part in the ocean rather than on land, although a small amount of oil was produced from organisms in fresh water lakes.

As indicated in Fig. 4.1, the rate at which organic matter is formed by photosynthesis in the ocean is comparable to the rate at which it is produced on land. A major difference, however, is that the organisms that inhabit the ocean are relatively short-lived. There is no great build-up of organic matter in the ocean comparable to what occurs for example in a tropical rainforest. Trees can live for hundreds of years: marine phytoplankton and zooplankton come and go in a matter of days. Large marine mammals, whales for example, may survive much longer (80 years or so) but there are not very many of them and their contribution (and the contribution of all fish) to the total abundance of marine biomass is relatively minor. The build-up of organic carbon in the ocean and sediments and the subsequent conversion of a small fraction of this carbon to crude oil are associated primarily with the lifecycles of microscopic organisms—phytoplankton and zooplankton and a variety of small marine animals going by names such as foraminifera and coccolithophoridae. Phytoplankton are the primary producers of organic matter in the sea, the ultimate source of the energy we harvest today as oil.

The first condition that had to be met for formation of an economically viable concentration of oil in the ancient ocean was a locally high rate of primary productivity. Supply of nutrients, specifically nitrate and phosphate, from deep waters to the surface is the primary factor limiting the production of biomass in the ocean. High rates of primary production are associated with regions of local upwelling (where nutrient-rich deep water rises to the surface) and with estuaries where high concentrations of nutrients enter the ocean from the land. Coastal areas are particularly productive (for reasons that have to do with an offshore flow of surface waters driven by the prevailing winds; surface waters moving offshore are replaced by nutrient-rich water rising from below). High latitude regions are also notably productive, a consequence in this case of the influence of winter storms drawing nutrients to the surface that can be taken up readily when the sun returns to the region in spring (a phenomenon known as the spring bloom).

A composite satellite image of ocean color is presented in Fig. 6.1. The wavelengths of the satellite instruments are selected to provide maximum sensitivity to chlorophyll, the primary pigment involved in photosynthesis. Note the presence of highly productive regions off the coasts of California, Peru, and northwest Africa, in the southern ocean near Antarctica, and over extensive regions of the North Atlantic and North Pacific. The interiors of the major ocean basins are characterized by downward flows of surface waters and stand out as deserts from the point of view of organic matter production in the ocean. For a more detailed discussion of the factors regulating biological productivity in the ocean, the reader is referred to McElroy (2002).

The second condition that must be satisfied to ensure formation of an economically viable oil reservoir is that the rate at which organic matter reaches the sediment should exceed the rate at which it can be consumed by bottom-dwelling organisms. The key factor here is that the accumulation of organic matter in sediments should exceed the rate at which oxygen can diffuse from the overlying water to satisfy respiratory demands of the aerobic (oxygen breathing) bacteria that would otherwise be effective in consuming the incoming organic matter. Decomposition proceeds much more slowly under anaerobic (oxygen deficient) conditions favoring accumulation of organic-rich matter in the sediments. Transformation of the sedimentary organic matter to the chemical building blocks of oil requires that the organic matter be exposed to temperatures in the range of 60 to about 160°C.

Temperatures increase with depth below the Earth's surface, typically by about 25°C per km, responding to what is known as the geothermal gradient (reflecting transfer of heat from higher temperature regions of the planetary interior).[2] Conditions for the formation of oil are realized therefore typically at depths of between about 2 and 6 km below the ocean bottom. In addition to organic matter, the oil-forming regions of the ocean sediment include typically also large numbers of fine grained particles composing of what is referred to collectively as shale. The porosity of shale (the gaps between the constituent particles) decreases steadily with depth as the shale is compacted under increasing pressure imposed by the overlying sediment. If the oil-producing organic matter is confined in place, it may be incorporated in the shale forming what is known as oil shale (we return later to the potential of oil shale as a potential future substitute for liquid petroleum). Alternatively, oil can migrate through permeable rock (sandstone for example) finding its way to an environment where it may be trapped (confined for example by an overlying layer of impermeable shale). Potential traps for oil can be identified by seismic probing of sediments. All too often, however, oil companies drill into promising regions only to come up dry. Finding oil today, especially in deeper coastal waters, is an expensive business, as much an art as a science.

6.4 HISTORY OF THE OIL USE

The history of the use of coal as a fuel dates back at least to the time of the Romans in England, as noted in the preceding chapter. The story of oil is similarly deep-rooted. Oil seeps (pools of oil emerging from below the ground accumulating at the surface) were exploited in Mesopotamia as early as 5000 BC to provide a source of asphalt and pitch used as mortar in constructing the walls and towers of Babylon.[3]

Genesis records God's instruction to Noah to "make yourself an ark of gopher wood; make rooms in the ark, and cover it inside and out with pitch." Yergin (1991) quotes the Roman naturalist Pliny extolling in the first century AD the pharmaceutical properties of oil, specifically bitumen or asphalt, highlighting among other attributes its role in checking bleeding, treating toothache, curing chronic coughing, treating diarrhea, and relieving problems of both rheumatism and fever. Oil wells were drilled to depths as great as 243 m in China as early as the fourth century AD using drill bits attached to bamboo poles. Oil produced from these wells was burned to evaporate brines and produce salt. By the tenth century, the Chinese had developed the means to transport oil from the source wells over significant distances through pipelines constructed of bamboo (http://en.wikipedia.org/wiki/Petroleum). The Greeks learned that oil could be employed as a fearsome new weapon of war—Greek fire or oleum incendiarum (fire oil). We noted earlier (Chapter 2) the critical role Greek fire played in the defense of Constantinople from marauding Vikings. As Yergin (1991) describes it, Greek fire was a weapon more terrible than gunpowder: means to produce and apply it were guarded by the Byzantines as a critical state secret.

The need for an inexpensive fuel that could be used for lighting provided an important stimulus for the development of the modern oil-based economy. The industrial revolution transformed the approach to production of goods, notably textiles, moving the locus of manufacture from the household to the factory. Absent efficient means for illumination, the machines that increased the efficiency of the manufacturing process (the quantity of goods that could be produced per unit of labor) were limited by the number of hours per day that they could be deployed in the absence of acceptable lighting. To operate these machines with lighting provided by candles would have been unacceptably hazardous to the operators of the machines. Whale oil provided a temporary respite but as whaling fleets were forced to travel further afield in search of whales this resource became prohibitively expensive, accessible only to the rich: the price of whale oil by the end of the first quarter of the nineteenth century had risen as high as $2.50 a gallon. Camphene, produced from turpentine (a resin extracted from pine trees and a variety of other conifers), offered an alternative but was dangerously flammable. Town gas, extracted from coal, was more acceptable but, again, expensive. The cost-effective solution was provided eventually by what came to be known as kerosene distilled from crude oil.

As noted earlier, kerosene refers to a mix of hydrocarbon molecules composed of between nine and approximately sixteen carbon atoms ($C_9 H_{20}$ to $C_{16} H_{34}$). The name (taken from the Greek words for wax and oil, keros and elaion) was coined first by the Canadian Abraham Gesner who developed a process to extract oil from asphalt converting it to a liquid with excellent illuminating properties (when combusted, kerosene decomposes into molecular fragments that radiate significant amounts of energy in the visible portion of the electromagnetic spectrum). As Yergin (1991) tells the story, Gesner applied for a patent in the United States in 1854 establishing his rights to "a new liquid hydrocarbon, which I denominate kerosene, and which may be used for illuminating or other purposes." He built a kerosene plant in New York and by 1859 his plant was producing five thousand gallons of kerosene a day. Similar facilities were established contemporaneously elsewhere in the U.S. Kerosene in these early plants was produced mainly from coal rather than oil which was much less available. All that would change when George Bissell had the vision to imagine that large amounts of feedstock for kerosene production might lie below the Earth's surface, accessible in liquid form as oil that could be brought to the surface inexpensively by drilling wells in appropriate locations.

George Bissell had an inspiration in 1856 when he saw an advertisement in a shop window in New York touting a patent medicine produced from oil. The oil depicted in the advertisement was obtained from a well drilled into a brine reservoir to access an inexpensive source of salt (salt was a valuable indispensable product employed extensively as a preservative for food stuffs prior to the development of refrigeration). Bissell succeeded in putting together the finances required to pursue his brainchild, hiring Edwin "Colonel" Drake to oversee the project. He was successful beyond his wildest dreams (confounding prevailing wisdom) when the blacksmith Drake hired to carry out the drilling, Uncle Billy Smith, struck oil at a depth of 69 ft in a well drilled in Titusville in northwestern Pennsylvania on August 27, 1859. News of the discovery spread like wildfire. In no time the problem was not to find a source of cheap feedstock for kerosene but a means to store it and transport it to market. Whiskey barrels provided a convenient temporary solution. Soon, Yergin relates, the price of oil was cheaper than the price of the barrels needed to store it. To this day, acknowledging this early history, oil is traded in units of barrels (as indicated earlier a standard barrel defines a volume of 42 gallons of oil).

The oil extracted from the initial well drilled by Smith had to be pumped to the surface. Less than 2 years later, in April 1861, drillers struck the first flowing well. Oil gushed to the surface spontaneously in this case establishing a sustained flow at a rate of three thousand barrels of oil a day. The beginning of this new phase of oil extraction, however, was less than auspicious. Gases in the gushing oil caught fire and the well burned for 3 days killing 19 people (Yergin, 1991). More than two hundred years later, Saddam Hussein would deliberately set fire to oil wells in Kuwait when forced to withdraw from that country at the end of the first Gulf war. But by that time the oil industry had learned how to cope with oil fires. The problem was solved when the larger than life Texas firefighter Red Adair and his band of intrepid oil firefighting specialists flew in, averting a problem some thought could have resulted in a regional, possibly a global, disaster: accumulation of a far flung layer of soot in the atmosphere with potential to obscure the sun, triggering a persistent change in global climate (cooling), a phenomenon referred to as nuclear winter, a term coined by the late Carl Sagan and associates in a paper published in 1983 (Turco et al., 1983) exploring the potential consequences of a major nuclear war.

Despite rapid up and down gyrations in the price of oil as markets sought to establish a sustainable balance between supply and demand, Bissell and his associates profited royally from their initial investments in the early Pennsylvania oil rush, as did many of the wildcatters who followed their lead. The most successful beneficiary of the new oil age, however, was John D. Rockefeller who emerged as founder and principal shareholder of one of the world's first great, vertically-integrated, oil companies, Standard Oil.

Rockefeller's entrepreneurial skills were expressed at a very early age. Born in 1839, he established a business selling turkeys at age seven. He went to work for a produce company in Cleveland at age sixteen forming a business partnership with Maurice Clark shipping produce at age twenty (in 1859). A railway link completed in 1863 provided an important opportunity for Cleveland to capitalize on the copious sources of oil available from Titusville and neighboring oil-producing regions of Pennsylvania. A series of refineries was constructed to convert this oil to kerosene. One of the largest and best capitalized was built by Clark and Rockefeller. Their partnership was dissolved amicably in 1865 when Rockefeller bought out Clark for $72,500. Within a few years, Rockefeller had built a second refinery and had turned his attention to the task of establishing a reliable market for his product, of controlling all aspects of its distribution from refinery to customer.

He acquired land to grow timber to make the barrels needed to store his kerosene. He made aggressive deals with railroads to ensure that his products could get to market at least cost, often to the disadvantage of his competitors. He acquired his own railroad tank cars, built warehouses to store his product in New York, and established a firm in New York (in 1866) under the direction of his brother William to oversee markets for his products on the East Coast and to explore opportunities for export to Europe.

The Standard Oil Company was established in 1870 with five shareholders. Rockefeller held a quarter of the stock in the new company, which by that time controlled a tenth of the total U.S. refining capacity. Ten years later, Standard would control more than 90% of total U.S. refining capacity with dominant positions not only in refining and marketing but also in transportation. Independent oil producers had sought to break Standard's stranglehold on the oil market by constructing a pipeline in 1879 that could carry oil 110 miles east from the oil-producing region to connect with the Pennsylvania and Reading Railroad, a technological achievement Yergin (1991) considers comparable to the building 4 years later of the Brooklyn Bridge. Standard responded to this challenge by building pipelines a few years later that connected the oil fields to Cleveland, New York, Buffalo, and Philadelphia.

Standard's success and what appeared to many its predatory business practices soon prompted calls for redress. The principals' response was to organize the disparate properties they controlled wholly or partially under a Trust, formalized as the Standard Oil Trust Agreement signed on January 2, 1882. A Board of Trustees was appointed. Shares of stock in all of the entities owned by Standard were transferred to the Trust. The Trust in turn issued stock—700,000 shares, 191,700 of which were allocated to Rockefeller. Separate organizations were set up in individual states to administer the affairs of the properties controlled by the Trust, answerable, however, ultimately to the Executive Committee of the Trust. In theory, the Trust was designed to head off charges of monopolistic practices by Standard Oil. The Trust did not directly control the operations of the various Standard Oil corporate entities: it merely administered the shares of the 41 shareholders who held stock in the component companies. In practice, though, nothing changed. The entire operation was tightly managed by the Officers of the Trust giving them what Yergin (1991) termed "the shield of legality and the administrative flexibility to operate effectively what had become virtually global properties."

The Trust would survive until May 1911, when the Supreme Court finally ruled on a suit filed by President Theodore Roosevelt's Justice Department in 1906 under the Sherman Antitrust Act of 1890 accusing Standard Oil of conspiring to restrain trade. The court that first heard the case in 1906 found that Standard Oil was guilty and that the company should be dissolved. Announcing the decision of the Supreme Court rejecting Standard Oil's appeal in May 1911, Chief Justice White stated: "no disinterested mind can survey the period in question [since 1870] without being irresistibly drawn to the conclusion that the very genius for commercial development and organization . . . soon begat an intent and purpose to exclude others . . . from their right to trade and thus accomplish the mastery which was the end in view." A few months later, in July, the company was broken up into separate units, the largest of which was Standard Oil of New Jersey, the forerunner of what is now Exxon Mobil. Stock in the individual companies was distributed to shareholders in the Trust in proportion to their estimated value at the time of breakup. Rockefeller chose to hold onto his stock. Within a year the value of his holdings had more than doubled, to what Yergin (1991) estimated as the equivalent of $9 billion in 1991 dollars. So much for the penalty of corporate guilt in capitalistic early twentieth century America!

Roosevelt's vigorous prosecution of Trusts under the Sherman Act played an important role in the subsequent success of American industry: he filed more than forty-five motions to dissolve perceived monopolies over the course of his term in office. One of the companies formed in the breakup of Standard Oil, Standard Oil of Indiana, was an immediate beneficiary. The market for oil products was changing towards the end of the nineteenth century. Electricity was beginning to replace kerosene as the preferred source of lighting following Thomas Edison's perfection of a long-lived incandescent light bulb in 1880 and his initiative in building the first distributed electrical system in Lower Manhattan in 1882 (more on this later in Chapter 11). Further contributing to this change was the increasing popularity of the automobile resulting from Henry Ford's success in introducing the affordable Model T in 1908. Demand for gasoline had outstripped demand for kerosene in the U.S. by 1910 and the oil industry was hard put to accommodate the rapidly changing dynamics of the market. A chemist working for Standard Oil of Indiana, William Burton, came up with a solution: technology allowing high molecular mass components of the oil to be broken down, cracked, to form lower mass molecular components suitable for incorporation in gasoline. Ironically, Burton had begun to work on this project prior to the breakup of Standard Oil, unknown to his bosses at the Trust headquarters in New York. The new cracking technology was patented by Standard Oil of Indiana and its competitors were obliged to pay royalties if they were to make use of it. Yergin (1991) summarized the situation thus: "the most galling thing the president of Jersey Standard [the largest of the breakup companies] had to do each month was to sign the fat royalty checks – made payable to Standard of Indiana."

With the growing market for oil products, concerns were raised as to whether the supply of oil might be able to keep up with demand. Was the industry, enjoying such a glorious beginning, destined soon to fizzle? It was clear early that the resource available in Pennsylvania was limited. Yergin (1991) quotes the State Geologist of Pennsylvania warning in 1885 that "the amazing exhibition of oil [was only] a temporary and vanishing phenomenon - one which young men will live to see come to its natural end." These fears would prove unfounded though they would recur frequently in the decades that followed. A new source of oil was discovered and exploited in the 1880s in northwestern Ohio straddling the border with Indiana (the Lima–Indiana field), followed soon by discoveries in Kansas, California, Texas, Louisiana, and Oklahoma. The Texas discovery was so large that by 1901 the price of oil had plunged to as little as three cents a barrel, less than the price of a cup of water (Yergin, 1991) but it was soon surpassed by Oklahoma (by the middle of the decade). And major discoveries were not confined to the U.S. Important sources of oil were discovered and developed in Baku on the western shores of the Caspian Sea in 1871–1872, in Sumatra in the Dutch East Indies (now Indonesia) in 1885, in Borneo in 1897, in Persia (now Iran) in 1908, in Mexico in 1910, in Venezuela in 1922, in Bahrain in 1932, in Kuwait in 1938, and finally, the largest strike of all, a few weeks later in 1938, in Saudi Arabia.

The dynamics of the global oil industry would change dramatically as a consequence of events that ensued during the course of World War I. Winston Churchill, serving as First Lord of the Admiralty, took steps prior to the outbreak of hostilities (in August 1914) to oversee conversion of the Royal Navy from a fleet that relied to a significant extent on coal to one that ran almost completely on oil. The German navy continued to depend almost exclusively on coal. Coal fired ships required large numbers of men engaged more or less full time stoking the ships' boilers. Replacing coal with oil allowed a greater fraction of the ship's complement to be employed in the primary business of a warship: the objective to outmaneuver and outgun potential enemies. Yergin (1991) summed up the situation thus: the advantage oil-fired Royal Navy ships enjoyed over the coal-fired German Navy was "greater range, greater speed, and faster mobility." The German fleet was confined to port as a consequence for much of the war.

Oil would play an important role also in determining the outcome of the conflict on land. Germany relied mainly on its network of rail lines (and coal fired locomotives) to move its troops and supplies to the front. Tactical deployment of its resources was limited thus to a significant extent by the location of the railroads. Plans to take Paris in the early days of the war were foiled when the French marshaled a fleet of gasoline-fueled taxis to transport thousands of troops rapidly and surreptitiously to the front, confronting the Germans where they were most vulnerable, forcing them to retreat. Later in the war, the Allies would introduce another powerful new oil-fueled weapon to their arsenal—the tank. Tanks would play a decisive role in the Battle of Amiens on August 8, 1918, breaking through the German lines forcing the Germans to conclude a few months later that victory was no longer possible.

World War I also saw the introduction of the airplane as a weapon of war, less than 15 years following the first successful powered flight by the Wright brothers at Kitty Hawk, North Carolina in 1903. The initial military application of the plane was for reconnaissance. Before too long, though, it came to be used not only for aerial combat but also as a strategic bomber. The Germans bombed targets in England in a forerunner of the World War II Battle of Britain. Before the war was over, the Allies were routinely bombing strategic targets in Germany. Oil again played a critical role in enabling this new military technology.

None of the major combatants in World War I (until the U.S. entered the fray on April 6, 1917) had the benefit of significant domestic sources of oil. Up to 80% of the Allies' wartime requirements were supplied by the U.S., which by 1917 accounted for 67% of total global oil production. The Germans made good use however of their limited supplies of oil developing a fleet of submarines powered by the diesel engine technology developed earlier by German engineer Rudolf Diesel (in 1893). German U boats exacted a devastating toll on Allied shipping in 1916 and 1917. The Allies met the challenge by grouping supply ships in convoys herded by fast moving destroyers and by the introduction of a powerful new weapon—the depth charge—that effectively neutralized the submarine threat.

In the early days of the oil era, individual companies played a dominant role motivated solely by the interests of their principal owners and stockholders to make money. By the end of World War I, however, oil had achieved the status of a strategic national asset, one that had to be managed to preserve not just the interests of individual companies but more importantly the strategic interests of the nations in which they were domiciled. The die was cast first when Winston Churchill convinced the British government in June 1914 to make a major investment in the Anglo-Persian Oil Company (which controlled the rights to development of oil in Persia) to ensure that the Royal Navy would have access to a reliable supply of oil. For the first time, a government was now a major player in the oil business. The importance of oil would only grow in the decades that followed. Today, nine of the ten largest corporations in the world are directly involved in some combination of production, processing or consumption of oil. Transportation fuels—gasoline, diesel, and jet fuel, all produced from oil—account for approximately 60% of oil consumed world-wide today. The balance is used to generate electricity, to heat our homes and offices, and to fabricate a myriad of the products we take for granted in modern society, including, but not confined to, fertilizers, pesticides, insecticides, lubricants, tires, specialized chemicals, dyes, cosmetics, pharmaceuticals, aspirin, toothpaste, detergents, clothing, building materials, insulating agents, and plastics.

A chronology of major events in the development of the modern oil industry is presented in Table 6.5. Historical data on oil production for the world and the U.S. are summarized in Fig. 6.2 indicating that the peak in U.S. domestic production was reached in the early 1970s. Consumption in the U.S. decreased briefly following the first oil crisis triggered by the OPEC boycott in 1973, to a greater extent following the Iranian hostage crisis in 1979. Trends in U.S. production, consumption, and imports since 1950 are summarized in Fig. 6.3 while historical data on the price of oil are presented in Fig. 6.4. It is interesting to note in the latter context that the price of oil in late 2008 ($110 a barrel in September of 2008, $55 a barrel in May of 2009, having reached a peak of close to $150 in July 2008), though higher than prices that prevailed (adjusting for inflation) through most of the twentieth century, is not without precedent. Expressed in 2005 dollars, prices were comparably high in the 1860s and at the peak of the disruption associated with the Iranian revolution.

6.5 ENVIRONMENTAL PROBLEMS

Environmental problems arise at all stages of the oil industry, from production, to transport, to refining, to eventual consumption of products. Many of the chemicals released to the environment by the industry are toxic. Some are carcinogenic. Others are responsible for the formation of photochemical smog. And almost all contribute to the build-up of CO_2 and a variety of other greenhouse gases including O_3 (ozone), CH_4 (methane), and N_2O (nitrous oxide).

Table 6.5 Chronology of major events in the history of the global oil industry

Date	Event
1854	Abraham Gesner applies for U.S. patent establishing his right to kerosene.
1859	Edwin "Colonel" Drake drills first oil well in Titusville, PA.
1861	First flowing well drilled in PA; accompanying fire burns for 3 days killing 19.
1870	John D. Rockefeller forms Standard Oil Company.
1873	Development of oil field in Baku.
1882	Standard Oil Trust formed.
1885	Oil in Sumatra. Development assigned to Royal Dutch Company.
1892	Marcus Samuel sends oil tanker through Suez Canal; beginning of Shell Oil Company.
1897	Oil find in Borneo.
1901	William D'Arcy granted concession to explore for oil in Persia (now Iran); beginning of British interests in the Middle East.
1901	Oil gusher in Spindletop, Texas; beginning of Sun, Texaco and Gulf Companies.
1905	Major discovery of oil in Oklahoma.
1907	Shell and Royal Dutch companies combined.
1908	Oil discovered in Persia (Iran); leads to formation of Anglo Persian Company (later British Petroleum).
1910	Major discovery of oil in Mexico.
1911	U.S. Supreme Court orders break-up of Standard Oil Trust.
1912	Standard Oil of Indiana patents cracking process developed by William Burton.
1913	British government buys 51% of Anglo-Persian Company.
1922	Discovery of oil in Venezuela.
1930	Major discovery of oil in East Texas.
1931	Discovery of oil in Bahrain.
1933	Standard Oil of California awarded concession to explore for oil in Saudi Arabia.
1934	Gulf and Anglo-Iranian awarded concession for Kuwait.
1938	Oil discoveries in Kuwait and Saudi Arabia.
1938	Mexico nationalizes foreign oil company.
1956	Suez Canal crisis.
1956	Oil discovered in Algeria and Nigeria.
1960	Formation of OPEC.
1968	Discovery of oil on Alaska North Slope.
1969	Discovery of oil in North Sea.
1969	Oil spill by Union Oil of California in Santa Barbara: important influence on formation of the U.S. Environmental Protection Agency.
1973	Yom Kippur War. Arab Oil embargo. Price of oil more than triples over 3-month period reaching level of $11.65 a barrel in December.
1974	First oil landed from North Sea.
1977	Alaskan oil comes to market; transported through pipeline constructed following approval in 1973.
1978	Hostage crisis in Iran. Oil climbs from $13 to $34 a barrel. Second oil crisis of the 1970s.
1989	Exxon Valdez accident in Alaska.
1990	Invasion of Kuwait by Iraq.
1991	Gulf War. Iraq sets fire to Kuwaiti oil fields before withdrawing.
2003	Invasion of Iraq by U.S. and coalition partners.
2006	Oil price rises to record level of $78.40 a barrel on July 13.
2008	Oil hits a record price of $147 a barrel in mid-July.

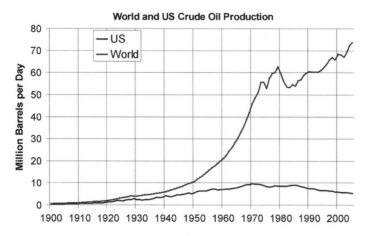

Figure 6.2 Data from EIA.(2007)

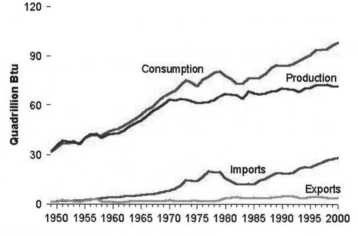

Figure 6.3 U.S. petroleum overview (from http://www.eia.doe.gov/emeu/aer/eh/
frame.html).

An impressive infrastructure is involved in the transport of oil from the
wellhead to where it is refined and eventually consumed. Most of this transport
proceeds through a combination of ships and pipelines, with an additional
contribution (mostly of refined products) from trucks, barges, and trains.
Transoceanic shipment engages to an increasing extent in recent years large
super tankers, ships with capacity to carry more than 250,000 tons of crude.
Tankers with dead weight tonnage (dwt) (a measure of carrying capacity) in
the range 240,000 to 320,000 tons are referred to as Very-large Crude Carriers
(VLCC); tankers with more than 350,000 dwt, ranging up to 500,000 dwt, are

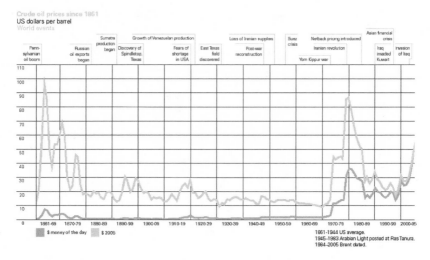

Figure 6.4 Crude oil prices since 1861 (BP Statistical Review of World Energy, 2006).

known as Ultra-large Crude Carriers (ULCC)). Box 6.1 provides an estimate of the quantity of crude transiting the high seas at any given time.

Oil spills are an almost inevitable consequence of the large quantities of oil present on the ocean at any given time. Table 6.6 presents a summary of ships, dates, locations, and quantities of oil spilled for the largest accidents over the past 40 years (ordered in terms of amounts of oil spilled). The quantity of oil released into the ocean is at best an imperfect indicator of potential environmental damage. In general, spills that occur in the open ocean are less serious than those that take place in coastal regions (there is a greater opportunity for oil to be dispersed). Spills in warm climates are generally less serious than those in cold climates (warmer temperatures promote evaporation of the lighter, generally more hazardous, components of the oil).

The Exxon Valdez ran aground on Bligh Reef in Prince William Sound, Alaska, on March 24, 1989 carrying 53 million gallons of oil, 10.8 million gallons of which were released into the ocean impacting eventually 1900 km of Alaskan coastline. Based on the quantity of oil spilled, the Exxon Valdez accident rates number 35 on the list of accidents included in Table 6.6. The accident, however, occurred in an ecologically sensitive coastal environment at just about the worst possible time of year, when biological activity is at a peak at the end of the long Arctic winter. The Exxon Valdez Oil Spill Trustee Council (a state and federal group set up to manage post spill restoration and research overseeing $900 million Exxon was obliged to pay in 1991 to settle a civil suit) estimates that the spill was responsible for the deaths of more than 250,000 birds, 2800 sea otters, 300 seals, 250 bald eagles, and billions of salmon and herring larvae. When clean-up efforts eventually got underway, up to 10,000

Box 6.1 _____

Estimate the total quantity of oil in transit on the ocean at any given time.

Solution:

Average daily production of oil globally amounted to 81 million barrels in 2005 (BP, 2006).

Approximately 60% of oil produced globally is transported to market by ship:

Shipment = (0.6) 81 million barrels/day = 49 million barrels/day

Assume an average shipment time of 20 days:

$$\text{Total oil in transit} = (49 \text{ million barrels/day}) (20 \text{ days})$$
$$= 980 \text{ million barrels}$$
$$= 133 \text{ million tons}$$

Approximately 3500 ships are involved in the global oil trade. Assuming that only about a third of these ships are carrying oil at any give time (the rest are either on their way back to acquire a new consignment, are loading or unloading, or undergoing repairs), estimate the load per ship:

$$\text{Load per ship} = \frac{980 \text{ million barrels}}{1200}$$
$$= 817,000 \text{ barrels}$$
$$= 110,000 \text{ tons}$$

Approximately one-third of the oil is carried by VLCCs with capacities of more than 240,000 tons (1.76 million barrels).

The quantity of oil in transit would be enough to cover an area of 16,000 km^2 with a 1 cm thick slick.

At $60 a barrel, the value of the oil in transit is equal to ($60/barrel) (980 million barrels) = $58.5 billion.

---

workers, 1000 boats, and 100 aircraft were mobilized at a cost of more than $3 billion to Exxon (Allen, 1999). Exxon was found guilty subsequently of reckless behavior in connection with the accident and ordered to pay $5 billion to parties judged to have been economically damaged by the spill (including 34000 fishermen), an amount subsequently reduced to $2.5 billion on December 22, 2006 as a result of the latest of a series of appeals filed by Exxon.

Following the Exxon Valdez accident, the U.S. Congress passed the Oil Pollution Act of 1990 requiring that all tankers entering territorial waters of the U.S. should be double-hulled by 2015. A similar measure was enacted by the European Union in response to the accident involving the Prestige in 2002 (spill number 20 in Table 6.6: 63,000 tons). All tankers carrying oil into European territory are required to be double-hulled by 2010. Despite these precautions, oil spills remain a serious environmental issue. Over 1.1 billion

Table 6.6 Major oils spills since 1967 (International Tanker Owners Pollution Federation Limited)

Position	Shipname	Year	Location	Spill size (tons)
1	Atlantic Empress	1979	Off Tobago, West Indies	287,000
2	ABT Summer	1991	700 nautical miles off Angola	260,000
3	Castillo de Bellver	1983	Off Saldanha Bay, South Africa	252,000
4	Amoco Cadiz	1978	Off Brittany, France	223,000
5	Haven	1991	Genoa, Italy	144,000
6	Odyssey	1988	700 nautical miles off Nova Scotia, Canada	132,000
7	Torrey Canyon	1967	Scilly Isles, UK	119,000
8	Sea Star	1972	Gulf of Oman	115,000
9	Irenes Serenade	1980	Navarino Bay, Greece	100,000
10	Urquiola	1976	La Coruna, Spain	100,000
11	Hawaiian Patriot	1977	300 nautical miles off Honolulu	95,000
12	Independenta	1979	Bosphorus, Turkey	95,000
13	Jakob Maersk	1975	Oporto, Portugal	88,000
14	Braer	1993	Shetland Islands, UK	85,000
15	Khark 5	1989	120 nautical miles off Atlantic coast of Morocco	80,000
16	Aegean Sea	1992	La Coruna, Spain	74,000
17	Sea Empress	1996	Milford Haven, UK	72,000
18	Katina P	1992	Off Maputo, Mozambique	72,000
19	Nova	1985	Off Kharg Island, Gulf of Iran	70,000
20	Prestige	2002	Off the Spanish coast	63,000
35	Exxon Valdez	1989	Prince William Sound, Alaska, USA	37,000

tons of oil were spilled in the decade of the 1990s with an additional 163 million tons released over the first 6 years of the 2000s (International Tanker Owners Pollution Federation Limited (2007)).

Transportation is responsible for about 67% of all oil consumed in the U.S. Industrial uses account for an additional 23% with the balance contributed by a combination of the residential/commercial and electric utility sectors. Gasoline used to drive automobiles and light trucks accounts for more than 44% of total oil consumed in the U.S. Gasoline consists of a complex mix of up to 200 different hydrocarbons including C_4–C_{12} paraffins (see Table 6.4), olefins (molecules in which two of the carbon atoms are linked by a double bond), aromatics (so-called because of their characteristic odor, examples of which include benzene and toluene), and selected oxygenated species supplied to enhance the octane content of the fuel (including most recently ethanol).[4]

A fraction of these chemicals is released to the atmosphere as a result either of evaporation (both from vehicles and from storage facilities employed to fuel them) or incomplete combustion. Combustion accounts for additional emissions of species including NO_x and CO.

The transportation sector is a major contributor to the problem of air pollution. A list of some of the more abundant hydrocarbon species observed in the atmosphere over Southern California during the summer of 1987 is presented in Table 6.7. Photochemical reactions involving anthropogenic sources of nitrogen oxides and hydrocarbons lead to the production of high regional levels of ozone with important negative impacts not only for public health but also for the environment more generally (including agriculture— high levels of ozone can damage the photosynthetic apparatus of plants). Oil and its various downstream uses are largely responsible for these problems. The transportation sector is also responsible for an important source of CO_2, 1.96 billion tons in the U.S. in 2005 out of a total of 6.00 billion tons (Energy Information Administration, 2005). Box 6.2 provides an estimate of the total amount of CO_2 produced by driving 20,000 miles with an average fuel efficiency of 20 mpg. Implications of CO_2 emissions for climate change are discussed further in Chapter 13.

6.6 PROSPECTS FOR OIL'S FUTURE

Future prospects for oil are unquestionably limited. It took approximately 300 million years to form the oil we have used to date and what remains still in the ground. Total consumption up to now amounts to about 900 billion barrels. BP (2006) estimates present worldwide reserves (resources still in the ground) of conventional oil (the type that can be pumped in liquid form from the ground) at 1200 billion barrels, suggesting that the total quantity of oil formed by natural agents over the past 300 million years amounted to 2100

Table 6.7 Mixing ratios (ppb carbon) for some of the most abundant non-methane hydrocarbons observed in Southern California air during the summer of 1987 (Lurmann and Main, 1992)

Species	Chemical formula	Mixing ratio (carbon content in ppb)
Propane	C_3H_8	56.0
i-Pentane[*]	C_5H_{12}	52.4
Toluene	C_7H_8	49.1
Butane	C_4H_{10}	42.0
Ethane	C_2H_6	27.1
Xylene	$C_6H_5(CH_3)_2$	25.2
Pentane	C_5H_{12}	24.0
Acetone	$(CH_3)CO(CH_3)$	22.4
Ethene	C_2H_4	22.3

[*]i-Pentane is one of three possible isomers (different skeleton forms) of pentane.

Box 6.2

Estimate the total quantity of CO_2 produced by driving a car 20,000 miles a year with a fuel efficiency of 20 mpg.

Solution:

The CO_2 produced by burning a typical gallon of U.S. gasoline is equal to 8.877 kg.

With our assumptions, driving 20, 000 miles would consume 1000 gallons of gasoline corresponding to emission of 8877 kg or 8.877 tons of CO_2.

Assuming an average cost per gallon of $3, you would have paid $3000 for this CO_2 corresponding to an average price of $338 per ton of CO_2 emitted. If you were required to pay a tax for carbon emitted and if this tax were levied at a level of $1 per gallon of gasoline, it would correspond to a tax of $113 per ton of CO_2.

Table 6.8 Reserves and R/P values for top 10 oil countries

Country	Reserves (billion barrels)	R/P (years)
Saudi Arabia	264.2	65.6
Iran	137.5	93.0
Iraq	115.0	>100
Kuwait	101.5	>100
United Arab Emirates	97.8	97.4
Venezuela	79.7	72.6
Russian Federation	74.4	21.4
Kazakhstan	39.6	79.6
Libya	39.1	63.0
Nigeria	35.9	38.1
Global Total	1200.7	40.6

billion barrels implying an average rate for formation of 7000 barrels per year, clearly negligible when compared with current consumption of close to 30 billion barrels a year (BP, 2006). The ratio of present identified reserves (R) to production by the global oil industry (P) provides a useful (if imperfect) measure of how much time we have left. The BP data implies a value for global R/P of 40.6 years. A summary of reported reserves for the ten largest oil-producing countries together with relevant R/P values is presented in Table 6.8 (BP, 2006).

Exploitation of a non-renewable resource such as oil follows a predictable pattern as discussed more than 50 years ago by Hubbert (1956). Utilization rates were low in the early years, increasing rapidly as the resource found a cooperative market, reaching a peak when approximately half was depleted. The overall behavior of the production function exhibits a bell-shaped curve.

The behavior on the declining portion of the curve depends on a number of factors including the price for extraction of the remaining resource and the price and availability of alternatives. Hubbert used his analysis to correctly predict the peak in oil production for the U.S. (which occurred in 1970) almost 20 years in advance. Applying his formalism to the global industry, accepting the BP figures for current reserves, suggests that we may be now close to the peak in global production. Various studies (with different assumptions with respect to reserves) place this peak (the Hubbert peak) somewhere between 2005 and 2015 (Hirsch et al., 2005; Bentley, 2002).

Estimates for amounts of recoverable oil are necessarily uncertain. They depend for example on the prevailing price of oil. The higher the price, the more profitable it is for producers to expend funds to enhance recovery, either by searching for new resources or by increasing recovery of oil from existing resources. During the early stages of exploitation of an oil reservoir, so-called primary recovery, oil flows to the well bore driven by ambient pressure in the reservoir. As little as 10% of the oil in the reservoir is brought to the surface, however, during this phase of production. Recovery rates can be enhanced by injecting water into the reservoir under pressure displacing an additional fraction of the less mobile oil. This practice, referred to as secondary recovery, has more than doubled the yield of oil from existing fields in the U.S. over the past 50 years. Even with secondary recovery, however, as much as 60 to 70% of the oil remains in the ground. There have been important investments in recent years in what is referred to as tertiary or enhanced oil recovery (EOR), encouraged by higher prices for oil. An increasingly important approach involves injecting CO_2 into the reservoir. The CO_2 acts as a solvent, increasing the mobility of the oil remaining in the reservoir, enhancing prospects for its recovery.

The CO_2 injection technique was applied first in Scurry County, Texas, in 1972 and is now widely used throughout the Permian Basin of West Texas and to a limited extent in a variety of other mature oil fields (DOE, 2006). Until recently most of the CO_2 used in this process came from geological sources (sub-surface gas reservoirs). Increasingly, though, companies are exploring opportunities to make use of industrial sources of CO_2, reducing in the process emissions to the atmosphere of this important greenhouse agent—an important added benefit. And there is promise that this initiative could be progressively more profitable in the future, both for the party supplying the CO_2 and for the party using it to enhance oil recovery. The Dakota Gasification Company operates a plant in Beulah, North Dakota (set up originally as a demonstration facility with funding from the U.S. government). The plant currently converts 18,000 tons of lignite coal daily to 170,000 cft of syngas. The CO_2 produced in this facility is piped 300 km to the Weyburn oil field operated by Encana in Southeastern Saskatchewan, Canada. There it is pumped 1500 m into the ground cutting the viscosity of the thick residual oil in the oil reservoir by a factor of 4. The investment is expected to result in production of 130 million barrels of oil that would otherwise have remained in the ground, providing at the same time an income for the Dakota Gasification Company of more than $150,000 a day for a commodity that would be regarded otherwise as a source of pollution (Fairley, 2005; DOE, 2006).

As supplies of conventional oil run down, there are opportunities to tap large reserves of non-conventional oil present for example in the tar sands of Alberta, Canada, or in the Orinoco Belt of Venezuela. The oil present in tar sands is not strictly oil but rather a thick heavy form of bitumen with the consistency of tar (hence the name). There are two ways to extract the tar from the sand. Either it can be mined and separated subsequently at the surface or it can be heated in situ causing it to flow, allowing it to be piped to the surface. The in situ technique makes use increasingly of what is known as steam assisted gravity drainage (SAGD). This involves drilling two wells into the tar deposit, one at the top, the other reaching a few meters below the bottom (the wells can extend over large distances in the horizontal direction). Steam, produced by burning natural gas, is injected into the upper well at temperatures as high as 300°C causing the bitumen to melt and flow under gravity to the lower well from which it can be pumped to the surface. The surface mining process involves some of the largest trucks and power shovels that can be found anywhere in the world. Fully loaded, one of these trucks can weigh more than the equivalent of two Boeing 747s and it takes two tons of tar sands to produce a single barrel of synthetic crude according to Blum (2005).

Production of oil from the Alberta tar sands has now reached a level of more than a million barrels a day accounting for 26% of Canada's oil production and is expected to double by 2010. The Alberta Energy and Utilities Board estimates the potential oil content of Alberta's tar sands at 1.6 trillion barrels of oil, of which approximately 11% (175 million barrels) can be extracted economically under present market conditions. The current cost for production of oil from the Albertan tar sands is about $30 a barrel. The breakeven price is likely to rise, however, in the future as companies are required to expend significant funds for environmental remediation: to clean up contaminated water produced in the combined extraction/production process and to restore vegetation to landscapes denuded by current extraction practices. The devastation exacted by current extraction practices is illustrated dramatically in Fig. 6.5.

Production of oil from tar sands is an energy intensive process: net emission of CO_2 per barrel of tar sands derived oil (accounting both for production and consumption) is reported to be more than two-and-a-half times greater than that associated with consumption of conventional oil (Sierra Club of Canada, 2005). Similar problems are associated with extraction of oil from oil shale, which, it is estimated, could contribute as much as 630 billion barrels to U.S. domestic reserves of oil (Holland and Petersen, 1995). Adding the potential sources of oil from tar sands (including both the resources in Canada and Venezuela) and oil shale could more than double current estimates for global oil reserves. It would do so, however, at a cost in terms of significantly increased emissions of CO_2.

Figure 6.5 Suncor's Millennium Mine (Kolbert, 2007).

6.7 CONCLUDING REMARKS

The health of the global economy is inextricably linked, at least in the near term, to prospects for a continuing reliable supply of affordable oil. Twice over the past 35 years—during the Arab embargo imposed in 1973 following the Yom Kippur war and then again during the Iranian hostage crisis in 1979—we experienced first hand our vulnerability to even modest interruptions of supply. Global demand for oil has increased by almost 25% since 1980 with important new players on the international stage including most notably China. The resources available to meet this increasing demand remain heavily concentrated, however, in a small number of potentially unstable parts of the world, the majority of them in the Middle East as indicated in Table 6.8.

The roots of the current instability in the Middle East may be attributed in no small measure to the vast sums of capital that have flowed into the region in response to the world's insatiable demand for oil. This newfound wealth has posed an important challenge to traditional values in the region. It has provided an opportunity for the region to modernize but forced it at the same time to select between different, often conflicting, paths for modernization. The overall response, as Huntington (1996) terms it, has been an Islamic Resurgence prompted in no small measure by a desire "to reverse the relations of domination and subordination that had existed [previously] with the West," and to revive "Islamic ideas, practices, and rhetoric and the rededication to Islam by

Muslim populations." Again quoting Huntington (1996), "the Resurgence is mainstream not extremist, pervasive not isolated," and "to ignore the impact of the Islamic Resurgence on eastern hemisphere politics in the late twentieth century is equivalent to ignoring the impact of the Protestant Reformation on European politics in the late sixteenth century."

The Resurgence, however, is not monolithic. The Islamic world is split between Sunni and Shiite, no less than the Christian world was divided between Roman Catholicism and Protestantism in the fifteenth century. It includes both moderates and extremists (western-style realists in contrast to suicide bombers and international terrorists). The West, specifically the United States, has been regrettably slow to grasp the significance and historical underpinnings of this Islamic renaissance, the extent to which it is based on a rejection of values accepted by many in the West as universal. We need to understand the background to these differences—and quickly—if we are to avoid becoming immersed in an impossible maelstrom of Middle Eastern political conflict. From a purely selfish point of view, the interests of the West are directed to forestall a potentially serious disruption of essential oil supplies. Adopting a longer range, more progressive, perspective, it is important that we assert essential core values, principles that reflect universal rather than simply parochial western values. If we are to assume the high ground, we need to ensure that these values are not subverted by overriding requirements to ensure a reliable continuing supply of oil. But, if we are to do so, it is essential that we wean ourselves off what President Bush described as our addiction to imported oil. The challenge is to do so in a manner that is not only economically advantageous but also constructive with respect to the environment.

NOTES

1 Carbon atoms in the interior of alkane molecules are linked by single bonds to two carbon neighbors and to two neighboring hydrogen atoms (the electronic structure of carbon is such that the atom can accommodate four single bonds). This accounts for the factor $2n$ in the formula defining the hydrogen composition of the alkane. The extra two hydrogen atoms are associated with the carbon atoms at the end of the molecule (carbon atoms that have only a single carbon neighbor and are able therefore to tie on to an extra hydrogen atom).

2 Decay of radioactive elements in the Earth's interior is primarily responsible for the outward transfer of heat reflected in the geothermal gradient. The upward flux of heat (energy) is higher in regions that are tectonically active (in the Basin and Range Province of the Western U.S. for example), lower elsewhere.

3 (http://en.wikipedia.org/wiki/Petroleum)

4 The octane number of a fuel is a measure of the extent to which the fuel–air mixture can be compressed without causing the fuel to ignite spontaneously causing the engine to knock or ping with resulting loss of power. A fuel consisting of pure isooctane is assigned an octane number of 100 with straight chain heptane, which ignites very easily is assigned an octane number of 0. Alternative fuels are rated for their antiknock characteristics based on this standard. The octane number for pure ethanol is 115.

REFERENCES

Allen, S. (1999). Worst oil spill in U.S. has lingering effects for Alaska industries. *Boston Globe*, March 7, 1999. http://www.jomiller.com/exxonvaldez/bostonglobe.html, read January 21, 2007.

Bentley, R.W. (2002). Oil and gas depletion: an overview. *Energy Policy*, **30**, 189.

Blum, J. (2005). Where oil is mined, not pumped. *Washington Post*, June 15, 2005. www.washingtonpost.com/wp-dyn/content/article/2005/06/14/AR2005061401533. html. Read January 27, 2007.

BP Statistical Review of World Energy (2006). Quantifying energy. http://www. bp.com/liveassets/bp_internet/globalbp/globalbp_uk_english/reports_and_pub-lications/statistical_energy_review_2006/STAGING/local_assets/downloads/pdf/ oil_section_2006.pdf. Read January 1, 2007.

DOE (2006). U.S. Department of Energy, Fossil Energy Office of Communications. Enhanced oil recovery/CO_2 injection. October 18, 2006. http://www.fossil.energy. gov/programs/oilgas/eor/index.ht, read January 22,2007.

EIA (2005). Energy Information Administration, U.S. Department of Energy, Emis-sions of greenhouse gases in the United States 2005, DOE/EIA-0573.

EIA (2006). Energy Information Administration, U.S. Department of Energy, Interna-tional Energy Annual 2004, Report released May-July 2006, http://www.eia.doe.gov/ iea/ Read January 1, 2007.

EIA (2007a). Energy Information Administration, U.S. Department of Energy, International Petroleum (Oil) Imports and Exports, http://www.eia.doe. gov/emeu/cabs/topworldtables1_2.html, http://www.eia.doe.gov/emeu/cabs/ topworldtables3_4.html, read on January 22, 2007.

EIA (2007b). Energy Information Administration, U.S. Department of Energy, Crude Oil and Total Petroleum Imports Top 15 Countries, November 2006 Import Highlights: Released on January 16, 2007.

http://www.eia.doe.gov/pub/oil_gas/petroleum/data_publications/company_level_ imports/current/import.html, read January 22, 2007.

Fairley, P. (2005). Carbon dioxide for sale. *MIT Technology Review*, July 2005. http:// www.technologyreview.com/article/16270/. Read January 25, 2007.

Hirsch, R.L, Bezdek, R., and Wendling, R. (2005). Peaking of world oil production: impact, mitigation and risk management. Paper prepared for the National Energy Technology Laboratory of the U.S. Department of Energy by Science Applications International Corporation. Paper quoted by Olah et al. (2006).

Holland, H.D. and Petersen, U. (1995). *Living Dangerously: The Earth, Its Resources and the Environment*. Princeton: Princeton University Press.

Hubbert, M.K. (1956). Nuclear energy and the fossil fuels, American Petroleum Institute Drilling and Production Practice. Proceedings of the Spring Meeting, San Antonio, March 7–9, 1956.

Huntington, S.P. (1996). *The Clash of Civilizations and the Remaking of World Order*. New York: Simon and Schuster.

International Tanker Owners Pollution Federation Limited (2007). http://www.itopf. com/stats.html. Read January 20, 2007.

Kolbert, E. (2007). Unconventional crude: Canada's synthetic-fuels boom. The New Yorker, November 12, p. 46.

Lurmann, F.W. and Main, H.H. (1992). Analysis of the ambient VOC data collected in the Southern California Air Quality Study. Final Report. ARB contract No.A832-130, California Air Resources Board, Sacramento, California.

McElroy, M.B. (2002). *The Atmospheric Environment: Effects of Human Activity.* Princeton: Princeton University Press.

Olah, G.A., Alain Goeppert, A., and Surya Prakash, G.K. (2006). *Beyond Oil and Gas: The Methanol Economy.* Wiley-VCH: Weinheim.

Turco, R.P., Toon, O.B., Ackerman, T.P., Pollack, J.B., and Sagan, C. (1983). Nuclear winter: global consequences of multiple nuclear explosions. *Science,* **222** (4630), December 23, 1983.

Sierra Club of Canada (2005). News release, April 22, 2005. www.sierraclub.ca/national/media/item.shtml?x=823 - 17k. Read January, 27, 2007.

Yergin, D. (1991). *The Prize: The Epic Quest for Oil, Money and Power.* New York: Free Press.

7

Natural Gas: Origin, History, and Prospects

7.1 INTRODUCTION

Natural gas, composed mainly of methane, is the cleanest of the fossil fuels—the environmentally most benign. Not surprising, with increasing recognition of the negative environmental impacts of coal and the volatility of oil prices (compounded by uncertainty with respect to the security of future supplies), use of natural gas has increased significantly in recent years. Natural gas accounted for 15.9% of total global primary energy supply in 1973. By 2004, it had risen to 20.7% and, at least according to one model, is projected to grow to more than 22% by 2030 (EIA, 2008).

The primary environmental issue relating to the use of natural gas involves its role as a source of CO_2. Natural gas accounted for 14.4% of global emissions of CO_2 from combustion of fossil fuels in 1970, as compared to 34.9% for coal and 50.7% for oil. The comparable percentages for 2004 were 19.9% for natural gas, 40% for coal, and 39.9% for oil (IEA, 2007). Trends in global emissions of CO_2 over the period 1973 to 2005 for the different fuel sources are illustrated in Fig. 7.1 (here the category other refers to minor sources relating to combustion of industrial and municipal wastes).

The energy content of natural gas, expressed in chemical form, was harnessed from the sun by photosynthesis over millions of years in the past. The origin of natural gas was basically similar to that of oil. The organisms that provided the feedstock transformed eventually into oil and gas lived primarily in the ocean. The life cycle of these organisms was relatively brief compared with the life cycle of the terrestrial plants that provided the source material for coal. When they died, a fraction of their body parts survived scavenging by other organisms to reach the underlying marine sediment. If the productivity

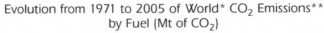

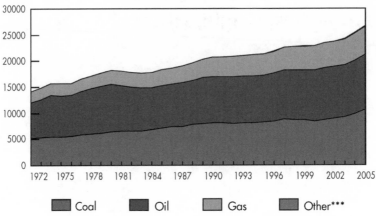

Figure 7.1 Evolution from 1971 to 2005 of world CO_2 emissions by fuel (Mt of CO_2) [IEA, 2007].

of the overlying water was high enough, the supply of organic material in the form of dead bodies reaching the sediment would have exceeded the capacity of organisms living in the sediment to consume it.

A portion of the chemically reduced carbon in the organic matter reaching the sediment would have been consumed near the sediment–water interface by oxygen breathing (aerobic) organisms, converted to CO_2, and released back into the water column where it would be transformed to HCO_3^-, the dominant form of inorganic carbon in the ocean. The supply of oxygen in high productivity regions would have been insufficient, however, to keep up with the metabolic requirements of these aerobic scavengers. Anaerobic bacteria would have taken over some distance below the water sediment interface. A fraction of the detrital organic carbon would have been converted in this case to methane (CH_4) but relatively little of this methane would have been conserved. Most of it would have been released to the water column where it would have been oxidized to CO_2 and transformed subsequently to HCO_3^- (CH_4 would have provided a ready source of food for the aerobic organisms living in the oxygen-rich ocean water). If productivity was high enough, however, a significant fraction of the organic matter reaching the sediment would have survived, however, to be buried under subsequent accumulations of sediment.

With increasing deposition of sediment, the surviving organic matter would have been subjected to steadily increasing temperatures and pressures (the rise in temperature with depth reflecting conduction of heat from the underlying mantle responsible for the so-called geothermal gradient). As temperatures rose to values in excess of about 60°C (typically at a depth of about 2 km in the accumulating sediment), the organic matter would have broken down to form the complex mix of hydrocarbons responsible for the formation of oil (see Chapter 6). At higher temperatures and greater depths (temperatures in excess of 160°C, depths greater than about 5 km), only the simplest of the hydrocarbon products would have survived. Methane and low molecular weight compounds such as ethane (C_2H_6) and butane (C_4H_{10}) would have represented the end products of chemical processing in this case. At even greater depths and higher temperatures (temperatures greater than about 230°C, depths in excess of 7 km), methane and the other light hydrocarbons would also have been unstable. The ultimate product of the carbon surviving to the elevated temperatures and pressures characteristic of this region would have been graphite (the soft black form of carbon employed as lead in pencils). The low molecular weight hydrocarbons that would eventually provide the dominant source of natural gas would be formed below the oil-producing region but above the zone favoring production of graphite.

The low molecular weight hydrocarbons responsible for the formation of natural gas are lighter—their density is lower—than the hydrocarbons implicated in the formation of oil. They will tend to migrate therefore, rising above the oil-forming compounds. In principle, they could find their way to the surface where they would be released either to the atmosphere or to the ocean and would be unavailable consequently as an economically harvestable source of commercial energy. If, however, along the way they encounter a region capped by impermeable rock (shale for example) they may be trapped, locked in place, providing thus an abundant energy source to be harvested by a technologically adept, energy hungry, society many millions of years subsequent to the time when the energy they contain was first captured from the sun by microscopic photosynthetic organisms living in an unusually favorable (nutrient rich) region of the world's ocean.

The composition of gas differs typically from reservoir to reservoir. Representative data are summarized in Table 7.1. Gas containing high concentrations of sulfurous compounds is referred to as sour. Gas brought to the surface is normally treated to remove these impurities prior to transmission to end-users. The ultimate product is essentially pure methane with trace quantities of ethane, propane, and butane. It is conventional though to add a small concentration of an odorous compound (t-butyl mercaptan for example) as an aid in leak detection (the smell serving to alert users to the presence of the leak).

The history of the development of natural gas is discussed in Section 7.2. The development of the infrastructure for its distribution is the subject of Section 7.3, which includes a discussion of the increasing importance of transport of the resource in the form of liquefied natural gas (LNG). The geographic distribution of natural gas and its capacity to serve anticipated future demand is treated in Section 7.4. Concluding summary remarks are presented in Section 7.5.

Table 7.1 Range of composition of gas recovered from different gas reservoirs

Component	Weight (%)
Methane (CH_4)	70–90
Ethane (C_2H_6)	5–15
Propane (C_3H_8) and butane (C_4H_{10})	<5
CO_2, N_2, H_2S, He, and other trace species	Balance

7.2 HISTORY

Humans were aware of the properties of natural gas long before they knew the nature and origin of the gas itself. Gas leaking out of the ground would frequently catch fire, ignited for example by a lightning stroke. One of the most celebrated of these leaks, on Mount Parnassus in Greece, prompted the ancient Greeks, more than three thousand years ago, to identify that site as the center of the Earth and Universe.[1] They erected a temple to Apollo on the spot and used the escaping gas to fuel a perpetual flame. Mystical properties were attached to this mysterious flame that could burn without need for an extraneous supply of fuel. The location became the home subsequently for the famous Oracle of Delphi.

The first recorded productive use of escaping gas was in China, dated at approximately 500 BC. The Chinese constructed a primitive pipeline using stems of bamboo. The gas was transported to a site where it was used to boil brine to provide a source of both economically valuable salt and potable water (Olah et al., 2006). Almost 2400 years would elapse before gas would first be tapped for productive use in the West.

A well drilled near Fredonia, New York, was used to provide an energy source for street lighting in 1821. The Fredonia Gas Light Company, formed in 1858, was the first commercial entity established specifically to market natural gas. The well drilled under the direction of Colonel Drake near Titusville, Pennsylvania, in 1859 (Chapter 6) was found to provide not only a source of oil but also of gas. A portion of the gas was transported from the wellhead to Titusville over a distance of about 9 km using a 5 cm diameter pipeline. Again the primary use was for lighting. J.N. Pew, an important figure in the early U.S. oil industry, formed a company that delivered gas to Pittsburgh in 1883. The gas was used there as a substitute for manufactured coal gas (so-called town gas). Appreciating the potential economic value of natural gas, previously regarded as an irritating side product of oil production (it had to be disposed of, usually by combusting it under controlled conditions on site, to avoid fires that might develop and burn out of control disrupting operations while endangering the lives of those working in the oil field), the Natural Gas Trust was formed in 1886 by Standard Oil (Rockefeller had a keen nose for all potential sources of profit). Pew sold out his gas interests to Standard Oil a few years later.

The Bunsen burner introduced in 1855 by the German chemist Robert Bunsen (1811–1899), a refinement of an earlier invention by Michael Faraday (whose contributions to the development of electricity are discussed in some detail in Chapter 11), provided the stimulus for development of a wider market for natural gas towards the end of the nineteenth century. Bunsen's burner employed a controlled, optimal, mixture of air and gas which when ignited produced a blue flame. His original motivation in developing the device was to use it in connection with his laboratory investigations of the spectroscopic properties of light emitted by bodies heated by the burner, work conducted later in collaboration with Gustav Kirchoff (1824–1887) best known for the association of his name with Kirchoff's rules in electric circuit theory. The Bunsen technology would be deployed later in a range of devices permitting gas to be consumed not just for lighting but also for cooking and heating.

The first long-distance pipeline was built in the U.S. in 1891. Some 120 miles in length, it was used to deliver natural gas to Chicago from a gas field in central Indiana. By modern standards, this pipeline was primitive: it relied on the pressure of gas at the wellhead to propel the gas to its ultimate destination. As discussed below, long-distance pipelines today employ compressors to maintain an optimal flow of gas along the transmission line. The Indiana–Chicago pipeline may be credited, however, with setting the stage for the more elaborate gas distribution system that would be established in the U.S. some 50 years later. Advances in welding techniques, pipe rolling, and metallurgy in the post World War II period played an important role in the development of the extensive infrastructure that exists today through which gas is transported securely and at relatively low cost over thousands of kilometers from source regions to accommodate demands in essentially all of the major metropolitan centers of the U.S.

The natural gas industry was established much later in Europe. The initial development was prompted by discoveries in the 1950s of gas deposits in Italy (on the Po Plain), France (Lacq), and the Netherlands (Groningen). The important expansion took place, however, in the 1960s exploiting the extensive reserves of gas (also of oil) found to be present in sediments underlying shallow waters of the North Sea.

7.3 INFRASTRUCTURE

The infrastructure for distribution of natural gas in the U.S. today involves more than 300,000 miles of pipeline, more than 1400 compressor stations employed to maintain a steady flow of gas in these pipelines, 394 underground gas storage facilities, 55 locations where gas can be imported or exported by pipeline (to and from both Canada and Mexico), and 5 facilities for the import of liquefied natural gas.[2]

A map of the interstate pipeline network is displayed in Fig. 7.2. Shaded regions in the figure indicate the 31 states (including the District of Columbia) that are more than 80% dependent on this network for their supplies of gas.

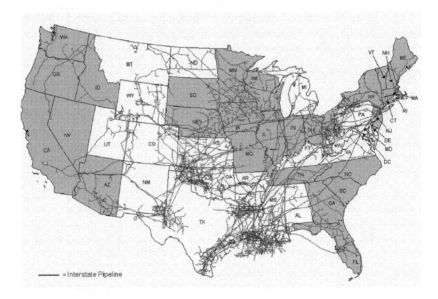

Figure 7.2 Interstate natural gas supply network (http://www.eia.doe.gov/pub/
           oil_gas/natural_gas/analysis_publications/ngpipeline/index.html, read
           November 4, 2007).

There are 29 hubs in the network at which gas can be directed to meet demands
in different parts of the country. In 2005, 85% of the 48 trillion cft of gas
transported throughout the U.S. moved through facilities owned by the major
interstate pipeline companies of which the thirty largest companies controlled
77% of the pipeline mileage and about 83% of its total capacity (EIA, 2007a, see
link above). Not all of the gas transported through the pipeline network is
delivered directly to end-users. A portion of the traffic is involved in moving
gas into and out of the various storage facilities.

As indicated in Chapter 3, retail customers in the U.S. are charged for natural
gas in units of therms, corresponding to a volume of 100 cft of gas delivered at
a temperature of 60°F at a pressure of 14.73 pounds per square inch, slightly
greater than the pressure of the atmosphere. The energy content of a therm is
equal to 10^5 Btu. Consumption of natural gas in the U.S. in 2005 amounted to
about 2.3×10^{16} Btu (2.3×10^{11} therms or 2.3×10^{13} cft), equal to about 23% of
total national energy use. The difference between these numbers and the
number quoted above for traffic through the pipeline system reflects presumably transfers of gas into and out of storage facilities.

Contracts for natural gas futures are traded on the New York Mercantile
Exchange based on prices for gas processed through the Henry Hub located in
Erath, Louisiana. Contracts are priced in units of dollars per million Btu, or
dollars per 10 therms. The closing spot price for natural gas quoted on the

Exchange on November 2, 2007, was $6.86 per contract, equivalent to 68.6 cents per therm, a decline of close to 10% with respect to the price that prevailed on the previous day (an indication of the considerable volatility of the market with the approach to winter with the related increased demand for natural gas in the U.S.). The comparable price for West Texas Intermediate crude oil was $10.88 per million Btu, almost 60% higher on an energy basis than the corresponding price for gas.

As indicated by the density of pipelines, there are three major source regions for natural gas in the lower 48 states of the U.S.: West Texas extending to Southeast New Mexico; the Gulf of Mexico from South Texas to Mississippi; and the Rocky Mountain region including portions of Utah, Colorado, and Wyoming. Gas is distributed from these source regions in trunk line pipes measuring up to 43 in. in diameter. More than 20 of the major interstate pipelines originate in the Southwest (including the Gulf region). Pipelines exiting this region have the capacity to carry up to 40.7 billion cft of gas per day of which 58% is delivered to the Southeast, 24% to the Central region of the country, 15% to the West, with the balance to Mexico (EIA, 2007a, cited above). The Rocky Mountain region supplies gas to the Western States, primarily California and Nevada with additional delivery to markets in Oregon and Washington, with further service to the Midwest—primarily to Iowa, Missouri, and eastern Kansas (EIA, 2007a). Two routes deliver gas from the Southwest to the Northeast. The first proceeds from East Texas and Louisiana through Mississippi, Tennessee, Kentucky, and Ohio, entering the Northeast either through West Virginia or Pennsylvania. The second extends northeastward from Mississippi by way of the Eastern Atlantic Coast States (EIA, 2007a). The pipelines servicing the Northeast are fully utilized in winter when demand for gas in the Northeast is greatest. Gas transported through these pipelines in summer is used mainly to replenish supplies withdrawn in winter from underground storage reservoirs located in West Virginia, Pennsylvania, Ohio, and New York (EIA, 2007a). More than 190 underground storage sites are located along these supply corridors (EIA, 2007a).

The U.S. also has significant proven reserves of natural gas in Alaska, estimated at 35 trillion cft, sufficient to accommodate about 18 months of current total U.S. national consumption, with potential for significant additional supplies that could be identified as a result of an aggressive future exploration program. Gross production of gas from the North Slope oil fields amounted to 3.15 trillion cft in 1999, of which 93% was reinserted into oil reservoirs to enhance secondary recovery of oil (Sherwood and Craig, 2001, http://www.mms.gov/alaska/re/natgas/akngases.htm, read November, 7, 2007). The energy content of gas contained in the Prudhoe Bay field is estimated to be comparable to that of the oil that remains in the ground (6 billion barrels out of the original content of 17 billion barrels as reported by Sherwood and Craig). Three options are under consideration to deliver the energy content of Alaskan gas to markets in the lower 48 states: one envisages converting the gas on site to liquid petroleum and transmitting it through the existing oil pipeline as current supplies of oil run down; the second contemplates construction of a pipeline to transfer the gas to a southern Alaskan port where it would be lique-fied (converted to LNG) for subsequent transshipment; the third contemplates construction of a pipeline to deliver the gas to the existing Canadian pipeline system through which it could be transferred to markets either in Canada or the U.S. The cost for option three (connection to the Canadian grid) is esti-mated at a minimum of $20 billion U.S. While the U.S. Congress has authorized a loan guarantee of $18 billion for the project (http://www.stateline.org/live/ViewPage.action?siteNodeId=136&languageId=1&contentId=58284read November 7, 2007), prospects for development of this new pipeline connection are presently unclear, reflecting in part concerns as to the possible damage its construction might cause to the fragile ecosystems over which it would inevitably have to pass.

Canada's gas resources are concentrated in two regions: portions of Alberta extending into northwestern British Columbia; and to a lesser extent in the east, specifically the region offshore of Sable Island. Imports accounted for 17% of natural gas consumed in the U.S. in 2006, an increase from 11% a decade earlier (EIA, 2007a). In terms of gas delivered by pipeline, Canada accounted for the bulk of these imports, 99.8% in 2006, with the balance from Mexico. There are four primary points at which gas is imported to the U.S. from Canada: Port of Morgan, Montana; Eastport, Idaho; Sherwood, North Dakota; and Noyes, Minnesota. The bulk of the imported gas is used to supplement require-ments in the West and Midwest. Gas from Sable Island, transported through the Canadian Maritimes and Northeast Pipeline connecting with the Portland Gas Transmission System in Wells, Maine, provides an important and growing source of supply for northern regions of New England including the Boston metropolitan area.

The European natural gas pipeline distribution system is illustrated in Fig. 7.3. Current demand for gas in the European Union (about 20 trillion cft per year) is comparable to that in the U.S. (about 23 trillion cft per year). Domestic production, primarily from the North Sea (which accounts for about 60% of total EU consumption today), is expected to decline however, by more than a third over the next several decades (Lochner et al., 2007). Supplies potentially available from neighboring regions, notably from countries of the former Soviet Union and from Iran and North Africa (including Nigeria), are considered adequate to meet increasing demand over at least the next several decades (Lochner et al., 2007) although there may be problems with the secu-rity of these supplies, particularly with those originating in the east and south-east (specifically from countries of the former Soviet Union and from Iran). Priority areas for expansion of the existing pipeline network, reflecting antici-pated future supplies, are indicated in the figure.

To an increasing extent, natural gas is being processed and delivered to mar-kets now in the form of LNG. Conversion to liquid has the advantage that the volume required to store a given content of energy can be reduced by a factor of 610 (the density of the liquid phase is 610 times higher than the density of

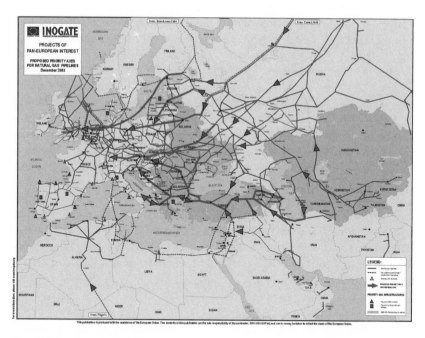

Figure 7.3 Europe's gas pipeline network (http://www.inogate.org/en/images/maps/
gas_map_big.gif, read Nov 8, 2007).

gas at a pressure of 1 atmosphere). Special facilities are required, however, to
effect the transition from gas to liquid. This takes place typically at a temperature of about –260°F (depending on the composition of the gas), with significant associated costs both in terms of capital to supply the necessary equipment
and energy to operate it (although the latter is usually small given the ready
availability and low cost of gas available in source regions where the liquefaction process is most likely to be implemented). Following liquefaction, the gas
is transported to market in specially designed container ships. Arriving at its
destination, the gas must be reconstituted prior to distribution to consumers
through the conventional pipeline system. Special facilities are required to
implement this regasification process. There are four sites at which LNG can be
received and regasified in the U.S.: Everett, Massachusetts; Lake Charles,
Louisiana; Elba Island, Georgia; and Cove Point, Maryland. EIA (2003) estimates
the breakdown of costs for the different elements of the LNG production/
transport/supply chain as follows: 15 to 20% for processing and delivery of gas
to the liquefaction plant; 30 to 45% for operations at the plant; 10 to 30% for
shipping; 15 to 25% for costs at the receiving station for unloading, storage,
regasification, and subsequent distribution of the product. With experience
and expansion of the LNG market, all of these costs have been on a downward
trajectory in recent years. The cost for a tanker capable of transporting 2.9 billion

cft of gas in the mid-1980s, for example, hit a peak of $280 million. The cost for a similar tanker delivered in November 2003 had dropped to about $155 million, reflecting primarily expansion of the number of shipyards available for construction of these tankers (EIA, 2003).

Asia and Oceania accounted for 65% of all worldwide imports of LNG in 2005: 4.5 trillion cft out a total world trade of 6.8 trillion cft. Europe ranked second with net imports of 1.7 trillion cubic feet (24% of the world total). Imports to North America, specifically the U.S., amounted to 630 billion cft (9% of the world total, less than 0.3% of gross national consumption). Indonesia and Malaysia accounted for 32% of total global exports, almost all of which were delivered to countries in East Asia, specifically to Japan (62%), South Korea (23%), and Taiwan (15%). European supplies of LNG came primarily from Africa (82% of the total), mainly from Algeria, Nigeria, and Egypt. The bulk of LNG imported into the U.S. was shipped from Trinidad and Tobago (70% of the total), with Algeria and Egypt accounting for 15% and 11% respectively. A summary of data for the 10 largest importers and exporters of LNG is presented in Table 7.2.

7.4 CONSUMPTION PRODUCTION AND RESERVES

On a national basis, the U.S. is the world's largest consumer of natural gas, accounting for 22% of total global consumption in 2006, with the Russian Federation occupying second position with 15.1% of the total (BP, 2008). From a regional perspective, Europe and Eurasia (excluding the Asia Pacific Region) accounted for 40.1% of global consumption in 2006, followed by North America (27.3%), Asia Pacific (15.3%), the Middle East (10.1), South and Central America (4.6%), and Africa (2.6%). Trends in regional consumption since 1981 are illustrated in Fig. 7.4. In relative terms, the growth in

Table 7.2 Major importers and exporters of LNG in 2005: data quoted in units of billion cft (http://www.eia.doe.gov/emeu/international/ LNGimp2005.html, read November 1, 2007)

Importers			Exporters		
Rank	Country	Imports	Rank	Country	Exports
1	Japan	2858	1	Indonesia	1123
2	South Korea	1075	2	Malaysia	1031
3	Spain	769	3	Qatar	987
4	United States	631	4	Algeria	886
5	France	453	5	Australia	581
6	Taiwan	340	6	Trinidad and Tobago	492
7	India	222	7	Nigeria	429
8	Belgium	95	8	Oman	336
9	Italy	88	9	Brunei	333
10	Portugal	60	10	United Arab Emirates	273

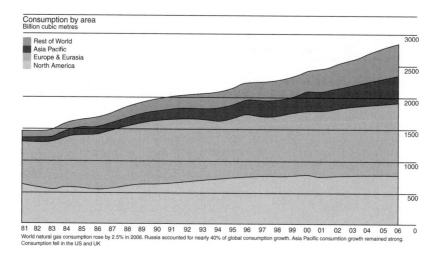

Figure 7.4   Trends in regional consumption of natural gas since 1981 (BP, 2007,
http://www.bp.com/productlanding.do?categoryId=6848&contentId=
7033471, read Nov 13, 2007).

consumption over the past decade was greatest in the Asia Pacific Region. This
region accounted for 10.6% of global consumption in 1996. By 2006, its share
had risen to 15.4%. Over the same period, Europe/Eurasia's share fell from
43.4% to 40%, while consumption in North America (as a fraction of the global
total) decreased from 33.5% to 27%.

Consumption data for the 10 largest consuming countries are summarized
in Table 7.3. This table includes information on changes in consumption for
these countries over the past decade with specific information on the relative
changes (expressed in percentage terms) for the period 2005 to 2006. Notable
in the latter case is the decrease reported for the U.S. in 2006 attributable to the
relatively warm conditions that prevailed in that country over the winter
months of November, December, and January, normally the period of peak
gas usage. In contrast, the summer months of July and August (for the same
year) were unusually warm. Demand for electricity for air conditioning
increased over this period with an associated rise in demand for gas by the
electric sector. Use of gas in the U.S. in the residential, commercial, and industrial sectors decreased in 2006 relative to 2005 by 9%, 6%, and 2% respectively.
Partially offsetting this reduction, consumption by the electric sector increased
by 6% (EIA, 2007b, http://www.eia.doe.gov/pub/oil_gas/natural_gas/feature_
articles/2007/ngyir2006/ngyir2006.pdf, read November 1, 2007).

Existing proven reserves of natural gas, as illustrated in Fig. 7.5, are concentrated to a large extent in two regions, the Middle East and Europe/Eurasia,
specifically countries of the former Soviet Union. The Russian Federation alone

Table 7.3 Major consumers of natural gas in 1996 and 2006 with percentage change from 2005 to 2006. Data quoted in units of billion cubic meters (1 cubic meter = 35.3 cubic feet). (from BP, 2007)

Rank	Country	Consumption (1996)	Consumption (2006)	Change 2005 to 2006 (%)
1	United States	640	620	−1.7%
2	Russian Federation	380	432	+6.7%
3	Iran	39	106	+2.7%
4	Canada	85	97	+5.7%
5	United Kingdom	83	91	−4.5%
6	Germany	84	87	+1.1%
7	Japan	64	85	+7.0%
8	Italy	52	77	−2.1%
9	Saudi Arabia	44	74	+3.5%
10	Ukraine	83	66	−8.8%

Table 7.4 Major producers of natural gas in 1996 and 2006 with percentage change from 2005 to 2006. Data quoted in units of billion cubic meters (1 cubic meter = 35.3 cubic feet). (from BP, 2007)

Rank (2006)	Country	Production (1996)	Production (2006)	Change 2005 to 2006 (%)
1	Russian Federation	561	612	+2.4%
2	United States	534	524	+2.3%
3	Canada	164	187	+0.6%
4	Iran	39	105	+4.1%
5	Norway	37	88	+3.1%
6	Algeria	62	85	−4.3%
7	Indonesia	68	74	+0.3%
8	Saudi Arabia	44	74	+3.5%
9	Turkmenistan	33	62	+5.9%
10	Malaysia	34	60	+0.4%

accounts for 26.3% of total world reserves. The published estimate (BP, 2008) for worldwide reserves (6168 trillion cft) is sufficient to accommodate current global demand (101 trillion cft per year) for a minimum of 61 years. Paralleling the situation discussed earlier for oil (Chapter 6), estimates of available reserves of natural gas have risen steadily, over the past 20 years, from 3800 trillion cft in 1986, to 5220 trillion cft in 1996, to 6168 trillion cft in 2006. They are likely to climb further in the years ahead with anticipated expanded investments in exploration, with advances in technology, and with expected increases in global demand. Consumption of natural gas in China, for example, amounted to a little less than 2 trillion cft in 2006, 9% of the quantity consumed the same year in the U.S. Demand in China increased, however, by more than a factor of 3 over the past decade, a trend that is likely to persist, probably to accelerate, in the future.

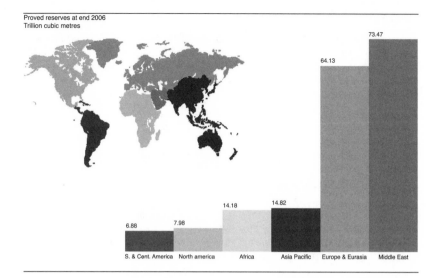

Proved reserves at end 2006
Trillion cubic metres

73.47

64.13

14.18   14.82

6.88   7.98

S. & Cent. America   North america   Africa   Asia Pacific   Europe & Eurasia   Middle East

Figure 7.5   Existing proven reserves of natural gas (BP, 2007, http://www.bp.com/ productlanding.do?categoryId=6848&contentId=7033471, read Nov 13, 2007).

The world is unlikely, we conclude, to run out of either gas or (as indicated in Chapter 6) oil in the foreseeable future. The key questions for the future of gas relate to the speed with which the world elects to invest in the infrastructure required to facilitate enhanced delivery of the product from source regions to markets (pipelines and LNG facilities) and, perhaps even more important, the security of regions and markets in which reserves are currently concentrated.

7.5 CONCLUDING REMARKS

Supplies of natural gas are sufficient to accommodate expected growth in the world's energy economy for the foreseeable future. Natural gas has the advantage that it is environmentally preferable to alternative fossil fuel sources of energy such as coal and oil. In particular, consumption of natural gas is associated with minimal emissions of particulates and acidic compounds such as SO_2. It has the disadvantage, however, that an unfettered, expanded, market in natural gas will inevitably contribute to an increase in emissions of CO_2 (less, though, than would be the case if the growth of the world's energy economy were to favor coal and oil rather than gas).

NOTES

1 (http://en.wikipedia.org/wiki/Delphi, read November 2, 2007).

2 (http://www.eia.doe.gov/pub/oil_gas/natural_gas/analysis_publications/ ngpipeline/index.html, read November 4, 2007).

REFERENCES

BP Statistical Review of World Energy (2008). June 2008. http://www.bp.com/liveassets/bp_internet/globalbp/globalbp_uk_english/reports_and_publications/statistical_energy_review_2008/STAGING/local_assets/downloads/pdf/statistical_review_of_world_energy_full_review_2008.pdf (read May 8, 2009).

EIA, Energy Information Administration, U.S. Department of Energy (2003). The global natural gas market: status and outlook. DOE/EIA-0637.

EIA, Energy Information Administration, U.S. Department of Energy (2007a). About U.S. natural gas pipelines.

EIA, Energy Information Administration, Office of Oil and Gas, U.S. Department of Energy (2007b). Natural gas year-in-review 2006.

EIA, Energy Information Administration, U.S. Department of Energy (2008) http://www.eia.doe.gov/oiaf/ieo/pdf/ieorefcase.pdf (read May 8, 2009).

Lochner, S., Bothe, D., and Lienert, M. (2007). Analyzing the sufficiency of European gas infrastructures—the Tiger Model. Conference Paper presented at ENERDAY 2007, Dresden, Germany, April 13, 2007. (http://www.ewi.uni-koeln.de/fileadmin/user/Veroeff/ENERDAY07_paper.pdf, read November 1, 2007).

Olah, G.A., Goeppert, A., and Surya Prakash, G.C. (2006). *Beyond Oil and Gas: The Methanol Economy.* Weinheim, Germany: Wiley-VCH.

Sherwood, K.W. and Cray, J.D. (2001). Prospects for development of Alaska natural gas: a review. U.S. Department of the Interior, Minerals Management Service, Alaska OCS Region.

8
Energy from Water and Wind

8.1 Introduction

For most of history, humans relied on a combination of personal muscle power and animals (horses, donkeys, oxen, and water buffalo) to do work. Only over the past few millennia did we develop the capacity to tap into the energy of running water. And applications of wind power came even later.

Energy in running water was harnessed first to turn millstones to grind grain. The earliest water mills employed a number of scoops immersed in running water, momentum transfer caused the scoops to rotate around a vertical axis. The vertical axle passed through the lower of the two stones that constituted the milling apparatus and was attached to the upper one. As the axle rotated responding to the force of the running water, the upper stone turned, grinding the grain inserted in between the two stones, the lower of which was held fixed in place. Mills such as this were widely deployed throughout Europe during medieval times and may have been introduced first in Northern Italy during the Roman period (Derry and Williams, 1960). There is no record of use of this form of waterpower in earlier civilizations, either in Mesopotamia or Egypt, presumably, Derry and Williams (1960) suggest, because the rivers there were either too sluggish or were subject to too frequent changes in surface elevation (which, given the design of the system, would have required the heavy milling apparatus to have been moved either up or down in response to the rise and fall of the river).

An alternate design placed the axle of the apparatus in the horizontal. The wheel could be made to turn then under the force of the water in the stream into which it was immersed, an approach referred to as the undershoot method. A more efficient strategy involved damming the stream and supplying water in a controlled fashion as needed (in a chute) to buckets at the top of the wheel. The wheel would be forced to turn then under the weight of the water and the buckets would empty their water into the stream when they reached the bottom of their trajectory. This approach is known as the overshoot method. Derry and Williams (1960) describe an overshoot Roman mill at Venafro (a Roman town situated in a hilly but fertile region on the border separating Campania, Abruzzo, and Lazio in east central Italy) with a wheel 7 ft in diameter that was capable of grinding 400 lb of corn an hour. By way of comparison, they state, two men using muscle power to turn a millstone would have been able to grind at most 10 lb an hour. Box 8.1 offers an estimate of the horsepower required to operate the mill at Venafro.

Waterpower came into wide use only in the later stages of the Roman Empire. There was little incentive earlier to invest capital in labor saving devices when labor, mostly slave, was widely available and cheap. Better, it was judged, to invest in monuments that would attest to the glory of the Roman state and culture and in roads and aqueducts—engineering marvels—that would ensure the continuing security of the Empire. By the fourth century AD, however, the situation had changed. Labor was then in short supply and priorities shifted to promote construction of large numbers of water-powered mills. Derry and Williams (1960) report that by 310 AD, as many as 16 overshoot wheels had been installed in the city of Barbegal (near Arles) alone with capacity to produce more than 3 tons of corn an hour, sufficient to feed a population

---

Box 8.1

Compute the horsepower (hp) generated by the overshoot mill at Venafro assuming that the work expended by two men was sufficient to grind 10 lb of corn in an hour. The mill at Venafro had an output of 400 lb per hour.

Solution:

A healthy man is able to put out work at average rate of about 100 W. Assume that half of this work can be deployed to rotate the millstone. It follows, according to our assumptions, that an expenditure of 100 W (two men working) is sufficient to produce 10 lb of ground corn.

1 hp is equivalent to 745.7 W. It follows that the horsepower required to grind 10 lb of corn is given by

$$(100 \text{ W}) \times (1/745.7 \text{ hp/W}) = 0.134 \text{ hp}$$
$$\text{Horsepower required to grind 400 pounds} = (40)(0.134) = 5.36 \text{ hp}$$

Derry and Williams (1960) quote a figure of about 3 hp for the work output of the Venafro mill, suggesting that the efficiency of the mill may have been somewhat higher than estimated here taking the output of useful work by the human agent as 50 W. We could reconcile the two estimates if we assumed an output of 60 W for the human agent.

of 80,000, almost eight times the population of Arles. Rome's late recognition of the advantages of mechanization would set the stage for extensive future developments of waterpower elsewhere in Europe. The Domesday Book records that by 1086 AD as many as 5624 water mills were in operation in England south of the River Trent. Waterpower was used not just to grind grain but for a myriad of other tasks including, but not confined to, sawing wood, hammering metals, crushing ore, and pumping the bellows of industrial furnaces (Derry and Williams, 1960).

It appears that Persians (Iranians) were first, during the seventh century AD, to appreciate the potential of wind as a source of mechanical power, although there may have been earlier applications of wind power to pump water in China. The Persian wind mills incorporated as many as 12 sails covered in either reed matting or cloth, rotating around a vertical axle similar to the design incorporated in the early water wheels. They were used both to grind grain and to pump water for irrigation (Derry and Williams, 1960). Windmills came into wide use in Europe later, during the twelfth and thirteenth centuries AD. The design that caught on in Europe was built around a horizontal rather than a vertical axle, a more efficient design in that a larger fraction of the available wind energy can be captured with this design and converted to rotary motion (more on this later). Windmills of this design were in common use in Northeastern Europe by the end of the thirteenth century. The earliest type, referred to as the post mill, was anchored to a strong vertical post around which the entire structure, sails and machinery, could be rotated to ensure that the sails could be pointed into the wind. Tower mills, a later innovation, included a fixed lower portion, constructed usually of either stone or brick, which contained the milling machinery. With this design, only the upper portion of the mill that carried the sails had to be rotated to catch the wind.

As was the case earlier with waterpower, wind power was deployed in a variety of applications, the most important of which arguably was its use to pump water to keep the fens and polders of Holland dry. At one time, as many as 8000 windmills were scattered across the Dutch countryside, the definitive feature of the Dutch landscape. Landes (1998) summed it up thus: "it was the windmill that made Holland." Wind power would play an important role later in the U.S. where it was used extensively to pump water from below ground (from aquifers) to satisfy the needs of early Midwestern farmers and to supply the copious quantities of water required by the steam powered locomotives of the early railroads (steam powered locomotives dominated rail transportation in the U.S. up to as recently as 1950).

We begin in Section 8.2 with an account of the role waterpower played in the early industrialization of New England. Today, wind and waterpower are employed mainly to generate electricity. The importance and future prospects of hydroelectricity are discussed in Section 8.3. The physics of wind, the fraction of solar energy converted ultimately to wind, is discussed in Section 8.4. The current status and future prospects of wind-generated electricity is treated in Section 8.5 with concluding remarks in Section 8.6.

8.2 WATERPOWER IN THE INDUSTRIALIZATION OF NEW ENGLAND

Precipitation in New England averages about 40 in. a year. A significant fraction of the water that falls on the region (as either rain or snow) is carried to the ocean by a relatively small number of rivers, the most important of which for Massachusetts and New Hampshire is the Merrimack. The Merrimack River drains an area of 5010 square miles (Fig. 8.1). The flow of water in the river, as measured in the lower regions of the catchment basin, averages annually about 6000 cft per second (cfs, the conventional unit used to describe the flow of water in rivers). As indicated in Box 8.2, approximately 41% of the precipitation that falls on the drainage basin of the Merrimac is carried off to the ocean by the river: the balance is returned to the atmosphere through transpiration and evaporation, mainly during the warm, dry, conditions of summer. Over

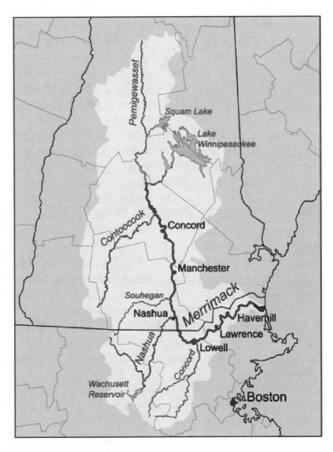

Figure 8.1 Merrimack River watershed. Created by Karl Musser based on USGS data (http://en.wikipedia.org/wiki/Image:Merrimackrivermap.png. Read on February 22, 2007).

Box 8.2

Given that precipitation averages 40 in. a year over the Merrimack drainage basin, that the drainage basin occupies an area of 5010 square miles, and that the flow of water in the river averages 6000 cfs, the question is: what fraction of precipitation reaches and is eventually removed by the river.

Solution:

Normally, we would prefer to carry out all calculations in the SI or mks system of units. But, since all relevant quantities are given here in English units, we'll use English units in this case.

$$\text{Area of the drainage basin} = 5.01 \times 10^3 \text{ (mile)}^2$$
$$1 \text{ mile} = 5.28 \times 10^3 \text{ ft}$$
$$1 \text{ (mile)}^2 = 2.79 \times 10^7 \text{ ft}^2$$
$$\text{Area of the drainage basin} = (5.01 \times 10^3)(2.79 \times 10^7) \text{ ft}^2$$
$$= 1.4 \times 10^{11} \text{ ft}^2$$
$$\text{Annual precipitation} = 4 \times 10^1 \text{ in.} = 3.3 \text{ ft}$$
$$\text{Volume of annual precipitation} = (1.4 \times 10^{11})(3.3) \text{ ft}^3 = 4.6 \times 10^{11} \text{ ft}^3$$
$$\text{Volume of precipitation per second} = 4.6 \times 10^{11}/3.15 \times 10^7 \text{ ft}^3 \text{ s}^{-1}$$
$$= 1.46 \times 10^4 \text{ ft}^3 \text{ s}^{-1}$$
$$\text{Fraction reaching the river} = 6 \times 10^3/1.46 \times 10^4 = 41\%$$

the course of its path from Lake Winnipesaukee in the north to the ocean in the southeast the Merrimack travels a total of about 200 miles dropping a modest average of 2.6 ft per mile. Much of this drop is concentrated, however, in six regions, three of which, near what are now the cities of Manchester, Lowell, and Lawrence, would evolve later as centers of the textile industry that would revolutionize life in nineteenth century New England, setting the stage for the mechanized industrial development (energized not just by falling water but later to a greater extent by coal and electricity) that would take hold subsequently in a variety of other locations in the United States with an expansive range of valuable industrial products (including but not confined to guns, steel, ships, railroad locomotives, and automobiles).

The Merrimack Valley offered an important source of agricultural and timber products for the growing cities to the south, notably for Boston. The frequent falls on the river posed a significant obstacle, however, to the ability of the residents of the valley to bring their goods to serve this market during the colonial era. To circumvent the difficulties a series of locks and canals was constructed in the latter half of the eighteenth century and by 1814 it was possible for river traffic to move freely all the way from Concord, New Hampshire to Boston (Steinberg, 1991). One of these canals, the Pawtucket Canal completed in 1796, designed to circumvent the falls in the river at what is now the city of

Lowell, the Pawtucket Falls, would play a critical role in the subsequent development of the region's, and indeed the nation's, textile industry.

Steinberg (1991) credits two Boston merchants, Francis Cabot Lowell and Nathan Appleton, with the original inspiration for the New England textile industry. They conceived this vision on a visit to Scotland in 1810. There, Appleton visited the cotton mills at New Lanark on the River Clyde. Steinberg quotes Appleton recording in his journal that "Lanark stands on the top of a hill and hence we proceed to the River where are a range of very extensive cotton mills through the windows of which we saw the wheels in full motion for 4 or 5 stories together." Returning to Boston, Lowell, with Appleton an initially hesitant participant in terms of investment, proceeded to acquire a charter from the Massachusetts legislature on behalf of the Boston Manufacturing Company (BMC) to "locate a factory in Suffolk County or within a fifteen mile radius of the city of Boston." With Patrick Tracy Jackson joining BMC and taking a major role, they settled on a site at Waltham on the Charles River, acquiring a preexisting mill used to produce paper taking advantage of an important fall in the river at that location. A critical challenge was to develop a reliable power loom. Jackson was successful in this endeavor and in the fall of 1813, wrote Lowell (Steinberg, 1991) informing him that "I have got our loom up and yesterday wove several yards by water." The key innovation of the BMC's Waltham factory was its success in integrating all elements of textile production—carding, spinning, and weaving—in a single facility and mechanizing all of these tasks using energy extracted from falling water.

By 1831, the BMC had more than 8000 spindles and 200 looms producing close to 2 million yards of cloth a year in its factories at Waltham. It enjoyed unprecedented economic success returning dividends of close to 19% a year to its shareholders over the period 1817 to 1826 (Steinberg, 1991). Eventually, though, its growth would be restricted by the relatively limited power that could be extracted from the Charles River and the principals, Appleton and Jackson (Lowell was dead by this time), turned their attention north, to where the potential supply of energy available from the Merrimack was almost twenty times greater than could be supplied by the Charles. Joined by Thomas M. Clark, a merchant from Newburyport (a town sited at the mouth of the Merrimack), they acquired a parcel of land between the Pawtucket Canal and the Merrimack River, taking over the company that had constructed the Pawtucket Canal in 1796, the Proprietors of Locks and Canals on Merrimack River (PLC), one of the earliest U.S. corporations and with it the rights to its water. They created a new company, the Merrimack Manufacturing Company, formed to manufacture and print cotton textiles. Kirk Boott, who had spent time at Harvard but failed to earn a degree, was appointed agent for this new Company. Boott, together with Jackson and Paul Moody, who had been involved earlier with the Waltham operation, were charged with the design, oversight, and management of the water distribution system that would eventually become the signature of the Company's investments in Lowell. They built a temporary dam across the Merrimack River upstream of the falls and

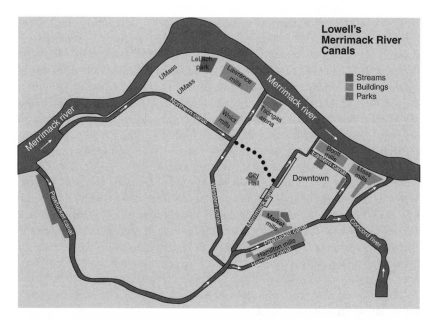

Figure 8.2 Lowell's Merrimack River canals (http://www.canalwaters.org/maps.
shtml. Read February 13, 2007).

proceeded to enlarge the Pawtucket Canal. In 1826, all of the property and
water rights owned by the Merrimack Company were transferred back to the
PLC which took charge of the management and expansion of the total water
distribution system operating much like a modern electric or gas utility. As
Steinberg (1991) described the subsequent developments: "by the mid-1830's
the Pawtucket Dam sent water into a maze of long waterways that provided
energy to twenty-six textile mills. Before the water found its way back to the
Merrimack, it performed some outstanding acrobatics." A map of contemporary Lowell depicting the complex network of canals is displayed in Fig. 8.2.
Milestones in construction of the canal system are summarized in Table 8.1.

Table 8.1 Milestones in development of the canals at Lowell

1796: Pawtucket Canal completed allowing shipping to bypass the 31 ft drop of the
   Merrimack River at Pawtucket Falls. The canal extended from upstream of the falls to
   the Concord River before the Concord joined the Merrimack below the falls.

1823: Merrimack Canal, Guard Locks, Swamp Locks, and Lower Locks constructed
   providing power to Merrimack Mills.

1828: Completion of Hamilton and Lowell Canals providing power to the Hamilton–
   Appleton Mill and the Lowell Manufacturing Company.

1836: Completion of the Western and Eastern Canals providing power to the Lawrence and
   Boott Mills respectively.

1848: Completion of the Northern Canal and the underground Moody Street Feeder.

The development of the textile industry in Lowell, and subsequently else-where in the Merrimack Valley, was orchestrated by a group of wealthy New England entrepreneurs known collectively as the Boston Associates. Theirs was an informal association, "a network of individuals and families joined by bonds of marriage, friendship, and, of course, finance," as defined by Steinberg (1991). By the mid-1830s they had extended their reach from Lowell upriver to Manchester and Concord in New Hampshire and a decade later downriver to Lawrence, Massachusetts. The Merrimack Valley assumed a dominant pres-ence in the textile industry, not just locally but also globally taking advantage of its status as the world's most efficient producer of cotton goods. The energy source that made all this possible was waterpower. The Associates took steps to control this resource all the way from the headwaters of the river to its mouth. They had a keen sense for profitable business investments. But they also had the wisdom to invest in the technological innovations that would enhance the overall efficiency and profitability of their enterprise. James B. Francis, an English born engineer, was appointed as Chief Engineer of the PLC in 1837. Francis would play a critical role in a wide range of activities sponsored by the PLC over the next 40 years while serving at the same time as general engineer-ing consultant to the Associates.

Francis is credited with introducing a scheme for monitoring the flow of water in the river and for devising the protocol used to regulate and charge for the efficient distribution of water to the factories owned by the Associates. In the early days of what became known as the Waltham–Lowell system, over-shoot water wheels were used to harness the power of falling water, technology not very different from that used more than a thousand years earlier in Italy as discussed in Section 8.1. Francis's most important and durable contribution, however, involved his role in perfecting the turbine that bears his name, and which continues in wide use today (converting waterpower to electricity in its modern usage)—the Francis turbine.

The blades of the Francis turbine were curved so that water entering the turbine was directed both inward and downward. Momentum transfer from the water to the blades as the water changed direction and speed passing through the turbine imparted a force to the blades causing them to rotate. The system was devised so that the greatest possible fraction of the energy of the water entering the turbine could be converted to rotational motion. This rotational energy was transferred to a massive flywheel capable of storing a significant fraction of the energy generated by the turbine, providing an essentially constant source of energy for the series of belts and gears that distributed this energy to the various floors of the factory where it could be applied to mechanize the multiple tasks involved in the conversion of cotton bolls to finished cloth. The efficiency of the Francis turbines was rated as high as 80%. By the mid-1850s turbines of the Francis design had replaced most of the water wheels installed initially in the Associates' factories, with close to a factor of two improvement in efficiency relative to the water wheels employed earlier.

The natural fall in the river at Lowell amounts to 31 ft. Assuming an average flow of water in the river of 6000 cfs, the potential power available from the river at Lowell (annually averaged) amounts to 21,100 horsepower (hp) (Box 8.3). Power was supplied by the PLC to the Associates' factories at Lowell in what was referred to as mill power units. One mill power unit corresponded to a supply of 25 cfs of water at a fall of 30 ft, equivalent to a maximum power output of 84 hp (Box 8.4). The power canal system in Lowell involved canals at two different levels with falls of 13 and 17 ft. Mill power units assigned to the factories on these two levels amounted to 60.5 cps and 45.5 cps respectively (Steinberg, 1991).

The investments of the Boston Associates would eventually exhaust the waterpower potential of the Merrimack Valley. Steam would supplant running (and falling) water as the dominant source of industrial power in the Valley. Coal from Eastern Pennsylvania would come to provide the energy source employed to generate this steam. The transition from waterpower to coal would eliminate eventually the competitive advantage enjoyed earlier in textile manufacture by the Merrimack factories. A particular beneficiary of the water to

---

Box 8.3

Assuming an average flow of water in the Merrimack River of 6000 cfs and a drop in elevation at Pawtucket Falls of 31 ft, estimate the power available potentially from the water as it drops over the Falls. Express answers both in units of kilowatts and horsepower.

Solution:

It is a good idea to carry out this calculation using the SI or mks system of units.

$$1 \text{ foot} = 3.048 \times 10^{-1} \text{ m}$$
$$1 \text{ cubic foot} = (3.048 \times 10^{-1})^3 \text{ m}^3 = 2.83 \times 10^{-2} \text{ m}^3$$
$$1 \text{ m}^3 \text{ of water corresponds to a mass of } 10^3 \text{ kg}$$
$$\text{Flow of water} = 6000 \text{ cfs} = (6 \times 10^3 \text{ cfs}) (2.83 \times 10^{-2} \text{ m}^3/\text{cft})$$
$$= 1.7 \times 10^2 \text{ m}^3 \text{ s}^{-1}$$
$$\text{Mass flow} = (1.7 \times 10^2 \text{ m}^3\text{s}^{-1}) (10^3 \text{ kg m}^{-3}) = 1.7 \times 10^5 \text{ kg s}^{-1}$$
$$\text{Fall of 31 ft is equivalent to } (3.1 \times 10^1) (3.048 \times 10^{-1}) \text{ m}$$
$$= 9.45 \text{ m}$$
$$\text{Power generated} = (\text{mass flow per second}) (g) (h)$$
$$\text{Power (Js}^{-1}) = (1.7 \times 10^5 \text{ kg s}^{-1}) (9.8 \text{ m s}^{-2}) (9.45 \text{ m})$$
$$= 1.57 \times 10^7 \text{ J s}^{-1}$$
$$= 1.57 \times 10^4 \text{ kW}$$
$$1 \text{ horsepower} = 7.45 \times 10^2 \text{ W}$$

Power released at the Falls is equivalent (transferred with 100% efficiency) to 1.57×10^4 kW or 21,100 horsepower.

Box 8.4

Estimate the power equivalent to 1 mill power unit expressing answer both in units of watts and horsepower.

Solution:

Mill power unit corresponds to 25 cfs with a fall of 30 ft.

$$\text{Mass of 1 cf of water (Box 8.3)} = 2.38 \times 10^1 \text{ kg}$$
$$30 \text{ ft} = (3 \times 10^1 \text{ ft})(3.048 \text{ m ft}^{-1}) = 9.144 \text{ m}$$
$$\text{Power (Js}^{-1}) = (2.5 \times 10^1 \times 2.83 \times 10^1 \text{ kg s}^{-1})(9.8 \text{ m s}^{-2})(9.144 \text{ m})$$
$$= (7.08 \times 10^2 \text{ kgs}^{-1})(9.8 \text{ m s}^{-2}) (9.144 \text{ m})$$
$$= 6.3 \times 10^4 \text{ W}$$
$$= 85 \text{ hp}$$

---

steam transition was the town of Fall River in southeastern Massachusetts. The mills in Fall River, steam powered, doubled the number of spindles employed in their factories between 1885 and 1865 and by 1875 could boast of almost twice the number of spindles in operation at the same time in Lowell.

At its peak, as many as 10 factories were busy in Lowell producing textiles 15 h a day powered by water. The mills are now idle but the turbines in the basements of the mills are still turning under waterpower supplied by the power canals constructed more than 150 years ago by the PLC. The product today is electricity rather than cotton fabric. As an interesting footnote, the system is still managed by the PLC, one of the oldest continually functioning companies in the U.S. The PLC, however, is no longer independent. It operates today as a subsidiary of the Italian energy company, Enel, sponsor recently (in 2007) of an environmental economics program at Harvard University.

8.3 HYDROELECTRICITY

Canada leads the world in present day production of electricity from waterpower followed by Brazil, China, the U.S., and Russia although given current trends it is likely that China will soon be number one. A list of the world's 13 largest producers of hydroelectric power in 2003 is presented in Table 8.2.

Waterpower accounted for 16.2% of electricity produced globally in 2004, 2827 billion kWh out of a total of 17,450 billion kWh (IEA, 2006) (countries included in Table 8.2 were responsible for about two-thirds of the total global hydroelectric production). The fraction of electricity produced from waterpower in the U.S. decreased from about 40% in the heyday following the major dam building period of the 1930s and early 1940s to less than 7% today.

Table 8.2 Largest national sources of hydroelectricity (EIA, 2003)

Country	Rank	Production (billions of kWh)
Canada	1	333
Brazil	2	303
China	3	279
United States	4	276
Russia	5	171
Norway	6	104
Japan	7	104
India	8	69
Venezuela	9	60
France	10	59
Sweden	11	52
Paraguay	12	51
Spain	13	41

Table 8.3 Largest U.S. state sources of hydroelectricity for 2006 through October (EIA, 2006)

State	Rank	Production (billions of kWh)
Washington	1	70.0
California	2	43.6
Oregon	3	32.0
New York	4	21.3
Idaho	5	10.0
Montana	6	8.5
Tennessee	7	6.1
Alabama	8	6.0
Arizona	9	5.8
Maine	10	3.5

Signature projects in the immediate pre-World War II era included the Hoover Dam on the Colorado River at the Arizona/Nevada border (operational in 1937), the Grand Coulee Dam on the Columbia River in Washington (1941), and the Bonneville Dam on the Columbia River at the Washington/Oregon border (1937). Table 8.3 presents a summary of electricity generated using hydropower for 2006 through October for the 10 largest producing states in the U.S. Note that the bulk of this production is sited in the west and northwest. Hydroelectricity accounts for about 60% of the power consumed in Washington, Oregon, Idaho, Montana, and Wyoming, and for about 20% of total electricity generated in California (Mapes, 2001). The hydroelectric generating capacity of the U.S. is not expected to increase significantly, however, in the future. Most of the best sites have already been developed and to an increasing extent environmental objections are being raised not only to potential new projects but also in opposition to existing installations.

The Glen Canyon dam on the Colorado River at Page, Arizona, completed in 1966, provides a case in point. As discussed in Wikipedia (2007), the dam

Box 8.5

Production of cement is associated with emission of CO_2. Manufacture of cement accounts for about 7% of CO_2 emitted annually by the U.S. Given that 4.9 million cubic yards of concrete were used in the building of the Glen Canyon Dam, estimate the corresponding emission of CO_2. Knowing that the dam provides 3.2×10^9 kWh of electricity annually, discuss how the savings in CO_2 emissions associated with the electricity generated by the dam compares with the CO_2 emitted during its construction.

Solution:

Using data from[1] we can calculate that 631 lb of CO_2 are produced in the manufacture of a cubic yard of concrete. This allows both for the energy expended in producing the concrete (1.7 million Btu per cubic yard) and the CO_2 emitted in converting limestone to the cement incorporated in the concrete. The source employed here assumes that cement accounts for about 12% by weight of the concrete used in building the dam with the balance composed of a mixture of crushed stone, sand, and water.

Total CO_2 emitted in conjunction with the manufacture of the concrete used in construction the Glen Canyon Dam:

$$\text{Total } CO_2 = (4.9 \times 10^6)(6.3 \times 10^2) \text{ lb} = 3.1 \times 10^9 \text{ lb} = 1.4 \times 10^6 \text{ tons}$$

If electricity is generated using coal, the emission of CO_2 per kWh is about 0.9602 kg. CO_2 saved by generating 3.2×10^9 kWh annually at Glen Canyon with hydropower rather than coal:

$$\text{Savings} = (3.2 \times 10^9)(0.9602) \text{ kg } CO_2 = 3.1 \times 10^9 \text{ kg} = 3.1 \times 10^6 \text{ tons } CO_2$$

Conclusion: it would take a little more than 5 months to reach the break-even point in terms of CO_2 emissions (assuming that the electricity consumed otherwise would have been generated using coal). If we were to account for all of the additional fossil energy expended in building the dam (transporting materials, operation of heavy equipment used in construction etc.), it is likely that the break-even point would be perhaps 3 times as long, say 15 months. Given the extended life of the dam, its hydroelectricity plant clearly contributes to an important reduction in emission of CO_2.

stands 710 ft tall with a crest width of 1560 ft. The thickness of the dam at the top is 25 ft expanding to 300 ft at the base. Its construction required 4.9 million cubic yards of concrete contributing to emission of more than 3 million tons of CO_2 (Box 8.5). The Glen Canyon hydroelectric plant incorporates eight 155, 500 horsepower Francis type turbines. Water is transported from the dam to the turbines through eight penstocks. The hydroelectric plant at Glen Canyon produced 3.2 billion kWh of electricity in 2005. Despite the fact that it is one of the most recent of the large dams constructed in the U.S. and that it provides not only an inexpensive source of electricity but also water (8.2 million acre-feet per year) for the residents of the neighboring states of California, Nevada, and

New Mexico in addition to Arizona, there has been significant opposition. The objections have centered for the most part on the impact the dam is having on the ecology of the region downstream, which includes the Grand Canyon. Under natural conditions, the flow of water in the Colorado River would be seasonally variable punctuated by episodic major floods. With construction of the dam and filling up of the major artificial lake behind it (Lake Powell), these floods have been largely eliminated. To offset the ecological problems perceived to be caused by the dam in the downstream region, environmental groups are calling in some cases for removal of the dam, in others, more moderately, for institution of a management protocol to regulate release of water from the dam in such a manner as to mimic as far as possible preexisting (natural) downstream flow conditions. Elsewhere there have been calls for elimination of four dams on the Snake River in Washington. The issue in this case relates to the negative impact these dams are perceived to have on migration of fish, specifically salmon, recalling problems of a similar nature encountered by the Boston Associates 150 years earlier as they sought to harness and control the flow of water on the Merrimack (Steinberg, 1991). Plus ca change, plus c'est la meme chose!

There are approximately 45,000 large dams in the world, almost half in China (World Commission on Dams, 2000). Only 25% of existing reservoirs are used to generate electricity, however. The primary purpose for most dams is to supply water for irrigation (IEA, 2007). The Three Gorges Dam on the Yangtze River when it goes into full operation in 2009 is an exception to this pattern. It is expected to have the capacity to generate 18.2 GW (billion watts) of electricity (Box 8.6) with an annual production of 85 billion kWh. The dam is designed to meet multiple objectives. By storing water during high flow conditions, it is expected to reduce the damage from floods in the downstream region. Second, proponents argued that when construction of the dam is complete, the river will be more easily navigable through the Gorges in the upstream region. Ships can be lifted or lowered through a series of ship locks allowing passage between the higher and lower levels of the water upstream and downstream of the dam. Third, the dam is expected to provide an important additional source of renewable energy. On the flip side, construction of the dam required displacement of close to two million people in the upstream region. There are concerns that sediment carried by the river will accumulate behind the dam requiring eventually expensive dredging to maintain the integrity of the dam. And there are ecological issues relating to the potential effects of the dam on a variety of fish and endangered species such as the Siberian Crane and the Yangtze dolphin. And concerns that pollution from upstream cities, notably Chongqing, given the longer residence time of water in the dam, could turn the Three Gorges region into a polluted, anoxic, open sewer. Finally, the electricity generated by the dam is not cheap. If we take the capital cost for construction of the dam as $25 billion as reported in some estimates and the electrical capacity as 18.2 G W, the cost per kW of production capacity would be close to $1400, almost 50% higher than comparable costs for conventional

Box 8.6

The Chinese government is nearing completion of what will eventually be the world's largest dam, the Three Gorges Dam on the Yangtze River. The dam will extend to a height of 181 m. The flow of water in the Yangtze River immediately upstream of the dam averages 60,000 m^3s^{-1}. Of this, 20,000 m^3s^{-1} will be used to drive turbines to generate electricity. The turbines are situated 125 m below the level of the water behind the dam. We refer to this as the hydraulic head for the water driving the turbines. Estimate the electrical power that would be realized if 100% of the potential energy of the water flowing through the turbines could be converted to electricity.

Solution:

Consider first the kinetic energy of a kg of water when it falls through 125 m before entering the turbines.

$$\text{Kinetic energy} = mgh$$
$$= (1 \text{ kg})(9.8 \text{ ms}^{-2})(1.25 \times 10^2 \text{ m})$$
$$= 1.23 \times 10^3 \text{ kg m}^2\text{s}^{-2}$$
$$= 1.23 \times 10^3 \text{ J}$$

The mass corresponding to 1 m^3 of water = 10^3 kg.

$$\text{The power generated} = (\text{kg s}^{-1} \text{ of water flow}) \times (\text{J kg}^{-1})$$
$$= (2 \times 10^7 \text{ kg s}^{-1}) \times (1.23 \times 10^3 \text{ J kg}^{-1})$$
$$= 2.46 \times 10^{10} \text{ J s}^{-1}$$
$$= 24.5 \text{ GW}$$

The dam has the potential to generate 24.5 gigawatts (GW) of electrical energy (a gigawatt = 10^9 W). The design objective is 18.2 GW, which can be realized by converting the potential energy of the water to electricity with an efficiency of about 75%, a realistic expectation for hydroelectric power generation using modern technology.

---

power generation. A fascinating account of the controversy concerning the Three Gorges Dam prior to its construction is presented in the book Yangtze! Yangtze! authored by Chinese dissident Dai Qing (1994).

The International Energy Agency's 2007 analysis of the future of renewable energy (IEA, 2007), projects in its Alternative Policy Scenario that the capacity for new hydro power could expand by as much as 70% worldwide by 2030, from the current level of 851 GW to about 1431 GW in 2030. Most of this new capacity would be established in developing countries, notably in Asia and South America. Capacity for hydroelectricity in China, it concluded, could increase from 105 GW in 2004 to 298 GW in 2030 with a comparable increase over the same period for India, from 31 GW to 105 GW. The Alternative Policy Scenario "shows how the global energy market could evolve if countries around the world were to adopt a set of policies and measures that they are now considering and might be expected to implement over the projection period."

Objectives in this scenario include "reductions in CO_2 emissions and improved security of supply." Despite the significant growth in hydropower envisaged in the Alternative Policy Scenario, 1.8% per year, the fraction of total electricity generated worldwide by hydropower would not increase significantly. It would remain at about 14% and fossil fuels would still account for 77% of total primary energy demand. Global emissions of CO_2 would be 8 GT per year higher in 2030 than today.

8.4 ENERGY: FROM SUN TO WIND

Energy in the atmosphere is apportioned among four principal modes: internal, potential, latent, and kinetic. The ultimate source for all of this energy (as for running water) is the sun. The energy absorbed from the sun, mainly in the form of visible light, is balanced by emission of infrared radiation to space (cf. Fig. 3.1). At present, the quantity of energy returned to space is slightly less than that absorbed from the sun. This imbalance is responsible for the phenomenon of global warming.

Internal energy (IE) refers to the energy the atmosphere possesses that is manifest on a molecular scale. At a temperature of 20°C, the average nitrogen molecule is moving between collisions at a speed of close to 300 m per second (about 670 miles per hour). It doesn't go very far, however, less than 10^{-7} m, before it runs into another molecule and is forced to change direction. If you could take a movie of the molecules that make up the atmosphere, what you would see is a lot of molecules moving rapidly back and forth in more or less random directions—furious activity but with no particular direction or purpose. They are not only moving but they are also spinning and internally vibrating. Internal energy is composed of a combination of the translational, rotational, and vibrational energy of the individual molecules that make up the atmosphere.

Latent energy (LE) reflects the potential of the atmosphere to gain or lose IE as a consequence of a change in the phase of water. Internal energy increases when water changes from vapor to liquid or ice; it decreases if the change in phase proceeds in the opposite direction. It is convenient for present purposes to include latent energy in an expanded definition of internal energy.

Potential energy (PE) refers to the energy the air has by virtue of its position relative to the ground. The concept is the same as discussed earlier for the case of a ball thrown or batted up in the air. As air rises, it must do work against the force of gravity. The energy required to accomplish this work is supplied by a transfer from the IE pool. Conversely, as air sinks, its potential energy decreases and its internal energy increases accordingly. The nature of this exchange between PE and IE is such that if we were to consider the total quantities of IE and PE contained in a column of air from the surface to the top of the atmosphere we would find that the ratio of IE to PE would be essentially the same wherever we choose to locate this column: IE is greater than PE by a factor of about 2.5 (5 units of IE for every 2 units of PE).[2]

Kinetic energy (KE) refers to the bulk (net translational) motion of the air, the property we experience as wind. To change the bulk motion of the air, we need to apply a force in the direction of its motion: work must be done to change the energy of the air mass. There are only two forces that can accomplish this task in the atmosphere. If air moves up, work must be done against the force of gravity and its vertical velocity and kinetic energy will decrease accordingly. As air moves down, it will gain energy from work supplied by the gravitational field and its speed and kinetic energy must increase in this case. To increase the speed of the air in the horizontal direction, air must move down a pressure gradient, it must move from a region of high pressure to one of low pressure. The kinetic energy of the air will increase accordingly. If the air is forced to move up the pressure gradient, from a region of low to one of high pressure, its kinetic energy will decrease. An increase or decrease of kinetic energy is accommodated by a corresponding change in what meteorologists refer to as available potential energy, the combination of IE and PE. Kinetic energy is dissipated ultimately by friction, primarily by interactions with the surface. Frictional dissipation is most effective when the air flows over an uneven, rough, surface.

The circulation of the atmosphere, the mean direction of air motion, proceeds for the most part along surfaces of constant pressure. The pressure force operates at right angles to the direction of motion (from high to low pressure) and is offset by what is known as the Coriolis force, a force reflecting the influence of the rotation of the Earth. The Coriolis force operates to the right of the direction of motion in the Northern Hemisphere (at right angles to the direction of motion), to the left in the Southern Hemisphere. As a consequence, air in the Northern Hemisphere tends to circulate in a clockwise sense around a region of high pressure (the Coriolis force acting to the right of the direction of motion pushes the air towards the region of high pressure opposing the tendency of the pressure force to push in the opposite direction). Motion proceeds in a counterclockwise sense around a low-pressure system in the Northern Hemisphere (the rule of thumb is that the air moves in such a direction as to keep the zone of high pressure to the right of the direction of motion). Directions for the flow of air around high and low pressures are opposite in the Southern Hemisphere (reflecting the fact that the Coriolis force is directed to the left of the direction of motion in this case). Responding to a balance between the Coriolis force and the pressure gradient, air tends to move along surfaces of constant pressure: the isobars (surfaces of constant pressure), at least some distance from the surface, tend to define surfaces of constant wind velocity. The magnitude of the Coriolis force is proportional to the magnitude of the wind velocity. The more densely packed the isobars, the greater the wind speed.

Wind directions deviate from isobaric surfaces mainly in the near surface region where air is subject not only to the influence of the pressure and Coriolis forces but is subject also by the effect of friction. Frictional effects cause air to deviate towards the center of a low-pressure system (air slows down and the Coriolis force is consequently less effective). The resulting convergence (tendency for net movement of air into a particular region) causes air to rise accounting for the fact that low-pressure systems are typically associated with inclement weather (it rains only when air rises, when the associated decrease in temperature causes the air to become saturated with respect to water vapor). Conversely, frictional effects tend to drive air out from the center of a high-pressure system (divergence) leading to descent. As a consequence, high-pressure systems are associated usually with clear skies and pleasant, often warm, conditions (air tends to heat up as it descends reflecting conversion of PE to IE).

Transfer of energy among the different modes in the atmosphere begins with the conversion of the energy in solar or infrared radiation to internal energy (with an important contribution from latent energy). Internal energy is converted subsequently to potential energy in the proportion discussed above as required to maintain hydrostatic balance in the vertical (with addition of IE, the atmosphere must expand converting the appropriate fraction of the new source of IE to PE). Kinetic energy is produced at the expense of IE. Again, PE must adjust to maintain the necessary IE to PE balance. Kinetic energy is both produced and consumed by work done by and against the gravitational and pressure gradient force fields. The net production of KE by these fields is offset by frictional dissipation of KE at the surface. The available data suggest that approximately 1% of the solar energy absorbed by the Earth is converted on a net basis to kinetic energy (the difference between sources and sinks), and is dissipated eventually at the surface, contributing a net frictional heat source of about 2.0 W m^{-2} averaged over the surface of the Earth (Lorenz, 1967; Peixoto and Oort, 1992).

If we assume that this energy is released uniformly over the entire surface area of the Earth (it isn't), this would imply a total potential rate of supply of energy equal to 2.97×10^{14} W or 8875 Quad per year. Doing the same calculation for the United States would indicate an available energy source of 1.83×10^{13} W or 546 Quad per year. These figures may be compared with current global and U.S. annual rates for consumption of commercial energy, 412 and 98 Quad respectively.

Wind speeds are typically low at the surface reflecting the influence of the frictional drag of air across the surface. They increase generally over the first few tens of meters above the surface, in the region of the atmosphere known as the boundary layer. To capture optimally the kinetic energy available in the wind (to justify the expense of construction of a high-standing tower vis-à-vis the potential return on the investment), turbines designed to generate electricity are located typically at the present time at elevations as much as 100 m above the surface (see below). The frictional boundary layer is lower over the ocean than over land (the land surface is generally rougher). The height of turbines required for economically viable production of electricity in offshore environments is consequently lower than those deployed to generate electricity on-shore. Costs for offshore wind farms are significantly greater, however, since the structures supporting the turbines must be grounded in this case at the sea-bottom (generally obviating the advantage of the fact that they need not extend so far above the surface) and additional costs must be incurred to bring the resulting electricity on-shore (usually through cables sunk on the sea bed). Costs for maintenance are also higher.

8.5 WIND TO ELECTRICITY: CURRENT STATUS AND FUTURE POTENTIAL

The U.S. pioneered in development of wind-powered generation of electricity in the 1980s and early 1990s but lost its lead to Europe in the late 1990s as cheap oil, coal, and gas reduced incentives for U.S. utilities to invest in alternative sources of energy. Germany assumed the position of number one in worldwide wind-powered generation of electricity in 1997 with 2.1 GW of capacity as compared to 1.6 GW in the U.S. By 2005, European countries, notably Germany (18.4 GW), Spain, (10.0 GW) and Denmark (3.1 GW), accounted for 53 % of total global capacity (up from 49 % in 1997) with the contribution of the U.S. dropping from 26 % (1.6 GW) in 1997 to 15% (9.1 GW) in 2005. A summary of historical data on global installed wind capacity is presented in Table 8.4.

Table 8.4 Historical data on wind generating capacity (MW) (http://www. earth-policy.org/Updates52_data.htm). Read March 1, 2007

Year	U.S.	Germany	Spain	Denmark	India
1980	8	0	0	5	0
1981	18	0	0	7	0
1982	84	0	0	12	0
1983	254	0	0	20	0
1984	653	0	0	27	0
1985	945	0	0	50	0
1986	1265	0	0	82	0
1987	1333	5	0	115	0
1988	1231	15	0	197	0
1989	1332	27	0	262	0
1990	1484	62	0	343	0
1991	1709	112	5	413	39
1992	1680	180	50	458	39
1993	1635	335	60	487	79
1994	1663	643	70	539	185
1995	1612	1130	140	637	576
1996	1614	1548	230	835	820
1997	1611	2080	512	1120	940
1998	1837	2870	830	1428	992
1999	2490	4445	1584	1718	1095
2000	2566	6113	2235	2300	1167
2001	4261	8754	3337	2417	1407
2002	4685	12001	4830	2889	1702
2003	6374	14609	6202	3110	2110
2004	6740	16629	8263	3117	3000
2005	10027	18428	10027	3122	4430

The increase in capacity globally from 1997 to 2005 amounts to 676%. Despite this impressive growth, wind power still accounts for less than 1% of total global production of electricity. The impressive increase in installed wind power capacity for the U.S. between 2002 and 2005 is illustrated in Fig. 8.3. The U.S. now, in 2009, ranks number 1 in installed windpower capacity having recently surpassed Germany; China ranks number 4 but its capacity is growing at a rate comparable to that of the U.S.

General Electric has emerged in recent years as a major player in the global wind turbine business following its acquisition of the wind division of Enron in May 2002. With over 1400 employees worldwide, GE Wind Energy designs, manufactures, installs, and operates a range of turbines with capacities rated from 900 kilowatts (KW) to 3.6 megawatts (MW). More than 3300 GE 1.5 MW systems have been installed worldwide over the past few years. All of these systems incorporate three blades mounted to a horizontal axle coupled through a gearing mechanism to the electrical generator. The blades on the 1.5s and 1.5se models are 35.25 m long sweeping out an area of 3904 m². The hub of the turbine is located almost 65 m above ground on a tower constructed of conical tubular steel. The rotor begins to turn at a wind speed of 4 ms^{-1} (9 mph) shutting down automatically at wind speeds higher than 25 ms^{-1} (56 mph) taking advantage of the fact that the system includes a fail-safe braking mechanism based on electromechanical controls regulating the pitch of the individual blades (pitch can be adjusted to ensure that the blades are feathered). The speed of the rotor varies between 12.0 and 22.2 revolutions per minute (rpm). At the higher revolution rate, the tips of the rotor blades are moving at speeds of 82 ms^{-1} (184 mph) (Box 8.7).

The power generated by a wind turbine depends on a number of factors, notably the wind speed and the area swept out by the rotor. The kinetic energy represented by unit volume of air moving at speed v is given by $\frac{1}{2}\rho v^2$ where ρ denotes the mass per unit volume and v defines the speed of the wind. The volume of air passing per unit time through a target of area A (oriented normal to the direction of the air flow) is given by vA. It follows that the energy intercepted by a target of area A in unit time is given by $\frac{1}{2}\rho v^3 A$. Only a fraction of this energy can be converted to electricity by the turbine, however. If all of the energy were captured, the wind speed would have to drop to zero leaving the rotor. The rotor would behave as though it were a solid disk in this case. Pressure would build up ahead of the disk and the incoming air would be diverted around the disk. The GE 1.5s turbine is designed to deliver 1.5 GW of electricity at a wind speed of 13 ms^{-1}. Incident wind energy is converted to electricity by this system with an efficiency of 28 % as indicated in Box 8.8. The electrical power generated by the GE 1.5s and 1.5se systems increases as approximately the cube of the wind speed for wind speeds between 4 ms^{-1} and 13 ms^{-1} reaching their maximum value of 1.5 MW for speeds higher than 12 ms^{-1} up to the cut-off limit of 25 ms^{-1} as indicated by the manufacturer's power curve reproduced in Fig. 8.4.

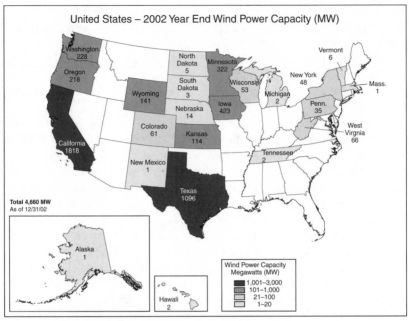

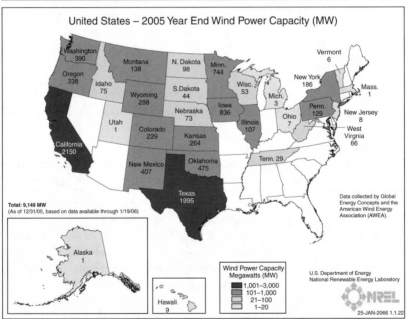

Figure 8.3 United States year end wind power capacity (MW) in 2002 and 2005.
[Source: http://www.eere.energy.gov/windandhydro/windpoweringamerica/
wind_installed_capacity.asp. Read March 9, 2007].

Box 8.7

_____

Assume that the blade of the wind turbine is rotating at a rate of 22.2 rpm. Calculate the speed of the blade at its tip.

Solution:

On each rotation, the tip of the blade moves through a distance equal to the circumference of the circle it tracks out on its rotation:

$$(2\pi)R = (2\pi)(35.25) \text{ m} = 221.5 \text{ m}$$

Distance traversed in a minute corresponding to a rotation rate of 22.2 rpm:

$$(22.2)(221.5) \text{ m} = 4917 \text{ m} = 4.917 \text{ km} = 3.06 \text{ miles}$$
$$\text{Speed} = 82 \text{ ms}^{-1} = 184 \text{ mph}$$

The modern wind turbine is an engineering masterpiece. The blades are shaped to ensure the maximum possible efficiency for conversion of the kinetic energy of the wind to rotational motion of the turbine. The wind is directed across the blades in such a manner as to ensure that wind speeds are significantly different on the upper and lower edges of the blades. According to

---

Box 8.8

_____

The GE 1.5s and 1.5se turbines are rated to deliver 1.5 GW of electricity at a wind speed of 13 m s^{-1}. Assuming that the area swept out by the rotor is equal to 3904 m^2, calculate the efficiency with which the kinetic energy of the wind is converted to electricity.

Solution:

The total energy intercepted by the rotor per unit time is given by:

$$P = \frac{1}{2}\rho v^3 A$$

where ρ is the mass density of air, 1.23 kg m^{-3} at sea level, v is the wind speed, and A is the area traced out by the rotor. Assuming values appropriate for the GE systems,

$$P = \frac{1}{2}(1.23 \text{ kg m}^{-3})(13 \text{ m s}^{-1})^3 (3904 \text{ m}^2)$$
$$= 5.27 \times 10^6 \text{ W}$$
$$= 5.27 \text{ MW}$$

The actual electrical power produced is 1.5 MW. It follows that the efficiency with which wind energy is converted to electricity is given by $P = \dfrac{1.5}{5.27} = 28\%$.

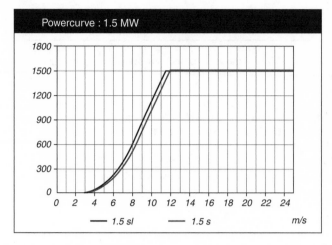

Figure 8.4 GE 1.5 MW wind turbine power curve. [http://www.gepower.com/prod_
serv/products/wind_turbines/en/15mw/tech_data.htm. Read
March 7, 2007].

Bernoulli's equation, the sum of pressure and kinetic energy per unit volume
(in the absence of friction) is a conserved quantity:

$$P + \frac{1}{2} \rho v^2 = \text{constant} \tag{8.1}$$

It follows that where the air is moving slowly the pressure is high. Conversely,
where the air is moving rapidly, the pressure is low. It is this differential in pressure that provides the torque that drives the rotational motion of the blades.
The underlying principle is similar to the physics that allows a heavy airplane
to offset the force of gravity and maintain a constant altitude. The wings of the
plane are shaped such that the flow of air under the wings is slower than the
flow above. The pressure of air under the wings is consequently greater than
the pressure above and it is this difference in pressure that accounts for the net
upward force, referred to in this case as lift, that counterbalances the downward directed force of gravity. It is of course the relative motion of air and the
plane that accounts for the lift. This depends primarily on the speed of the
plane (the plane's speed is typically much higher than the speed of the wind).
The lift in the case of the turbine depends on the velocity of the air relative to
the (rapidly) rotating blades of the turbine.

The wind power potential of a particular location is defined in terms of the
annual average speed of the wind and the annual average power available in the
wind (recall that power depends on the cube of the speed and is weighted
accordingly towards the higher wind speeds available at the location). Seven classes

of wind conditions are identified as indicated in Table 8.5 for elevations of 10 m and 50 m above the local surface. A map of wind power classes prepared for the lower 48 states of the U.S. by the U.S. Department of Energy Wind and Hydropower Technologies Program is presented in Fig. 8.5. Data with higher spatial resolution are displayed for California and Massachusetts in Figs. 8.6 and 8.7 respectively and there are plans to complete similar surveys for the entire U.S. The data for California and Massachusetts underscore the potential importance of offshore wind resources for these states. A major wind facility has been proposed for construction near the center of Nantucket Sound, off the coast of Cape Cod in Massachusetts. If approved, and implemented, this would represent the first major project for development of offshore wind power in the U.S.

The Cape Wind project would involve 130 turbines embedded in the sandy sea bottom in a region of shallow water known as Horseshoe Shoal. The turbines would be arranged in a grid of parallel rows separated by about 1 km. Individual turbines would be spaced by about 630 m and each turbine would be rated to produce 3.6 MW of power at peak performance. The entire project is expected to generate electricity at an average rate of 170 MW (allowing for the variability of wind conditions), sufficient to supply approximately 75% of current electrical needs of the Cape and Islands. The towers would extend vertically approximately 80 m (262 ft) above the sea surface and would be mounted on hollow steel pipes driven 24.4 m (80 ft) into the sea bottom. Power would be carried onshore by cables buried several feet below the sea bottom. For further details of the project, refer to the developer's web site at http://www. capewind.org/ (read March 7, 2007).

Despite the obvious merits of the proposal (notably the reduction in CO_2 emissions occasioned by exploitation of a renewable non-fossil energy resource),

Table 8.5 Classes of wind power

Class	Power density[1] (Wm⁻², 10 m)	Power density[1] (Wm⁻², 50 m)	Wind speed[2] (ms⁻¹, 10 m)	Wind speed[2] (ms⁻¹, 50 m)
1	<100	<200	<4.4(<9.8)	<5.6 (<12.5)
2	100–150	200–300	4.4–5.1(9.8–11.5)	5.6–6.4(12.5–14.3)
3	150–200	300–400	5.1–5.6(11.5–12.5)	6.4–7.0(14.3–15.7)
4	200–250	400–500	5.6–6.0(12.5–13.4)	7.0–7.5(15.7–16.8)
5	250–300	500–600	6.0–6.4(13.4–14.3)	7.5–8.0(16.8–17.9)
6	300–400	600–800	6.4–7.0(14.3–15.7)	8.0–8.8(17.9–19.7)
7	>400	>800	>7.0(15.7)	78.8(>19.7)

(1) Power densities quoted here assume a specific distribution of wind speeds around the stated mean. Since power density depends on the cube of the wind speed, densities are about a factor of 2 greater than would be implied if the wind speed was assumed to be constant equal to the mean.

(2) Numbers in brackets refer to wind speeds quoted in units of mph.

the Cape Wind project has been the subject of considerable controversy. Concerns have been raised as to the potential impact on birds and fish and shipping and tourism. Some of the issues clearly involve questions of aesthetics. The project would be sited in a region of unusual natural beauty, a popular summer vacation destination with expensive ocean front property. How would the development affect the value of this property? Would it result in irreversible damage to an otherwise pristine natural environment? Proponents view some of these objections as classic NIMBY (not in my back yard). But influential voices have been raised objecting to the project. Robert Kennedy Jr., son of the late Senator Robert Kennedy, in an op-ed published by the *New York Times* on December 16, 2006 concluded his article thus: "if Cape Wind were to place its project further offshore, it could build not just 130 but thousands of windmills—where they can make a real difference in the battle against global warming without endangering the birds or impoverishing the experience of millions of tourists and residents and fishing families who rely on the sound's unspoiled bounties." The company points out, though, that the economics of the project would be totally different if it had to be moved further offshore.

The current Governor of Massachusetts, Deval Patrick, is on record supporting the project while his predecessor, Mitt Romney, was an opponent. Congress, under the Energy Policy Act of 2005, designated the Minerals Management Service (MMS) of the Department of Interior as the entity of the U.S. government charged with granting easements for development in Federal waters on the Outer Continental Shelf. The Commonwealth of Massachusetts reported positively on an environmental assessment of Cape Wind on March 30, 2007. The preliminary report issued by MMS on January 14, 2008 was also positive but MMS has yet to issue the lease that would authorize the project to proceed. Construction could begin as early as 2010 assuming a positive outcome to the ongoing deliberations and the wind farm could be operational in this case by late 2011. The future of U.S. offshore wind power development could well depend on the outcome of this complicated, drawn-out process. As a footnote, it is interesting to note that offshore wind power, despite early opposition, is now widely accepted in Europe, with major facilities currently installed and operating in a number of countries including Denmark, the Netherlands, Sweden, the United Kingdom, and the Republic of Ireland.

As indicated in Table 8.4, growth of wind-powered generation of electricity (exclusively on-shore) in the U.S. has proceeded in spurts with major increases between 1980 and 1985, between 1997 and 1999, and most notably since 2002. Federal policy is largely responsible for this uneven pattern. The Energy Tax Act of 1978 (Public Law 95-618) provided tax credits for both residential and business installations of wind power. These credits expired, however, in 1985. The Energy Policy Act of 1992 (Public Law 102-486) instituted an income tax credit of 1.5 cents per kWh (available to eligible companies) for electricity generated by wind power. The impact of this initiative was mitigated by low prices for oil and fossil fuels that prevailed over the decade of the 1990s (Chapter 6). The Job Creation and Worker Assistance Act of 2002 (Public Law 107–147) extended the tax credit for wind power to apply through December 31, 2005. Credits were extended subsequently, initially through the end of calendar year 2007, at present through 2012. There can be little doubt that a longer-term commitment to wind power by the U.S. Congress would enhance prospects for the future of the industry although initiatives by individual states offer reasons for optimism.

Sales tax relief for wind power investments have been instituted in no fewer than 13 states. For details of incentives for renewable energy in individual states check the Database of State Incentives for Renewables and Efficiency (DSIRE) at http://www.dsireusa.org/ (read March 6, 2007). Koplow (2006) estimates the cost of current federal subsidies for all forms of renewable energy at $6 billion a year. To place this in context, subsidies for oil and gas run at a level of $39 billion a year with additional supports for coal, nuclear, and ethanol of $8 billion, $9 billion, and $6 billion respectively. And these costs do not include the much greater outlays for military forces required to ensure a continuing supply of oil from vulnerable regions of the world such as the Middle East.

Wind power has proved a boon for landowners in the U.S. who are receiving rents ranging from $2000 to $6000 annually for permission to site large turbines on their property. These rents are by no means exorbitant. Consider for example a commercial scale turbine generating electricity at an average power output of 1 MW. Over the course of a year the turbine could produce $(8.76 \times 10^3 \text{ h})(1 \times 10^3 \text{ KW}) = 8.76 \times 10^6 \text{ KWh}$ of electricity. At an average price of 5 cents per KWh this would provide an annual income to the producer of $438,000, more than sufficient to pay rents at the stated level to the landowner. Moreover, since the footprint of the turbine would be relatively minor, it need not seriously impact the ability of the landowner to continue to make use of his or her holding—to graze animals or plant crops for example.

Archer and Jacobsen (2005) estimated that the wind power resources of the world are capable of supplying electricity at a power level of as much as 14 TW ($1 \text{TW} = 10^{12}$ W), This would correspond to an annual production of electrical energy of 1.23×10^{14} or 420 Quad, comparable to the world's total current use of energy in all forms, or more than seven times present global consumption of electricity. They based their study on an analysis of data from 7753 surface meteorological stations complemented by data from 446 stations for which vertical soundings were available. They restricted their attention to power that could be generated using a network of 1.5 MW turbines tapping wind resources with annually averaged wind speeds in excess of 6.9 ms $^{-1}$ (class 3 or better) at an elevation of 80 m assuming that this power could be captured with an efficiency of 20%. They argued that their estimate for available power was conservative given that the stations employed in their analysis were not optimally located to define best possible sites for wind development (mountain tops or ridges or example). Regions identified as particularly promising for wind powered generation of electricity included the northeastern and northwestern coasts of North America, the region of the Great Lakes (shared by the U.S. and Canada), northern Europe along the North Sea, the southern tip of

South America, and the island of Tasmania in Australia. Short et al. (2003), using the National renewable Energy Laboratory's WinDS model, which accounts not only for the potential production of electricity from wind but also for its delivery to the market through present and potential future transmission systems, concluded that wind power could accommodate as much as 25% of the demand for electricity projected for the U.S. in 2050. The total wind power potential of the U.S. is estimated at 8 TW. To put this in context, wind resources available in North Dakota today could account for as much as 30% of current U.S. demand for electricity. With wind power continuing to supply only about 1% of U.S. consumption of electricity, there is obviously significant scope for growth. We shall return in the final chapter of this volume to a more extensive discussion of the potential of wind power to meet future demands for electricity not only for the U.S. but also for the larger global community.

8.6 CONCLUDING REMARKS

As we have seen, hydropower accounts for 18% of global electricity production today. The capacity for hydropower is expected to increase by about 1.8 % per year between now and 2030 with most of this growth taking place in developing countries, mainly in Asia and South America, according to projections by the International Energy Agency (2007). Despite this impressive growth, the fraction of total global electricity contributed by hydropower is not expected to increase significantly over the foreseeable future (International Energy Agency, 2007).

As discussed, hydropower, especially when developed on a large scale, is not a free good. It is associated with a variety of problems, both sociological and environmental. Large dams on major rivers can result in major changes of the environment both upstream and downstream of the dams. In the case of the Three Gorges Dam on the Yangtze River in China, construction of the dam required forced relocation of more than two million people in the upstream region. The Aswan High Dam in Egypt, completed in 1970, provides Egypt with an important source of electricity (2.1 GW). But the benefits are offset by loss of nutrient-rich silt carried downstream previously by the river during its annual flood stage. Farmers are forced now to rely increasingly on expensive chemical fertilizer to grow their crops. And there are problems with public health (increased incidence of malaria and schistosomiasis) and with subsidence and saltwater intrusion in the delta associated with the altered flow pattern of the river. It is important that major hydro projects in the future be informed by comprehensive prior assessments ensuring an equable balance of benefits and risks for affected communities.

Wind power is the fastest growing component of the world's renewable energy portfolio. The International Energy Agency (2007) in its Alternative Policy Scenario projects the possibility of an eighteen-fold increase in wind-powered generation of electricity globally by 2030. This would represent, however, an increase from a relatively modest base, 82 TWh in 2004, to 1440 TWh in 2030. The most impressive growth is projected for Europe where, it is suggested, 30% of total electricity could be contributed by renewables by 2030 (mainly wind). Europe's commitment to sustainable energy is based in large measure on a conviction that the threat of future climate change is real and urgent. The prior (Bush) administration in Washington had not seen fit to share this view. It remains to be seen whether future U.S. administrations will adopt a more aggressive stance (this statement written in early 2008: there are reasons for optimism with the initial performance of the Obama administration). Should they do so, it is clear that wind power could make an important contribution to a reduction in future emission of greenhouse gases in the U.S., significantly more than envisaged in the current assessment of energy futures by the International Energy Agency (2007). Wind energy is currently enjoying a boom in the U.S. A recent article in the Wall Street journal (Ball, 2007) begins with the tease: "Deep in the heart of Texas, multinational giants are gambling on a new supply of energy. The prize isn't oil. It's wind." For the boom to endure, it will be important for governments at both state and federal levels to demonstrate that their commitments are for the long term, not transitory, and for regulators to approve and fast track authorization for construction of the transmission systems needed to connect these new sources of power to the existing electrical grid enhancing the capacity of the grid where necessary.

NOTES

1 http://www.buildinggreen.com/auth/article.cfm?fileName=020201b.xml

2 This result reflects the fact that the vertical motions in the atmosphere, except possibly in a violent hurricane or tornado, are generally very gentle. The vertical structure of the atmosphere reflects primarily a balance between vertical forces of pressure and gravity (pressure pushing up, gravity pulling down). The net vertical force experienced by air at any given altitude is negligibly small; pressure and gravitational forces are essentially in balance, a condition referred to by atmospheric scientists as hydrostatic equilibrium.

REFERENCES

Archer, C. L. and Jacobson, M. Z. (2005) Evaluation of global wind power *J. Geophys. Res.* **110.**

Ball, J. (2007). Breezy talk. The Texas wind powers a big energy gamble. The *Wall Street Journal*, Page A1, March 13, 2007.

Dai, Q. (1994). *Yangtze! Yangtze!* Edited in English by Patricia Adams and John Thibodeau. London: Earth Scan.

Derry, T.K. and Williams, T.I. (1960). *A Short History of Technology: From the Earliest Times to A.D. 1900.* Oxford: Oxford University Press.

Energy Information Administration, U.S. Department of Energy (2003).

Energy Information Administration, U.S. Department of Energy (2006).

International Energy Agency (2006). Key world energy statistics.

International Energy Agency (2007). Renewables in global energy supply: an IEA fact sheet, January 2007.

Koplow, D. (2007). Subsidies in the U.S. energy sector: magnitude, causes, and options for reform. www.earthtrack.net, read March 1, 2007.

Landes, D.S. (1998). *The Wealth and Poverty of Nations: Why Some are Rich and Some are so Poor*. New York: W.W. Norton.

Lorenz, E.N. (1967). *The Nature and Theory of the General Circulation of the Atmosphere*. WMO Publication, 218, World Meteorological Organization, Geneva, Switzerland. Geneva.

Mapes, L.V. (2001). BPA officials planning huge wind-power buy. The *Seattle Times*, February 22, 2001.

Peixoto, J.P. and Oort, A. H. (1992). *Physics of Climate*. New York: American Institute of Physics.

Short, W., Blair, N., and Donna Heimiller, D. (2003). The long-term potential of wind power in the United States. Solar Today, November-December issue, 2003. http://www.nrel.gov/docs/gen/fy04/34871.pdf. Read, March 10, 2007.

Steinberg, T. (1991). *Nature Incorporated: Industrialization and the Waters of New England*. Cambridge: Cambridge University Press.

Wikipedia, The Free Encyclopedia, Glen Canyon Dam, http://en.wikipedia.org/wiki/Glen-Canyyon_Dam. Read February 21, 2007.

World Commission on Dams, http://www.dams.org/. Read February 22, 2007.

9

Nuclear Power

9.1 INTRODUCTION

An atom consists of a positively charged nucleus surrounded by one or more negatively charged electrons. Most of the mass of the atom is contained in the nucleus. Hydrogen is the simplest of the atoms. Its nucleus is represented by a single proton and is accompanied by a single orbiting electron. The mass of the proton is 1.673×10^{-27} kg: the mass of the electron is 9.11×10^{-31} kg. Nuclei of more complex atoms include not only protons but also electrically neutral particles known as neutrons. The mass of a neutron is similar to, though slightly larger than, the mass of a proton: 1.675×10^{-27} kg. The nucleus of the most common form of oxygen, for example, contains 8 protons and 8 neutrons. Offsetting the charge carried by the 8 protons are 8 electrons such that the net charge on the atom as a whole is zero. Atoms of oxygen exist also with nuclei that contain additional neutrons, a total of 9 and 10 with the 10-neutron form more common.

The chemical properties of an electrically neutral atom are determined by the structure of the orbital electrons, specifically by the properties of the electrons in the outermost region of the atom (the electrons that are least tightly bound to the parent nucleus). The number of protons in the nucleus (and the equivalent number of accompanying electrons) identifies what we refer to as the atomic number of an element. Oxygen and hydrogen are elements. Atoms of the same element with differing numbers of neutrons in their nuclei are referred to as isotopes of the element. The three stable isotopes of oxygen are identified by the symbols ^{16}O, ^{17}O, and ^{18}O where the superscripts denote the sum of the number of protons and neutrons in the nucleus. The heaviest isotope of oxygen contains 8 protons and 10 neutrons corresponding to an atomic mass number of 18. An alternate convention includes the number of protons in the nucleus (the atomic number designated by Z) as a preceding subscript accompanying the superscripted mass number. In this case, the isotopes of oxygen are indicated as $^{16}_{8}O$ $^{17}_{8}O$, and $^{18}_{8}O$. The ability of atoms to combine to form molecules depends on their electronic structure (the atomic number Z). Chemical species can react to form other chemical species. When energy is released in the process, the reaction is said to be exothermic. If energy must be supplied to accommodate the reaction, the reaction is said to be endothermic. In a chemical reaction, the nature and number of atoms involved in the reaction remain constant: they are simply organized in different chemical forms. In burning methane (CH_4), the primary component of natural gas, the net reaction may be written as

$$CH_4 + 2O_2 \rightarrow CO_2 + 2 H_2O \tag{9.1}$$

All of the atoms present on the left hand side of this reaction equation appear also on the right, but in different chemical form. Energy is released in the reaction, contributing, as noted earlier to an important source of atmospheric CO_2. This energy corresponds to the difference in the binding energy of the elements present in the compounds on the left as compared to the right (the net binding energy of the elements on the right is greater than the binding energy of the elements on the left allowing the excess to be released as heat or kinetic energy of the products).

Just as the elements composing a molecule may be considered in a lower energy state than the elements present separately in free form, so also with the protons and neutrons composing the nucleus. One of the great achievements of modern physics involved recognition of the basic equivalence of mass and energy expressed through the equation

$$E = mc^2 \tag{9.2}$$

where E denotes energy, m is mass, and c is the speed of light. The relationship was first derived by Henri Poincare in 1900 but is now more generally attributed to Albert Einstein who presented a more rigorous formulation of the law in 1905. The mass of the nucleus is similar to but not quite the same as the mass of the component protons and neutrons. The difference reflects the energy with which the protons and neutrons are bound together in the small volume occupied by the nucleus. If the mass of the nucleus is denoted by M with the masses of the protons and neutrons given by m_p and m_n respectively, and if we assume that the nucleus includes n_p protons and n_n neutrons, it follows that

$$M = n_p m_p + n_n m_n - B/c^2 \tag{9.3}$$

where B defines the energy with which the protons and neutrons are bound in the nucleus (the negative sign reflects the fact that since these particles are bound they must be present in a lower energy state than that corresponding to their unbound, free, form). The net binding energy of the protons and neutrons composing the nucleus provides a measure of the stability of the nucleus. The energy with which these particles are bound in a stable nucleus is typically orders of magnitude greater (a factor of 10^6 or more) than the energy with which atoms are combined in a molecule. This ultimately is the source of the vast quantities of energy that can be released by rearranging the components of the nuclei of atoms, the energy responsible both for nuclear bombs and nuclear power.

We begin in Section 9.2 with a discussion of the physics of the fission process by which this energy is released both in bombs and civilian nuclear power plants. Fission defines the process by which a nucleus is split into two or more components, in contrast to fusion in which two nuclei are combined to form a third. Fusion is the process responsible for the luminosity of the sun. Fission is the basis of all current civilian uses of nuclear power although large sums of money have been expended and will continue to be spent in hopes of developing a commercially viable fusion process. The problem is that in order to trigger the fusion process, nuclear particles must be squeezed into very close proximity. The objective of current research is to develop an energy efficient means to accomplish this task, which if successful could provide the world with an essentially unlimited source of clean energy for the future (our own star under our own control!).

The history of the development of nuclear power is recounted in Section 9.3. Problems that have plagued the development of nuclear power in recent years are addressed in Section 9.4, challenges that have rendered moot the ambitious prediction of Lewis L. Strauss, then chairman of the U.S. Atomic Energy Commission (AEC), when he stated in a speech quoted the following day in the New York Times of September 17, 1954, and frequently thereafter, that: "It is not too much to expect that our children will enjoy electrical energy in their homes too cheap to meter." Future prospects for nuclear power are discussed in Section 9.5 with concluding remarks in Section 9.6.

9.2 NUCLEAR FISSION

The primary fuel for nuclear power is uranium. Uranium is present in natural environments on Earth in three main isotopic forms, ^{234}U, ^{235}U, and ^{238}U with relative abundances (expressed in terms of numbers of atoms) of 0.0055%, 0.72%, and 99.275% respectively (Bodansky, 2004). The fissile isotope is ^{235}U (fissile refers to the property that the nucleus can be induced to break apart, undergo fission, when colliding with a neutron). A collision between a slow moving (thermal) neutron and ^{235}U can cause the nucleus of ^{235}U to splinter into two distinct components of unequal mass with release of additional neutrons.

A typical reaction (Bodansky, 2004) might involve production of isotopes of barium and krypton, ^{144}Ba and ^{89}Kr respectively. Two additional neutrons are produced in this case. The overall reaction may be summarized as:

$$n + {}^{235}U \rightarrow {}^{236}U \rightarrow {}^{144}Ba + {}^{89}Kr + 3n \qquad (9.4)$$

The numbers of protons (92) and neutrons (44) (referred to collectively as nucleons) are conserved in this process, a necessary condition for all such reactions. The combined mass of the species indicated as the final products of the reaction equation is less than the combined mass of the species involved in the initial reaction. The difference is reflected in the kinetic energies of the products. The isotope of uranium indicated as an intermediate in the reaction sequence, ${}^{236}U$, has a mass number that is a multiple of 4: isotopes with this property are particularly subject to fission.

Energies in atomic and nuclear physics are quoted normally in units of electron volts (abbreviated as eV). An electron volt defines the energy acquired by an electron when accelerated through a potential difference of 1 V:

$$1 \text{ eV} = 1.6022 \times 10^{-19} \text{J} = 1.519 \times 10^{-22} \text{Btu} \qquad (9.5)$$

The energies associated with the formation of chemical bonds (the linkage of atoms in molecules) are expressed typically in units of eV: the strong triple bond linking the atoms of nitrogen in the nitrogen molecule (N_2), for example, has an energy of 9.76 eV. Energies implicated in nuclear reactions are typically much larger than energies involved in atomic and molecular processes and are expressed in this case more commonly in units of millions of electron volts (MeV). The energy released in (9.4) (based on mass differences) amounts to 173 MeV. The bulk of this energy, 167 MeV, is communicated to the fission fragments, with the balance, 6 MeV, appearing as kinetic energy of the product neutrons.

Reaction (9.4) reflects but one of a large number of possible paths for fission of ${}^{135}U$. It exemplifies another common feature of the fission process. The barium and krypton isotopes produced in (9.4) are unstable. Initial decay of ${}^{144}Ba$ (on a time scale of less than a minute) leads to production of ${}^{144}Ce$ (cerium-144) with emission of energy in a form known as beta radiation. Beta radiation is associated with conversion of a proton to a neutron, or visa versa. Conversion of a neutron to a proton results in emission of an energetic electron with a companion particle known as a neutrino. Conversion of a neutron to a proton is accompanied by emission of a neutrino and a positively charged analog of the electron known as a positron. Neutrinos have very small cross-sections for interaction with matter: the cross-section is sufficiently small that neutrinos can pass without impact through masses as large or even larger than that of the Earth. The energetic electrons and positrons, however, interact readily with matter and are important contributors to the problems posed by exposure of humans to nuclear radiation (although exposure to gamma radiation is even more consequential). Initial decay of ${}^{144}Ba$ to ${}^{144}Ce$ takes place in the reactor. The lifetime of ${}^{144}Ce$ is sufficiently long (285 days) such that important quantities of the element are still present when the spent fuel is removed from the reactor. It decays eventually to form neodymium-144, ^{144}Nd, with additional emission of beta radiation. The problem posed by the persistence of long-lived radioactive elements formed as by-products of nuclear power generation lies at the root of the challenge faced by society seeking to identify a safe depository for long-term disposal of nuclear waste. Some of the products of uranium fission have lifetimes measured in tens or even hundreds of thousands of years, underscoring the complexity of the challenge.

Averaging over all possible products, fission of ^{235}U leads to production of approximately 2.5 neutrons per fission event, sufficient in principle to allow for a self-sustaining chain of further fission reactions (this allows for up to 1.5 neutrons to either escape from the system or to be captured in non-fission processes). If every fission reaction results in a single additional fission process, the state of the system is said to be critical (this is the condition that applies in a stable, functioning, power generating, nuclear reactor). If the number of successive fission events increases in time, the system is said to be supercritical (the situation that pertains for the case of nuclear weapons). The average energy of neutrons produced by fission of ^{235}U lies in the range 1 to 2 MeV. The problem is that neutrons in this energy range interact more readily with ^{238}U than with ^{235}U resulting in capture of the neutron by ^{238}U with production of plutonium-239 (^{239}Pu) rather than fission of ^{235}U. Plutonium-239 is also fissile and could be used potentially as the fuel for a nuclear bomb. Much of the concern about nuclear proliferation relates to the possibility that ^{239}Pu produced by civilian nuclear power reactors could fall into the hands of either rogue nations or individuals.

Conditions for reaction of neutrons with ^{235}U rather than ^{238}U are significantly more favorable at low neutron energies. At thermal energies (close to room temperature), the cross-section for reaction of a neutron with ^{235}U leading to fission exceeds that for capture by ^{238}U by a factor of more than 200. The abundance of ^{238}U is much greater, however, than that of ^{235}U in natural sources of uranium, by about a factor of 140 as indicated above. To enhance the conditions for sustained release of energy by fission of ^{235}U, nuclear power reactors normally adopt a two-pronged strategy: first, they incorporate what is referred to as a moderator, the function of which is to lower the energy of the prompt MeV neutrons to the thermal range thus reducing the possibility the neutrons may be absorbed by ^{238}U; second they use sources of uranium that are enriched to develop a higher relative abundance of ^{235}U, increasing again the chance for a fission reaction with ^{235}U rather than capture by ^{238}U. The exception is the CANDU reactor (so called in recognition of its Canadian origin and the nature of the key ingredients employed in its operation: D and U), which uses heavy water (water composed mainly of D_2O) as a moderator and is able to function using natural sources of uranium.

Bodansky (2004) provides an account of the fuel loading characteristics of a reactor based on the Westinghouse Corporation Pressurized Water Reactor (PWR) design, noting that the gross features of all large Light Water Reactors

(LWRs) (see below) are basically similar. Fuel is supplied to the reactor in the form of cylindrical pellets of uranium oxide (UO_2). The pellets measure 1.35 cm in length, 0.8 cm in diameter, and are placed in tubes composed of material selected on the basis of their structural strength and low cross-section for absorption of neutrons. The tubes are referred to as fuel rods. Individual fuel rods are composed of zircaloy, an alloy of zirconium (98% zirconium, 1.5% tin with small concentrations of other metals), and measure 3.7 m (12 ft) in length, 1.0 cm in diameter. They are packaged in units referred to as bundles or assemblies. A typical single assembly includes an array of 17×17 fuel rods. The reactor is loaded with 193 assemblies containing 50,952 fuel rods. A number of the assemblies incorporate control rods, the function of which is to provide a measure of control on the operation of the reactor. These control rods are composed of a silver–indium–cadmium alloy, which serves to limit the reactivity of the fuel by controlling the flux of neutrons available for fission. As the ^{235}U content of the fuel is depleted, the control rods can be withdrawn ensuring a relatively constant release of energy for power generation. A typical fuel assembly remains in the reactor for approximately 3 years (one-third of the assemblies are withdrawn and replaced on an annual basis).

Materials employed as moderators include normal (light) water (H_2O), heavy water, as noted above, and graphite (carbon). Light water suffers from the problem that low energy neutrons are absorbed efficiently by H (which is converted in the process to D). It has the advantage that it is readily available and cheap. Neutrons are absorbed by D_2O less efficiently than H_2O by about a factor of 80 (Bodansky, 2004). The ability of heavy water to serve as an effective moderator is limited, however, by the presence in the heavy water of trace quantities of H. Graphite is more effective than light water as a moderator though less effective than heavy water. Use of water (light and heavy) as a moderator has the advantage that the water can be employed also as a coolant. Light water reactors account for 88% of electricity generated by nuclear power plants currently on a worldwide basis with contributions from heavy water and graphite reactors trailing at 5% and 7% respectively.

Civilian reactors employ mildly enriched uranium, relative abundances of ^{235}U in the range 0.71–20%. Bombs require higher levels of enrichment, as much as 90% or even greater. A number of approaches are available to produce this enriched fuel. The method in most common use today, referred to as gaseous diffusion, exploits the fact that compounds of lower molecular weight diffuse more readily than those of higher molecular weight (the mean thermal velocity varies as the square root of the mass of the compound). To take advantage of this difference, the uranium must be converted to gaseous form. The compound of choice (the only one) is uranium hexafluoride, UF_6, present as a gas at room temperature. The problem is that the difference between the thermal velocities of the ^{238}U and ^{235}U forms of UF_6 is very small, less than 0.5%. Enrichment is accomplished by passing a stream of UF_6 through a barrier with small apertures, taking advantage of the fact that the lighter molecules pass through this obstacle more readily than the heavy resulting in an increase in

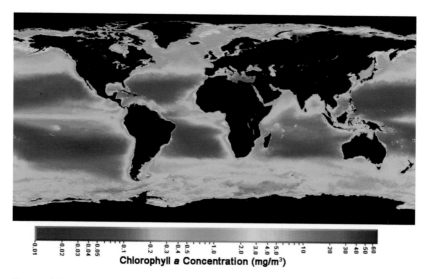

Figure 6.1 SeaWiFS average chlorophyll-a concentration (Oct 1997–April 2002).

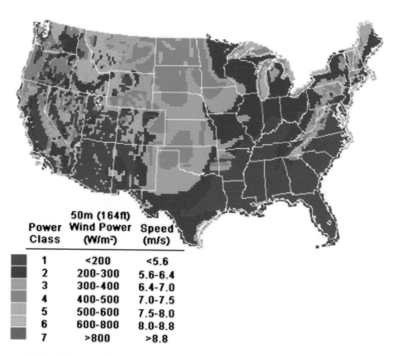

Power Class	50m (164ft) Wind Power (W/m²)	Speed (m/s)
1	<200	<5.6
2	200-300	5.6-6.4
3	300-400	6.4-7.0
4	400-500	7.0-7.5
5	500-600	7.5-8.0
6	600-800	8.0-8.8
7	>800	>8.8

Figure 8.5. U.S. annual wind power resource and wind power classes—Contiguous U.S. States. [Source: http://www1.eere.energy.gov/windandhydro/wind_potential.html. Read March 8, 2007.]

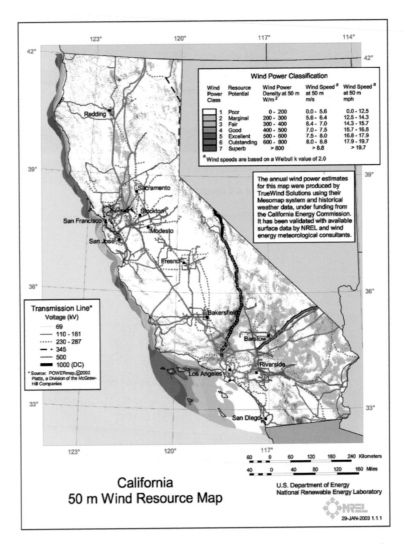

Figure 8.6. California wind resource map. [Source: http://www.eere.energy.gov/ windandhydro/windpoweringamerica/wind_maps.asp. Read March 8, 2007.]

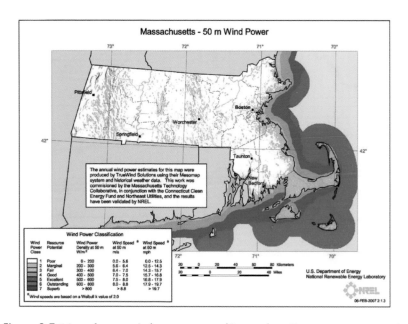

Figure 8.7 Massachusetts wind resource map. [Source: http://www.eere.energy.gov/
windandhydro/windpoweringamerica/wind_maps.asp. Read
March 8, 2007.]

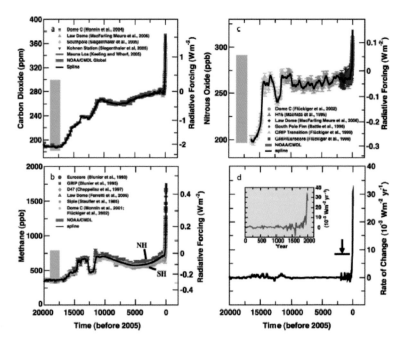

Figure 13.2 Concentrations and contribution to radiative forcing over the past 20,000 years (20 kyears) for (a) CO$_2$, (b) CH$_4$, and (c) N$_2$O. Contributions from the combination of all three gases to the rate of change of radiative forcing over the past 20 kyears is displayed in panel (d). Data reflect analyses of gas extracted from ice and firn samples from both Greenland and Antarctica complemented for the most recent period by direct measurements of the atmosphere. The range of variability observed over the past 650 kyears, excepting the recent post-industrial period, is indicated by the grey vertical bars in panels (a–c). From IPCC (2007).

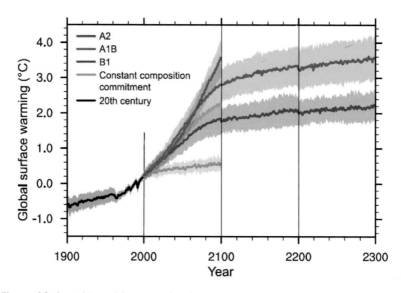

Figure 13.6 Multi-model means of surface warming (relative to 1980–1999) for the scenarios A2, A1B, and B1 as described in the text. From IPCC (2007).

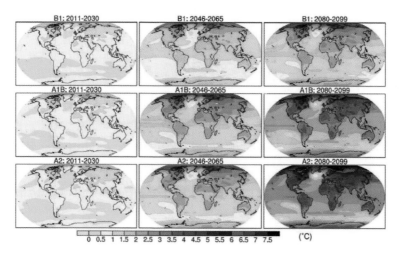

Figure 13.7 Multi-model means of annual mean surface warming (surface air temperature change, °C) for the scenarios B1 (top), A1B (middle), and A2 (bottom), for three time periods, 2011 to 2030 (left), 2046 to 2065 (middle), and 2080 to 2099 (right). Anomalies are relative to the average for the period 1980 to 1999. From IPCC (2007).

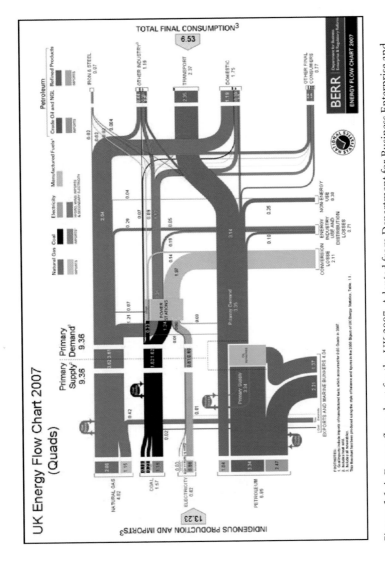

Figure 16.4 Energy flow chart for the UK 2007. Adapted from Department for Business Enterprise and Regulatory Reform (BERR), http://www.berr.gov.uk/whatwedo/energy/statistics/publications/flowchart/page37716.html, read October 8, 2008.

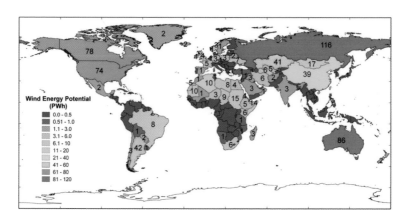

Figure 17.1 Annual onshore wind energy potential (in units of 10^{15} Watt-hour, PWh) country by country, restricted to installations on suitable areas with capacity factors greater than 20% (from Lu et al., 2009).

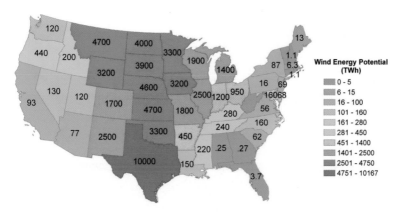

Figure 17.2 Annual onshore wind energy potential (in units of 10^{12} Watt-hour, TWh) on a state-by-state basis for the contiguous U.S. (from Lu et al., 2009).

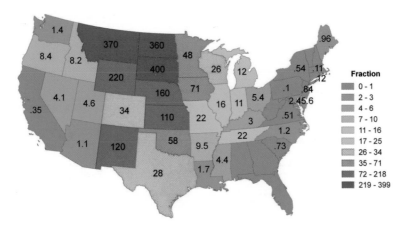

Figure 17.3 Annual potential generation of electricity from wind on a state-by-state basis for the contiguous U.S. expressed as fraction of total electricity retail sales in the State (2006). For example: the potential source for North Dakota exceeds current total electricity retail sales in the State by a factor of 414 (from Lu et al., 2009).(Note: Data source for total electricity retail sales: http://www.eia.doe.gov, read October 19, 2008.)

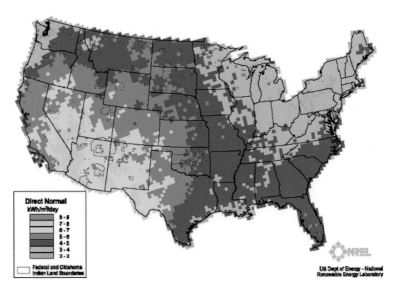

Figure 17.5 Available concentrated solar power (CSP) for the contiguous states of the U.S.

their relative abundance downstream. To accomplish the enrichment needed to supply the needs of a conventional nuclear reactor—to achieve an enrichment of say 4%—it is necessary to repeat this process as many as 1200 times (Bodansky, 2004). The procedure is expensive in terms of both equipment and manpower. It is also energy intensive. Future enrichment plants are expected to make use of an energy-more-efficient strategy based on gaseous centrifuge.

The centrifuge approach operates by spinning the uranium gas at high speed in a cylinder. The high mass components are concentrated by the centrifugal force in the outer regions of the cylinder, at greatest radial separations from the rotational axis of the cylinder. The small mass difference requires that multiple steps must be implemented to achieve the desired degree of enrichment. The slightly enriched uranium from one step is used as input to the next. The advantage of the centrifuge approach is that the power consumption is much less than that involved with gaseous diffusion. The current debate over enrichment activities in Iran involves a dispute as to whether the authorities there are engaged, as they claim, in producing a grade of uranium suitable for use as a fuel in civilian reactors or whether their real intent is to develop a fuel that could be employed to produce a nuclear bomb (or bombs).

The amount of ^{239}Pu in a typical thermal nuclear reactor increases with time as the ^{235}U content of the original fuel is depleted increasing the probability of capture of neutrons by ^{238}U rather than ^{235}U. Capture by ^{238}U results in production of ^{239}Pu. Capture of an additional neutron by ^{239}Pu results in production of an excited state of ^{240}Pu. Since the mass of this isotope is a multiple of 4, ^{240}Pu is susceptible to additional fission as noted above. For light water reactors, the yield is about 0.6 (loss of 10 atoms of ^{235}U is associated with production of 6 atoms of fissile ^{239}Pu). Yields of up to 0.8 are obtained readily with heavy water and graphite moderated reactors reflecting the higher concentration of thermal neutrons present in these systems (a consequence of the diminished role of the sink due to capture of neutrons by H). A yield of 1 would correspond to a condition in which rate for production of fresh fissile material would equal the rate at which preexisting fissile material was eliminated. Reactors for which this condition is met are referred to as breeder reactors. There are only two breeder reactors currently in operation, one in Russia, one in France. They rely on reactions involving energetic neutrons (for which the cross-section for capture by ^{238}U is large) rather than low energy neutrons produced as a consequence of energy loss due to interactions with the moderating material. They are referred to as fast breeder reactors underlining the distinction with the thermal reactors emphasized here, operation of which depends on the presence of low energy (thermal) neutrons to enhance the probability that fission of ^{235}U can compete with capture of neutrons by the more abundant ^{238}U.

In addition to the fuel rods that supply the initial neutrons and the moderating agent whose function is to reduce the energy of these initially fast neutrons to the thermal range where they can be captured most efficiently by ^{235}U,

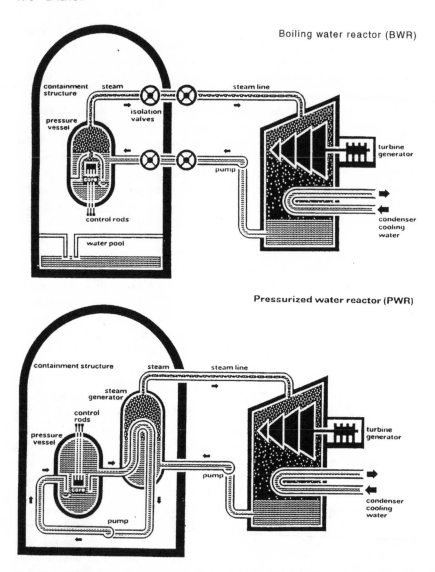

Figure 9.1 Schematic representation of BWR and PWR systems, emphasizing the difference in the means for providing steam to the steam turbine. [Bodansky, p. 182].

conventional nuclear reactors are equipped with control rods that can be inserted as required to turn off the reactor for routine maintenance or in response to an emergency. These control rods are fashioned typically of either cadmium or boron, elements with high cross-sections for neutron capture.

A schematic of the two main types of light water reactors is presented in Fig. 9.1. In both cases, the key components of the system—the fuel rods, control rods, and moderator (light water)—are contained in a large cylindrical steel tank, the reactor pressure vessel. This vessel stands approximately 40 ft in height with a diameter of 15 ft and is enclosed by thick walls capable in the case of the pressurized water reactor of withstanding pressures of up to 170 atmospheres (Bodansky, 2004). The heat released in the fission process is converted directly to steam in the boiling water reactor. The steam is used then to generate electricity by driving a turbine located exterior to the pressure vessel. Following this, the steam is cooled and pumped back into the reaction vessel. The water that serves as the moderating agent in the pressurized water reactor is maintained at pressures high enough so that it remains in liquid form as it processes heat received from the reactor. The steam used to generate electricity in this case is produced in a separate steam-generating tank. The pressurized water, having given up its heat to produce the steam, is pumped back into the reactor pressure vessel. The pressure vessel and steam generators (in the case of the pressurized water reactor) in reactors of western design are enclosed in a massive containment building which provides an additional level of protection in the case of an accident. This additional level of security proved to be important in preventing release of radioactive materials to the external environment in the case of the Three Mile Island accident in the U.S. in 1979. It was absent, however, in the case of the Chernobyl reactor in the Soviet Union, which blew up in 1986 as a result of an uncontrolled build-up of the nuclear chain reaction followed by a steam explosion and release of radioactive elements that spread subsequently over a large region of Europe. Chernobyl is the only case to date in which the generation of electricity using nuclear power is known to have resulted in loss of human life.

We introduced of necessity a large number of technical terms in the course of the discussion in this section. Although the terms were defined when first introduced, it would be too much, we believe, to expect the non-technical reader to commit all of this to memory. Table 9.1 is intended to provide a convenient summary, a ready reference, to the more important terms and concepts introduced here.

9.3 HISTORY OF NUCLEAR POWER

In detailing the key discoveries that led ultimately to the development of nuclear power, we noted the early importance of the contributions of Poincare and Einstein arguing for the equivalence of mass and energy (the celebrated $E = mc^2$ equation). The English physicist James Chadwick, a protégé of Ernest Rutherford, is usually credited with the discovery of the neutron for which he was subsequently awarded the Nobel Prize. In the understated language of the day, Chadwick began the summary of the work in his landmark paper (Chadwick, 1932) thus: "The properties of the penetrating radiation emitted from beryllium (and boron) when bombarded by the alpha particles of polonium have been examined. It is concluded that the radiation consists, not of quanta as hitherto supposed, but of neutrons, particles of mass 1 and

Table 9.1 Summary of terms used in discussion of nuclear power

Nucleus:	Positively charged central component of an atom that includes the bulk of its mass.
Nucleons:	Particles composing the nucleus consisting of protons and neutrons.
Protons:	Positively charged components of the nucleus.
Neutrons:	Electrically neutral components of the nucleus; mass of a neutron is comparable to, though slightly greater than, the mass of the proton.
Electrons:	Low mass, negatively charged, particles orbiting the nucleus of an atom.
Nuclear mass number:	Number of protons plus neutrons contained in the nucleus.
Z number:	Number of protons contained in the nucleus or equivalently the number of electrons orbiting the nucleus of an atom in its electrically neutral state.
Fission:	Process by which the nucleus of an atom splits into two or more smaller nuclei with emission of a combination of neutrons, photons, beta particles, and alpha particles.
Fusion:	Process by which several nuclei combine to form a heavier nucleus with emission of energy.
Photons:	The basic components of light.
Beta particles:	High energy particles—electrons and positrons—emitted in conjunction with decay of unstable nuclei.
Positron:	Positively charged counterpart to the electron with same mass and same charge (though opposite sign).
Alpha particle:	Combination of 2 protons and 2 neutrons representing the nucleus of the ^{4}He atom.
Neutrino:	Elementary electrically neutral particle traveling at close to the speed of light with very small cross-section for interaction with matter.
Electron volt:	Energy acquired by an electron when accelerated through a potential of 1 V.
Thermal reactor	Reactor in which fission is induced by interaction of fissile material with low energy (thermal) neutrons.
Fast reactor:	Reactor in which fission is induced by interaction of fissile material with high energy neutrons.
Breeder reactor:	Reactor in which as much fissile material is produced as is consumed.
Fuel rod:	Tube containing the fuel for the reactor.
Control rod:	Fashioned from materials with high cross-sections for neutron capture, the function of the control rod is to turn off the reactor either for routine maintenance or in case of an accident.
Moderator:	The component of a nuclear reactor whose function is to reduce the energy of neutrons to the thermal range where they are more effective in triggering fission of ^{235}U.
Light water reactor:	Reactor in which natural water serves as the moderator.
Heavy water reactor:	Reactor in which water enriched in deuterium relative to hydrogen serves as the moderator.
Pressurized water reactor:	Reactor in which water remains in liquid form under the influence of high pressure; steam to drive the turbine is produced by exchange of heat between the pressurized water circulating in the reactor and an external source of water.

charge 0. Evidence is given to show that the mass of the neutron is probably between 1.005 and 1.008." Enrico Fermi, working at the time in his native Italy, demonstrated (in 1934) that bombardment of uranium by neutrons resulted in production of a wide range of radioactive elements. For this work he also was awarded the Nobel Prize (in 1938). Experiments by German scientists Otto Hahn and Fritz Strassman in 1938, interpreted by Lise Meitner and her nephew Otto Frisch, provided the first conclusive proof for fission of uranium and for the vast quantities of energy released in the process. Meitner was born into a Jewish family in 1878 in Austria. Although she later became a Protestant, she was forced to flee Germany (with Hahn's help) in July 1938 as Hitler moved to crack down on individuals of Jewish heritage. Meitner relocated to Stockholm where she continued a clandestine collaboration with Hahn. Hahn was awarded the 1944 Nobel Prize in Chemistry (which he received finally in 1946), an honor many felt should have been shared with Meitner. The importance of the contributions of Hahn, Meitner, and Strassman to nuclear science was recognized by President Lyndon Johnson and the U.S. Atomic Energy Commission in 1966 with award of the Enrico Fermi Medal, the only time this honor was conferred on scientists of foreign nationality.

Learning of the experiments by Hahn and Strassman and the related work by Meitner and Frisch, it was immediately clear to scientists in the U.S. (and to their colleagues in France and the UK) that fission of uranium could provide an important new source of energy that could be used potentially to produce a devastating new weapon. Einstein wrote to President Roosevelt on August 2, 1939 informing him that: "In the course of the last four months it has been made probable through the work of Joliot in France as well as Fermi and Szilard in America—that it may be possible to set up a nuclear chain reaction in a large mass of uranium, by which vast amounts of power and large quantities of new radium-like elements would be generated. Now it appears almost certain that this could be achieved in the immediate future. This new phenomenon would also lead to the construction of bombs, and it is conceivable—though much less certain—that extremely powerful bombs of this type may thus be constructed." Einstein's letter was not immediately effective. But it was influential eventually in the establishment of what came to be known as the Manhattan Project. The Manhattan Project at its peak engaged more than 130,000 individuals with a budget of close to $2 billion (more than $20 billion in current dollars). It culminated in the development and explosion of the first nuclear bombs. Research conducted under the project played an important role later, however, in development of the first nuclear reactors constructed specifically for the purpose of peaceful generation of electricity.

The newly launched U.S. nuclear program achieved its first notable success on December 2, 1942 with demonstration of the first self-sustained nuclear chain reaction. The site for this historic occasion was a facility that served formerly as a squash court under the West Stands of Stagg Field at the University of Chicago. The experiment was conducted under the direction of Fermi who had emigrated to the United Sates by way of London after receiving his Nobel

Prize in Sweden in 1938 (Fermi's wife was Jewish; he was criticized by the Italian Fascist press for not wearing a Fascist uniform and for not giving the Fascist salute when he received his award). Word of the success of the experiment was communicated in a coded message by telephone from Chicago physicist Arthur H. Compton to James B. Conant (Professor of Chemistry, President of Harvard University and Chairman of the National Defense Research Committee). Compton's message was: "The Italian navigator has landed in the new World," to which Conant responded: "How were the natives?" "Very friendly" replied Compton.

The first nuclear bomb was detonated in a test (known as Trinity) that took place on July 16, 1945 at what is now the White Sands Missile Range near Alamogordo, New Mexico. The associated energy release was estimated at 20 kilotons of TNT (equivalent to about 8.4×10^{13} J or 8.0×10^{10} Btu). The fissile material used in this case, and in the bomb dropped later (August 9, 1945) on Nagasaki, Japan, was ^{239}Pu. The bomb that destroyed Hiroshima, Japan, on August 6, 1945, used ^{135}U. The energy yield of these early bombs was modest compared with the energy released by the weapons developed later which combined fission and fusion with yields up to a 1000 times greater than the bombs dropped on Hiroshima and Nagasaki. The bombs that exploded over Hiroshima and Nagasaki devastated both cities resulting in more than 100,000 deaths immediately with an uncertain additional number of casualties later as a result of exposure to radiation released by the bombs. There are eight countries now that admit possession of nuclear weapons. In order of their acquisition of these weapons, they are: the United States, Russia, the United Kingdom, France, China, India, Pakistan, and, prospectively, North Korea. Israel is most likely an eighth member of the nuclear club although they have never officially acknowledged this fact. The horrors of Hiroshima and Nagasaki, and the accidents that occurred later at Three Mile Island and Chernobyl, have contributed in no small measure to the unease the general public feels as to the safety of nuclear power. There are important differences, however, between nuclear power deployed as a weapon of war in contrast to its use for peaceful generation of electricity. The accident at Chernobyl resulted in a major release of radioactive material. The release was triggered, though, not by a nuclear explosion, but by a build-up of uncontrollable steam pressure followed by an intense fire fueled by hydrogen and carbon monoxide coming in contact with air. The operational conditions of a nuclear power plant are such as to absolutely preclude the possibility of a nuclear explosion such as occurs in a nuclear bomb.

The transition from military to civilian uses of nuclear power was slow to occur following the end of World War II. Research in the U.S., and elsewhere, continued to be cloaked under a veil of secrecy. President Truman established the Atomic Energy Commission in the U.S. on August 1, 1946 with responsibility to oversee both civilian and military applications of nuclear energy. The first pressurized water reactors were constructed by the U.S. Navy under the direction of Admiral Hyman Rickover in 1953 and were deployed subsequently to provide the power source for the new U.S. nuclear submarine fleet. The first civilian reactor, sited at Shippingport, Pennsylvania, with a capacity of 60 MW, did not become operational until the end of 1957. Orders for a total of 14 reactors were placed by utilities in the U.S. over the period 1953 to 1960. Only three of these reactors had capacities for generation of electricity greater than 100 MW: Indian Point 1 in New York (265 MW), Dresden 1 in Illinois (207 MW), and Yankee Rowe in Massachusetts (175 MW). All three were of the light water reactor type and all three have since been shut down, Yankee Rowe the most recent to go out of commission, in 1991 (Bodansky, 2004).

Nuclear power enjoyed a period of rapid growth in the U.S. between 1965 and 1974. By the end of 1974, 55 reactors were in operation with a total capacity of about 32 GWe [1] accounting for 6% of U.S. electricity generation (Bodansky, 2004). The largest of these reactors had a capacity of 1.25 GWe with an average capacity for reactors constructed after 1970 in excess of 1 GWe. The industry then went into a state of suspended animation. Only 13 reactors were ordered in the U.S. after 1975 and all of these orders were subsequently cancelled. There have been no new orders for nuclear reactors in the U.S. since 1978. What were the circumstances that led to this change of attitude? There were two important influences according to Bodansky (2004). One was the decrease in the growth of demand for electricity in the U.S. resulting from the economic shock occasioned by the oil embargo of 1973 (although it may be interesting to note that construction of coal-fired plants continued with minimal interruption). Second, arguably more important, was the decrease in public enthusiasm for nuclear power that developed following the accident at Three Mile Island in 1979, exacerbated by reactions to The China Syndrome movie released just 12 days prior to the accident. The movie, starring Jane Fonda, Jack Lemmon, and Michael Douglas, dramatizes the near meltdown of a nuclear reactor that takes place near Los Angeles as a result of deficiencies in design compounded by operator errors and irresponsible management. Followed so closely by the Three Mile Island accident, the movie had an immediate (if unwarranted) impact on public attitudes towards the safety of nuclear power, confidence undermined further by the subsequent accident at Chernobyl.

Worldwide trends in generation of electricity from nuclear power are summarized in Fig. 9.2. There are 435 commercial reactors operational today in 30 countries with a total capacity of 370 GWe supplying 16% of the world's demand for electricity (World Nuclear Association, 2007). Nuclear power accounted for 780.5×10^9 kWh of electricity produced in the U.S. in 2005—30% of the total power generated by nuclear energy worldwide. A summary of the top ten countries in terms of electricity generated by nuclear power is presented in Table 9.2. The table includes also information on the fraction of total electricity produced in these countries that was contributed by nuclear energy (19% for the U.S., a much larger fraction (79 %) for the second largest producer, France) also data on the number of operational reactors and total related capacities (World Nuclear Association, 2007).

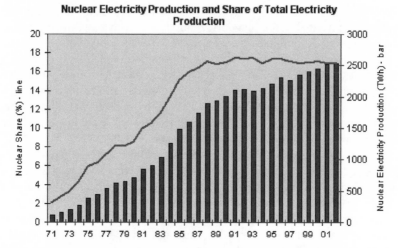

Figure 9.2   Increase in electricity generated worldwide by nuclear power since 1971. The solid line (referenced to the left vertical axis indicates the fraction of total electricity contributed by nuclear power. [World Nuclear Association, 2007].

Table 9.2   Nuclear powered generation of electricity: summary of conditions for the top ten producing countries (2005)

Rank	Country	Production (Billion kWh)	Percent[1]	Numbers of reactors (January 07)	Total capacity (MWe)
1	United States	780.5	19%	103	98,254
2	France	430.9	79%	59	63,473
3	Japan	280.7	29%	55	47,700
4	Germany	154.6	31%	17	20,303
5	South Korea	139.3	45%	20	17,533
6	Russia	137.3	31%	31	21,743
7	Canada	86.8	15%	18	12,595
8	Ukraine	83.3	49%	15	13,168
9	United Kingdom	75.2	20%	19	10,982
10	Sweden	69.5	45%	10	8975

[1]Indicates the fraction of electricity produced in the country that was contributed by nuclear power in 2005.

9.4 CHALLENGES FOR THE FUTURE OF NUCLEAR POWER

There are three primary challenges that must be confronted by nuclear power if it is to play an important role in meeting future demands for energy. First, there is the need for management protocols and regulatory oversight to ensure safety at all stages of the nuclear fuel cycle: from plant design and construction, to plant operation, to the treatment and handling of spent fuels. Second, there is need for a credible and publicly acceptable strategy for treatment and eventually for disposal of radioactive wastes. And third, there is a need for security measures to ensure that the by-products of nuclear industry should not be subverted to be used as fuel for illicit production of nuclear bombs—to guard against the risk of what is referred to as nuclear proliferation.

On the first of these counts, the nuclear industry, at least in the west, has an effectively unblemished record. The accident at Three Mile Island was contained. As of the end of 2003, reactors in the non-Soviet world had enjoyed 10,100 reactor years of successful operation (2870 in the U.S.) without a single documented loss of life as a result of exposure to nuclear radiation (Bodansky, 2004). On a worldwide basis, the only black spot has been Chernobyl, the result in this case of a combination of deficient design and inefficient management. There is every reason to expect this enviable record of success to continue in the future as the industry builds on past experience and moves increasingly to standardized designs for new reactors. Contrast this record with the large numbers of casualties resulting from accidents in the coal mining industry, premature deaths on the part of coal miners suffering from black lung disease, the toll exerted as a consequence of the exposure of large populations to fossil fuel related air pollution, and the thousands of deaths resulting over the years from accidents in oil refineries and chemical factories. Release of 40 tons of methyl isocyanate from a pesticide factory owned by Union Carbide in Bhopal, India, on December 3, 1984, for example, resulted in an immediate death toll of close to 3000 with more than 15,000 suffering from serious subsequent illnesses, as reported by the BBC, in what was arguably the world's worst single industrial accident.[2]

An estimate of the quantity of uranium required to produce a given amount of electricity is presented in Box 9.1. A little less than 18,000 metric tons of uranium per year are required to satisfy current U.S. demand. Global requirements amount to about 60,000 metric tons per year. As indicated earlier, fuel is supplied to reactors in the form of pellets of uranium oxide contained in tubes of zirconium packaged in units referred to as assemblies. When it is time to replace the fuel, the assemblies are withdrawn intact from the reactor. The combination of spent fuel, fission products, and associated packaging represents the source of high-level radioactive waste that must be treated and prepared for eventual disposal. The mass of this waste is typically about 75% greater than the mass of uranium contained in the original fuel (recall that the uranium is supplied in the form of UO_2 rather than pure elemental U). The total mass of high-level radioactive waste produced annually by civilian U.S. nuclear industry amounts to about 4500 metric tons, 15,000 metric tons for the world. This may be compared with the quantity of CO_2 that would be emitted to the atmosphere if electricity generated using nuclear power were replaced using electricity generated using coal: 760 million metric tons per year for the U.S., 2500 million metric tons per year for the world. While the mass of waste produced by nuclear industry is obviously significant, it is clearly much less

---

Box 9.1 Quantity of Uranium required to supply current U.S. production of electricity by nuclear power

---

Quantity of electricity produced using 1 metric ton of ^{235}U = 8.76 × 10^9 kWh (Bodansky, 2004).

Amount of uranium required to provide 1 metric ton of ^{235}U assuming an isotopic abundance for ^{235}U in natural uranium of 0.72% = 137 metric tons.

Electricity produced per metric ton of U

$$= \frac{8.76 \times 10^9 \text{ kWh}}{1.39 \times 10^2 \text{ tons}}$$

$$= 6.3 \times 10^7 \text{ kWh ton}^{-1}$$

Total U.S. production of electricity by nuclear power (2005, Table 9.2) = 7.805 × 10^{11} kWh

$$\text{Uranium required} = \frac{7.80 \times 10^{11} \text{ kWh}}{6.4 \times 10^7 \text{ kWh ton}^{-1}} = 12,389 \text{ metric tons}$$

Allowing for losses of ^{235}U in conjunction with production of enriched fuel for reactors gross annual demand for U in the U.S. is about 18,000 metric tons. Global annual demand for uranium used in power generation is about 60,000 metric tons U year^{-1}.

(by a factor of about 1.5×10^5) than the mass of waste associated with conventional power generation.

The spent fuel when removed from the reactor is not only radioactive but also hot. The radiation emitted and the heat released decrease by about a factor of 75 during the first year following removal of the spent fuel from the reactor. A year after discharge, a metric ton of nuclear waste is responsible for emission of radiation at a level of about 1.5 megacuries (MCi)[3] and for a source of heat at a power output of about 6 kW (roughly equal to the rate at which energy from the sun is incident on an area of 20 m^2 of the Earth's surface). Both radiation and thermal energy outputs decrease rapidly with time following removal of the spent fuel from the reactor, by about a factor 75 in year 1 as noted above and by a further factor of 5 between year 1 and year 10 as indicated in Fig. 9.3. The spent fuel assemblies are placed initially and allowed to cool in a pond of water from which they are transferred subsequently to steel canisters emplaced in turn in heavy casks with shielding sufficient to essentially eliminate risk of significant release of radiation. The casks employed typically in the U.S. have 29-in. thick outer concrete walls with steel inner liners with a minimum thickness of at least 1.5 in. The transfer operation is handled robotically, eliminating any possible risk of human exposure to radiation. The casks are

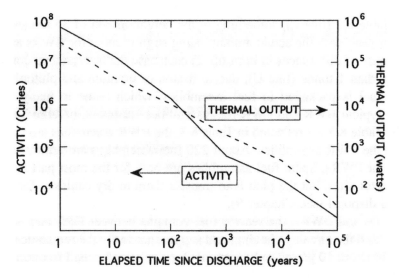

Figure 9.3 Emission of radiation (curies) and heat (watts) resulting from decay of spent fuel produced from 1 GWyear of operation of a pressurized water reactor: 1 GWyear is equivalent to 8.76×10^9 kWh (from Bodansky, 2004).

stored under dry conditions, cooled by natural convective flow of air through a gap between the canister and the outer wall of the cask (cool air enters at the bottom of the gap, moving upward in response to buoyancy provided by the heat absorbed from the canister, venting to the exterior of the cask at the top of the cavity).

Plans in the U.S. called originally for spent fuel to be transferred to a central location or locations for reprocessing after an initial period of storage and cooling on site at the reactors. These plans were abandoned in 1977, largely in response to concerns about nuclear proliferation, concerns that the plutonium in reprocessed fuel might fall into the wrong hands and could be used by a sub-national terrorist group to fabricate a crude nuclear bomb or bombs. Reprocessing facilities are currently operational in five countries: in France, the United Kingdom, Japan, India, and Russia. The major product of reprocessing is a mixture of uranium and plutonium oxides referred to as mixed oxide fuel or MOX, a significant fraction of which can be recycled and used as a source of fuel for reactors reducing the requirements for a fresh source of enriched uranium. The mass of waste is reduced significantly by reprocessing (uranium constitutes approximately 98% of the mass in the original spent fuel). Reprocessing does little, however, in the short term to reduce the radioactivity of the waste, contributed primarily by the original fission products. In the longer term, when most of the fission products have decayed (in about

30 years), reprocessing is significantly effective in reducing the residual radio-activity. It remains a challenge to identify a safe, acceptable, strategy to dispose of this material.

U.S. plans for long-term disposal of nuclear waste have focused on identifying a suitable geological medium in which the waste could be stored safely, ideally for hundreds of thousands of years or even longer (note that the activity of the waste, while it decreases by about a factor of 10^4 over the first 1000 years declines by less than a factor of 10^2 over the subsequent 100,000 years). After much study and controversy, Yucca Mountain in Nevada was selected in 1987 as the first site for construction of a depository for long-term storage of nuclear waste in the U.S., authorized under the Nuclear Waste Amendments Act of 1987. The schedule announced by the Secretary of Energy in 1989, called for "iterative scientific investigations of the site to examine its suitability," anticipating that the site would be open to receive waste by 2010. The site would be licensed to receive 105,000 metric tons of heavy metal (MTHM).[4] Approximately 90% of this storage would be allocated to wastes produced by the civilian power sector with the balance committed to wastes generated earlier by the military (by-products of the production of nuclear bombs). As of mid-2007, the schedule had been pushed back. The earliest the depository is anticipated to open now is 2017 (a 7-year delay with respect to the original schedule) but given the complex politics of the situation (the U.S. Senate Majority Leader, Harry Reid, hails from Nevada and is strongly opposed to the project) further slippage would seem likely. It the meantime, wastes will accumulate in interim storage facilities (in the future presumably mainly dry), most likely on site at reactors though conceivably in the future at central facilities set up specifically to receive, store, and manage these wastes. Bunn et al. (2001) concluded that: "interim storage, designed to last perhaps 30–50 years (though with flexibility to shorten that time to match the progress of permanent solutions) should be pursued as the best near-term approach to managing a large fraction of the world's spent fuel." They argued that this strategy should not only be cost-effective but that it should also be "secure and proliferation-resistant."

The International Atomic Energy Agency (IAEA) was set up in 1957 with the dual aim of promoting peaceful uses of nuclear energy while providing safeguards, including inspections, to minimize the risk of proliferation. An international agreement, the Treaty on the Non-Proliferation of Nuclear Weapons, designed to discourage military applications of nuclear power while at the same time promoting peaceful uses, was developed and went into effect in 1970. The Treaty made an important distinction between countries that had "manufactured and exploded a nuclear weapon or other explosive device prior to January 1, 1967" and those that had not done so. According to this distinction, the nuclear weapon states (NWSs) were identified as China, France, the USSR, the United Kingdom and the United States, all, as it happens, permanent members of the United Nations Security Council. The NWSs were obligated under the treaty to do nothing that would contribute to the development of nuclear weapons by non-NWSs. The non-NWSs in turn were committed neither to manufacture nor receive nuclear weapons. IAEA was charged with monitoring compliance of states under terms of the Treaty. India, Israel, and Pakistan declined to sign the Treaty and, as noted earlier, all opted subsequently to join the nuclear (weapons) club. The success of IAEA was acknowledged in October 2005 with award of the Nobel Peace Prize conferred jointly on the organization and its Director General, Mohamed ElBaradei, in recognition of "their efforts to prevent nuclear energy from being used for military purposes and to ensure that nuclear energy for peaceful purposes is used in the safest possible way."

Many view the distinction between NWSs and non-NWSs in the Non-Proliferation Treaty as unfortunate. Why should those who chose to develop a nuclear weapons' capability prior to January 1, 1967 be treated specially? Ultimately there is need for a comprehensive plan to phase out all nuclear weapons. In the absence of such an initiative, it is likely that the nuclear club will continue to grow. The Cold War established the precedent that if potential antagonists each possess nuclear weapons, each will be deterred from their use—the doctrine that came to be referred to as Mutually Assured Destruction (MAD). The same regrettable logic may be used to justify the nuclear weapons programs of India and Pakistan and could be invoked to rationalize development of a weapons capability in Iran (an offset to Israel) and potentially also in Japan (an offset to China and North Korea). As more countries proceed to adopt nuclear power as an energy option, the risk of proliferation increases. Plutonium formed as a by-product of peaceful nuclear power generation can easily be subverted and used to produce nuclear weapons. Further, facilities designed to enrich uranium to concentrations required for civilian reactors can readily be used to produce uranium that could be employed to make bombs. There is an urgent need for a new international non-proliferation agreement that would engage not only the current cohort of NWSs but also the additional members of the nuclear (weapons) club who declined to ratify the original Treaty. Given the authority, IAEA could play an important role in overseeing compliance of nations with terms of any such agreement.

The risk of proliferation due to interception of spent fuel by non-state terrorists is considered unlikely: the high level of radiation associated with the fuel would appear to argue against the possibility. Interception of the plutonium in reprocessed fuel could pose a more serious problem and it is clear that stringent security measures should be in place to guard against that possibility. On the whole, though, the risk of proliferation at the hands of non-state terrorists is considered small. The greater danger exists at the state level as exemplified by the current concerns with respect to events taking place in Iran and North Korea.

9.5 PROSPECTS FOR THE FUTURE OF NUCLEAR POWER

Political considerations rather than technological or economic factors are likely to determine the future of nuclear power, at least in the U.S. There are obvious

Table 9.3 Costs (cents per kWh) for generation of electricity with options potentially available by 2010: estimates from the U.S. Energy Information Administration (EIA) and the Near Term Deployment Group (NTDG). Costs for EIA data are quoted in 2001 dollars; costs for NTDG data are quoted in 2000 dollars (from Bodansky, 2004)

	Capital	Operation and maintenance	Fuel	Total
Advanced coal (EIA)	3.51	0.45	1.04	5.00
Advanced gas (EIA)	1.23	0.13	3.22	4.59
Reference nuclear (EIA)	5.00	0.74	0.46	6.20
Advanced nuclear (NTDG)	3.16	0.50	0.50	4.16
Wind (EIA)	4.06	0.82	0.00	4.88

advantages to nuclear power. In contrast to power generated using fossil fuels, nuclear power does not contribute to air pollution. There are no related emissions of greenhouse gases. Nuclear power is economically competitive with other sources of power (coal, oil, gas, wind) as indicated in Table 9.3.

The bulk of the costs involved in the generation of electricity using nuclear power relate to the capital expenditures required for the design, licensing, and construction of the plant: fuel costs are relatively minor. The entry in Table 9.3 for the Advanced Nuclear Option refers to an Advanced Light Water Reactor (ALWR), which, it is assumed, could be operational by 2010. The cost quoted reflects an estimate by the Near-Term Deployment Group (NTDG), a panel appointed by the U.S. Department of Energy to develop a road map for deployment of new nuclear power plants in the U.S. by 2010 (Near Term Deployment Group, 2001). The panel assumed a base construction cost of $1000 per kW with a discount rate of 12% (their estimate for the interest rate that might be required to finance the investment). The relatively high value assumed for the discount rate (U.S. 30-year Treasury Bills currently earn less than 5%) reflects the panel's view of the current high level of skepticism that pertains as to the future of the U.S. nuclear industry. This skepticism is due in no small measure to experience with the Shoreham nuclear plant constructed for the Long Island Lighting Company (LILCO) in the 1980s. The plant was completed in the mid-1980s and received a license in 1989 from the Nuclear Regulatory Commission to operate at full power. Local opposition, however, prevented the plant from going into operation and it was eventually dismantled at significant expense to LILCO, its customers, and New York State.

Costs for nuclear power as indicated in Table 9.3 are already competitive with costs for alternative sources of electrical energy. The costs indicated for coal and gas do not allow for possible surcharges in the future to discourage emissions of greenhouse gases. A tax of $50 on emissions of a metric ton of carbon, for example, would increase the price of coal-generated electricity by as much as 1.3 cents per kWh, 0.45 per kWh for electricity generated using natural gas. If carbon has to be captured and sequestered, it is estimated that associated costs could run as high as $100 per metric ton of carbon.

Bodansky (2004) quotes Larry Foulke, then Vice-President Elect, subsequently President, of the American Nuclear Society, advocating three measures required to ensure a revival of the American nuclear industry (Foulke, 2003): "help in meeting the high costs of the first plants, a guarantee of government purchase of some of the power output if private demand is less than expected, and recognition of nuclear energy's national benefits such as environmental quality, energy security, and the burn-up of weapons-grade fissile material."

Experience with nuclear power in Japan and elsewhere in Asia has been very different from that in the U.S. A number of Advanced Boiling Water Reactors (AWBRs) designed by General Electric in collaboration with Japanese partners have been installed in Japan and additional units are scheduled for Taiwan. The time from beginning of construction to completion and operation of these units has been as short as 4 years. General Electric projects construction costs for future units well below the costs for recently completed Light Water Reactors.

Worldwide demand for electricity is expected to increase by between a factor of 2 and 3 by 2050: to between 2600 and 4800 GWyear as compared to 1700 GWyear in 2001 (Nakicenovic et al. (1998)). Is it reasonable to expect that a significant fraction of this increased capacity could be provided by nuclear power? An output of 2600 GWyear from nuclear power would require an expansion of the world's current nuclear power generating capacity by about a factor of 7, to a level equal to about 40 times the current nuclear generating capacity of France. Addition of 2000 GWe to current capacity would require a capital investment of 2×10^{12}, assuming a cost per kW capacity of $1000 as considered by NTDG (Near Term Deployment Group, 2001). Given that this would amount to less than 3% of the value estimated for the world gross domestic product in 2020 (expressed in current dollars), an investment of this magnitude over a 40-year period would appear not to be unreasonable (Bodansky, 2004).

The population of the U.S. is projected to grow at an annual rate of about 0.8% over the next 20 years (DOE, 2003). Assuming that demand for electricity tracks the increase in population (a conservative assumption) and that the increase in population continues to mid-century, the U.S. will require an additional electricity generating capacity of about 200 GW over this period. If 50% of this additional capacity were to be met by nuclear power, this would require an approximate doubling of current nuclear capacity, addition of approximately 100 1 GW reactors at a capital cost of about $100 billion in current dollars. Since this investment would be spread over a number of years, the burden on the U.S. economy would clearly be acceptable (U.S. GDP in 2005 amounted to $12.5 trillion). A major commitment to nuclear power would increase, however, the urgency of finding a long-term solution to the nuclear waste disposal problem. As currently planned, Yucca Mountain will have a storage capacity of 70,000 MTHM. Waste generated by 200 GW of nuclear capacity would amount to about 6000 MTHM per year requiring the equivalent of approximately 11 additional Yucca Mountains. The global requirements for storage would be even more demanding (offset to some extent by the reduction in the mass of waste due to reprocessing of spent fuel).

9.6 CONCLUDING REMARKS

As noted above, the future of nuclear power will depend most likely, less on considerations of technical feasibility and economics, rather, more on how the public perceives the security of the nuclear option. There is an urgent need for public education and discourse to identify both the risks and benefits of nuclear power and for policies that can enhance the latter while reducing the potential of the former.

Risks of proliferation are real. They are most likely, as discussed, to involve rogue nations rather than non-state terrorist groups. To safeguard against these hazards, there is an urgent need for international oversight, for agreements involving all nuclear nations (including specifically the countries that do not subscribe to current international agreements notably India, Israel, and Pakistan) to limit the possibility of future nuclear proliferation.

The special status in the existing Treaty on the Non-Proliferation of Nuclear Weapons accorded the original members of the nuclear weapons club (coincidentally, as noted earlier, the permanent members of the UN Security Council) poses a problem. Why should these nations be permitted to retain nuclear weapons while others should be denied the right to do so? If the current members of the nuclear weapons club (old and new) were to commit to a phased elimination of their stock of weapons, they would be in a better position to exercise moral leadership to reduce the possibility that others should elect to proceed down a similar path.

Use of nuclear weapons to settle political, economic, or ideological disputes is unthinkable. It would involve what Max Born, the Nobel Prize winning physicist, referred to as the "unacceptability of a war degenerated to mass murder of the defenseless." Born concluded (Born, 1968) that "we must not tire of fighting the immorality and unreasonableness which today still governs the world." His advice was prescient in 1965 at the height of the Cold War. It is even more urgent that we heed his message today. Mutually Assured Destruction is not a recipe for a sustainable future. A strong international commitment, including a fully funded and empowered IAEA, can play an important role in minimizing future risks of proliferation, fostering at the same time the environment for safe, economically competitive, development of civilian nuclear power.

In addition to dealing with the risks of proliferation, it will be important to formulate also viable plans for both interim storage and longer-term disposal of nuclear waste. From a purely technical point of view both of these challenges appear manageable. Arguably, the latter is more difficult. To the extent that the policy process should require that a depository selected for long-term storage should be stable to a high degree of confidence for as long as a million years, the challenge is formidable. There is a growing recognition, however, that such a requirement may be unduly restrictive and unnecessary. Better, it would appear, to adopt a more flexible, more gradual, approach (Bodansky, 2004). An international committee convened under the auspices of the U.S. National Research Council (National Research Council, 2001) recommended as follows: "For both scientific and social reasons, national programs should proceed in a phased or stepwise manner, supported by dialogue and analysis . . . Decision makers, particularly those in national programs, should recognize the public's reluctance to accept irreversible actions and emphasize monitoring and retriev-ability. Demonstrated reversibility of actions in general, and retrievability of wastes in particular, are highly desirable because of public reluctance to accept irreversible actions." This would appear to provide useful guidance for devel-opment of a politically acceptable long-term strategy to deal with the waste problem. It is essential that such a strategy be in place if nuclear power is to live up to its potential.

Currently established supplies of uranium are sufficient (even allowing for significant growth in nuclear power as envisaged here) to accommodate antici-pated demand for more than 50 years. The price of uranium (traded as U_3O_8) was as low as $10.75 a pound in 2003. It has risen recently (in early 2007) to more than $100 a pound reflecting bullish sentiment in the financial commu-nity as to the potential future of nuclear industry (but has fallen since back to close to $30 a pound). As noted earlier, though, the price of fuel is not a signifi-cant factor in determining the price of electricity produced by nuclear power: capital costs are much more important. Supplies of uranium at reasonable cost are more than adequate to accommodate anticipated demand for the foresee-able future. A transition from a once through fuel cycle to breeder technology would only enhance this conviction although, by providing a potentially large source of fissile plutonium, it would increase the risk of proliferation.

Paradoxically, while substitution of nuclear for fossil fuel generated power could play an important role in reducing the risk of unacceptable future cli-mate change, the possibility of significant warming could complicate plans for expansion of the nuclear option. To operate safely, nuclear power plants require a reliable source of cold water to supply their cooling needs. Restrictions on the supply of cooling water during an episode of extreme heat in France during the summer of 2003 forced 17 nuclear power plants either to turn off or to operate at reduced capacity (Kanter, 2007). The price of electricity increased by as much as a factor of 10 and the French electricity company, Electricite de France, was obliged to purchase power from neighboring countries incurring an addi-tional expense of 300 million euro. The experience indicates that it may be important to consider climate change as a factor in selection of sites for future nuclear power plants.

Subject to the reservations expressed here, we conclude that nuclear power can make an important contribution to future national and international demands for energy, while at the same time reducing dependence on unstable supplies of fossil alternatives, enhancing national security, and reducing the risk of unacceptable future climate change, with additional benefits to be reaped in terms of improvements in local and regional air quality.

NOTES

1 The symbol e included in capacity values quoted here is intended to emphasize that the number refers to capacity for generation of electricity. The efficiency for conversion to electricity of the thermal energy produced by fission in a nuclear reactor is about 32%.

2 (http://news.bbc.co.uk/onthisday/hi/dates/stories/december/3/newsid_2698000/2698709.stm: read April 5, 2007).

3 The curie, named in honor of Marie and Jean Curie who shared the Nobel Prize in physics with Henri Becquerel in 1903, is a measure of the radioactivity of a substance: 1 curie corresponds to 3.7×10^{10} disintegrations per second, the approximate level of radiation emitted by 1 g of radium-226 (^{226}Ra). Marie Curie was the first woman to receive a Nobel Prize. She was awarded a second Nobel Prize in 1911, this time in Chemistry.

4 The unit MTHM measures the mass of uranium employed as input to the nuclear fuel cycle. Allowing for packaging, the mass of waste produced is slightly larger than this as discussed above.

REFERENCES

Born, M.(1968). *My Life and My Views*. New York: Charles Scribner's Sons.

Bunn, M., Holdren, J.P., Macfarlane, A., Pickett, S.E., Suzuki, A., Suzuki, T., and Weeks, J. (2001). Interim storage of spent nuclear fuel: a safe flexible, and cost-effective near-term approach to spent fuel management. A Joint Report from the Harvard University Project on Managing the Atom and the University of Tokyo Project on Sociotechnics of Nuclear Energy, June 2001.

Bodansky, D.(2004). *Nuclear Energy: Principles, Practices, and Prospects*, Second edition. New York: Springer.

Chadwick, J. (1932). The existence of the neutron. *Proceedings of the Royal Society A*, **136**, 692–708.

Foulke, l.R. (2003). The status and future of nuclear power in the United States. *Nuclear News*. 46, no.2, February, 2003, pp 34–38.

Kanter, J. (2007). Climate change puts nuclear energy into hot water. *International Herald Tribune*, May 20, 2007.

Nakicenovic, N., Grubler, A., and McDonald, A. (eds.) (1998). *Global Energy Perspectives*. Cambridge: Cambridge University Press.

National Research Council (2001). *Disposition of High-level Waste and Spent Nuclear Fuel: The Continuing Societal and Technical Challenges*. Washington, DC: National Academy Press.

Near Term Deployment Group (2001). A roadmap to deploy new nuclear power plants in the United States by 2010. Volume II, Main Report, U.S. Department of Energy.

World Nuclear Association, Nuclear Power in the World Today (2007). http://www.world-nuclear.org/info/inf01.html. Read April 2, 2007.

10

Steam Power

10.1 INTRODUCTION

For most of human history humans relied on personal muscle power to do work. As indicated earlier, a healthy human is capable of doing work, of expending energy, at a rate roughly equivalent to the energy consumed by a typical incandescent light bulb, about 100 W. When our ancestors learned to domesticate animals—about 12,000 years ago—they increased their capacity to perform work by a factor of at least two or three (Chapter 2). Cattle employed as draft animals were capable of working at rates equivalent to several hundred watts. Large horses could put out work at rates as great as 800 W. The unit of horsepower was introduced by James Watt—more on Watt later—to provide a measure of the rate at which Watt's steam engines could perform work as compared to the rate at which work could be extracted from strong horses: 1 horsepower (hp) is equal to 746 W, the output of energy required to lift 33,000 lb through 1 ft in a minute. As discussed in Chapter 8, the medieval over-shoot water wheel at Venafro in Italy was capable of harnessing a little more than 2.2 kW (3 hp) of waterpower and to use this power to grind 400 lb of grain per hour. Reflecting continuing progress, the Boston Associates (Chapter 8) had the capacity, by the middle of the nineteenth century, to garner close to 16,000 kW of power from the falls on the Merrimack River near Lowell (more than 21,000 hp). They used this power eventually with close to 80% efficiency to mechanize all elements of cotton cloth production in more than 20 mills at Lowell. Power was supplied to their mills through an elaborate canal-based distribution system, delivered, and charged for in quantities known as mill power units, each equivalent to 63 kW or 84 hp. Soon, though, as we learned, the capacity of the Merrimack River to supply the needs of the

New England textile industry was exhausted. Steam replaced running (and fall-ing) water as the agent of preference to run the machines of the American tex-tile industry, building on the steam revolution that had taken hold earlier in England. The ability to harness the chemical energy of coal, to convert it to heat, and to use this heat to produce steam that could perform mechanical work truly revolutionized the world. It continues to play a critical role in our electrically powered economy today.

We begin in Section 10.2 with an account of the physical properties of water. Heat water and it can be transformed to vapor (steam). Raise the temperature of water above the boiling point (100°C at atmospheric pressure) and the resulting steam can be used to drive a piston against the opposing pressure of the atmosphere. Cool steam, by bringing it in contact with cold liquid water for example, and one can create a near vacuum. The pressure of the atmosphere can be employed then to do work, to drive a piston into the region of lower pressure.

The early history of the steam age is discussed in Section 10.3, with a specific account of the contributions of Thomas Savery, Thomas Newcomen, and James Watt. The initial motivation for all of these pioneers was to develop an efficient means to pump water from underground mines. In this, Watt's pri-mary achievement was to improve on the earlier innovation by Thomas Newcomen. But where Newcomen failed to profit much from his invention, Watt was remarkably successful. He was recognized early in life as an accom-plished inventor and entrepreneur. He received more lasting fame in later life when he was elected, at age 47 (in 1785), to the Royal Society of London. Fifteen years after his death in 1819, a statue honoring him was installed in Westminster Abbey.

The inscription on Watt's statue reads as follows: "The King, his ministers, and many of the nobles and commoners of the realm raised this monument to James Watt who directing the force of an original genius early exercised in philosophic research to the improvement of the steam-engine enlarged the resources of his country increased the power of man and rose to an eminent place among the most illustrious followers of science and the real benefactors of the world." The monument was relocated to the crypt of Saint Paul's in 1966, and was moved subsequently to Scotland. Marsden (2002) speculates as to the significance of the decision to remove Watt from the Pantheon of British heroes at Westminster Abbey. Was it a sop to the Scots to return their distin-guished countryman to the land of his birth or did it in fact reflect "a manifes-tation of British ambivalence about the practical arts in which they consider themselves to have excelled?" In any event, Watt's place in history is assured. His name will be forever associated with the basic unit of power in the interna-tional system of scientific units, a decision taken by representatives of the inter-national scientific community meeting in Paris in 1889.

Later developments of steam power are discussed in Sections 10.4 and 10.5. Steam power made possible the railroad (Section 10.4) and steam powered ships (Section 10.5). It continues to play an important role in modern society as the primary agent for generation of electricity. Summary remarks are presented in Section 10.6.

10.2 PROPERTIES OF WATER

Water is commonly present on the Earth in all three phases: ice, liquid, and gas. The three phases coexist in equilibrium, however, only under special conditions, for a unique combination of pressure and temperature defining what is known as the triple point. The triple point of water at atmospheric pressure occurs at a temperature close to 0°C or 273 K. With an increase in pressure, the triple point shifts to slightly lower temperatures. At temperatures below the triple point, water, in equilibrium, is present in only two phases, vapor and ice. At temperatures above the triple point, it exists as either liquid or vapor.

That only two phases can coexist over most of the temperature range may come as somewhat of a surprise. After all, if you fill an ice cube tray with liquid water and place it in the freezer, it doesn't immediately turn to ice. Similarly, if you take ice out of the freezer, it will take some time for it to melt and for at least part of the time you clearly have a combination of all three phases. These situations, however, are unstable. The concept of equilibrium assumes that sufficient time has elapsed for the system to settle down to a steady state. Below 0°C, in equilibrium, water molecules leave the ice and return at the same rate. Any liquid that might have been present initially will inevitably, given time, turn to ice. The transition from ice to vapor is referred to as sublimation. Similarly, at temperatures above 0°C, ice will melt and when equilibrium is attained, water molecules will escape from and return to the liquid at precisely the same rate. Transfer of molecules from the liquid to the gas phase is referred to as evaporation. The reverse process, return of molecules from the vapor to either the liquid or ice phase, is known as condensation.

The higher the temperature, the greater the kinetic energy of the molecules in the liquid and thus the easier it is for them to escape the forces with which they are bound to neighboring molecules. It follows that a higher concentration of molecules will be required in the vapor phase at higher temperature to ensure that the rate at which gas phase molecules return to the liquid should be large enough to balance the rate at which they are released. Similar considerations apply to the equilibrium between ice and vapor. The abundance of molecules in the vapor phase required to balance evaporation from liquid or sublimation from ice at a particular temperature is expressed in terms of a quantity known as the equilibrium vapor pressure. Equilibrium vapor pressures over ice increase with temperature more rapidly than over liquid reflecting the fact that molecules are more tightly bound to neighbors in the ice phase as compared to the liquid. The variation of equilibrium vapor pressure with temperature is summarized for a range of temperatures in Fig. 10.1. Variations of the density of the liquid and vapor phases as functions of temperature are illustrated in Fig. 10.2.

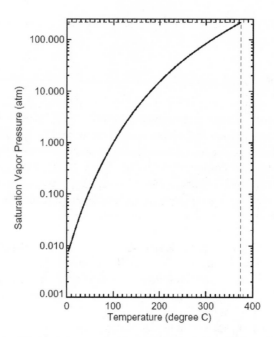

Figure 10.1 Equilibrium vapor pressure of water as a function of temperature.

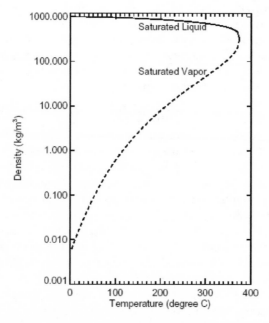

Figure 10.2 The density of the liquid and vapor phases as functions of temperature. The solid line refers to the liquid phase; the dashed line to the vapor phase.

As indicated in Fig. 10.2, the density of water in the liquid phase decreases as a function of temperature. The decrease in the density of ocean water over the past century arising as a result of the observed increase in ocean temperature (an increase detected to depths as great as 3 km) has been responsible for a rise in global sea level of about 17 cm (6.7 in.). The rate of rise has accelerated over the past decade—an increase of 3.1 cm between 1992 and 2003—and the Intergovernmental Panel on Climate Change (IPCC, 2007) projects that, absent aggressive efforts to arrest the rise in the concentration of atmospheric greenhouse gases, sea level is likely to increase by at least an additional 18 cm (7.1 in.), perhaps by as much as 59 cm (almost 2 ft), over the next century. The increase could be much greater—a meter or more—if the warming projected should result in destabilization of the ice sheets currently present on Greenland and Antarctica (IPCC, 2007).

The temperature at which the equilibrium vapor pressure of a liquid is equal to the pressure of the atmosphere defines what is known as the boiling point. For water, this occurs at a temperature of 100°C (373 K). Bubbles of gas form in the liquid at the boiling point. Lighter than the liquid they displace, the bubbles rise to the surface. If the surface is open to the atmosphere, the bubbles will break releasing molecules of water vapor to the overlying atmosphere. Since the pressure of the bubbles is comparable to the pressure of the atmosphere, the gas contained in the bubbles can displace the molecules of the air into which they are released. If the temperature of the liquid is maintained at the boiling point, transfer of liquid to gas will continue until the liquid is exhausted.

Suppose the water is contained in a closed container. In this case the pressure of the vapor in the headspace, and consequently the density of water molecules, will increase as the temperature of the liquid is raised above the boiling point (more water molecules in the same volume). At the same time, as indicated in Fig. 10.2, the density of the liquid will decrease (the liquid expands with increasing temperature). The density of liquid and gas are equal at a temperature of 374°C (647 K). This defines what is known as the critical temperature. The distinction between liquid and gas is moot for temperatures higher than the critical value. The pressure of water vapor at the critical point is equal to 218 times the pressure of the atmosphere. At temperatures above the critical temperature water exists only in the gas phase.

The energy or heat required to raise the temperature of a substance by 1 K is known as the heat capacity. Heat energies were expressed originally in units of calories. The heat required to raise the temperature of 1 g of liquid water by 1 K is equal to 1 calorie (cal) or 4.184 J. As we discussed earlier, the energy content of food is quoted customarily in units of calories. This is actually a misnomer: the calorie unit quoted by nutritionists (usually written as Calories, abbreviated to C) is actually equal to 1000 cal or 1 kcal. To raise the temperature of 1 kg of liquid water by 1 K requires an input of heat equal to 1 kilocalorie (kcal) or 4.184 kJ. The heat capacity of liquid water is equal therefore to 1 kcal kg^{-1} K^{-1} or 4.184 kJ kg^{-1} K^{-1}. A pound of butter contains fat with an energy content of 100 kcal (indicated as 100 calories on the package) and has punch sufficient to increase the temperature of 1 kg of liquid water from 0°C to 100°C!

Altering the phase of a substance requires an input of heat significantly greater than that required to raise its temperature. Heat equivalent to 333.5 kJ is required to melt 1 kg of water ice. A much larger heat input, 2257 kJ is needed to turn 1 kg of liquid into vapor. The heat required to transform a substance from the solid to the liquid phase is known as the latent heat of fusion, written as L_f. The heat consumed in converting a given mass of liquid to vapor is referred to as the latent heat of vaporization, written as L_v. Values of L_f and L_v for water at 1 atmosphere pressure are equal to 333.5 kJ kg^{-1} and 2257 kJ kg^{-1} respectively. The energies involved in melting ice and vaporizing liquid water are unusually large. For example, the latent heat required to melt 1 kg of lead (24.7 kJ) is less than 10% of that required to melt a similar mass of ice (of course, in order to reach the melting point of lead it would be necessary to raise the temperature of the solid phase to a much higher temperature than that of ice, 600 K as compared to 273 K). Box 10.1 presents a calculation of the energy required to convert 1 kg of ice at a temperature of −2°C (271 K) to steam at a temperature of 100°C (283 K). The effect of adding 20 kg of liquid water at a

---

Box 10.1

A household freezer is maintained typically at a temperature of about −2°C. Estimate the heat required to convert 1 kg of ice at −2°C to steam.

Solution:

We will do this in four steps. We first estimate the heat required to raise the temperature of the ice to the melting point. We add to this the heat required to convert the ice to liquid, the heat required to raise the temperature of the liquid to the boiling point, 100°C, and then finally the heat required to transform the hot liquid to vapor (steam).

The heat capacity of ice is 2.05 kJ kg^{-1} K^{-1}.

Raising the temperature of 1 kg of ice from −2°C to 0°C requires $(2) \times (2.05)$ kJ = 4.1 kJ.

Melting 1 kg of ice requires 333.5 kJ.

Raising the temperature of 1 kg of water by 100°C requires $(100) \times (4.18)$ kJ = 418 kJ.

Vaporizing 1 kg of water requires 2257 kJ.

Total heat = 4.1 + 333.5 + 418 + 2257 kJ = 3012.6 kJ.

Note that the bulk of the total heat supplied (75%) is used to convert liquid to steam. Only 14% of the heat is used to raise the temperature of the ice from −2°C to 0°C and the temperature of the liquid from 0°C to 100°C.

Box 10.2

A closed vessel contains 1 kg of 100°C steam produced by boiling water at atmospheric pressure. The pressure of the vapor is equal to the pressure of the atmosphere, 1 atm. Calculate the effect on the pressure of gas in the vessel introduced by adding 20 kg of liquid water at a temperature of 10°C.

Solution:

Direct equilibration of 1 kg of a substance at 100°C with 20 kg of another substance at 10°C would cause the temperature of the mixture to adjust to a common temperature of 14.3°C: $(1 \times 100 + 20 \times 10)/21$.

The mixture, however, would then be supersaturated with respect to the vapor. Vapor would condense with associated release of latent heat resulting in a further rise in the temperature of the liquid. The energy released by condensation of 1 kg of vapor is equal to 2257 kJ. Given the heat capacity of water, $4.184 \text{ kJ kg}^{-1}°C^{-1}$, this additional energy would cause the temperature of the 21 kg mixture to increase by an additional 25.7°C, to 40°C. The equilibrium vapor pressure of water at a temperature of 40°C is equal to 73.7 millibar (mb) or 0.074 atm. The pressure of the vessel would be lowered accordingly by a factor of close to 14.

The latent heat content of 1 kg of steam is sufficient to raise the temperature of 5.5 kg of liquid water from a temperature of 0°C to the boiling point. The trick in the Newcomen fire engine was to inject a sufficient quantity of cold water into the hot steam so that the increase in the temperature of the water resulting from release of the latent heat of the steam was insufficient to raise the temperature of the water to the point where the resulting equilibrium vapor pressure would significantly limit the ability of the atmosphere to depress the piston in the cylinder. As we shall see, the need to cool the walls of the cylinder markedly increased the quantity of cold water required to accomplish this task. Conversely, in Newcomen's case, the need to heat up the walls of the container when steam was injected required a significant increase in the necessary supply of steam. This, as we shall see, was the inefficiency Watt set out to eliminate.

temperature of 10°C to 1 kg of 100°C steam at a pressure of 1 atmosphere is computed in Box 10.2.

10.3 FROM SAVERY TO WATT

The story is told how James Watt as a boy would spend hours taking the lid off a kettle of boiling water, observing how the steam could be used to lift a spoon or a cup while pondering the mysterious influence responsible for the apparent force supplied by the steam. Whether the story is true or not, or whether the experience had any influence on his subsequent career, it is reasonable to conclude that Watt was not the first to observe this or similar phenomena. Hero of Alexandria is reported to have built a primitive steam engine almost 1000 years ago demonstrating how steam could be used to pump water from one container to another at higher elevation (Wikipedia, 2007). Others through the ages undoubtedly played around with a variety of the properties of steam. Thomas Savery, however, was arguably the first to envisage a serious commercial application and to take steps to ensure his prerogative.

Savery (1650–1715) received a patent in 1698 for "the sole exercise of a new invention by him invented, for raising water, and occasioning motion to all sorts of mill works, by the important force of fire, which will be of great use for draining mines, serving towns with water, and for the working of all sorts of mills, when they have not the benefit of water nor constant winds; to hold for 14 years; with usual clauses." The life of the patent was extended in 1699 to 1733. That pretty much took care of any potential competition for 35 years. When Thomas Newcomen (1663–1729), a Dartmouth ironmonger, began to experiment with what became known as a fire engine, he was obliged to enter into partnership with Savery. Newcomen's first full scale model was delivered to the colliery of the Earl of Dudley in 1712. It was a massive structure standing 50 ft high as illustrated in Fig. 10.3.

The primary purpose of Newcomen's machines was to pump water from mines, mainly from coal mines in the Midlands but also from tin mines in the south west of England, in Cornwall. The key element of Newcomen's invention involved a vertical cast iron cylinder open at the top accommodating a piston connected to a rod attached in turn to a chain strung from one end of a heavy wooden beam pivoted at its center on the top of a 50 ft high brick wall. The piston was free to move up and down in the cylinder. A second chain on the

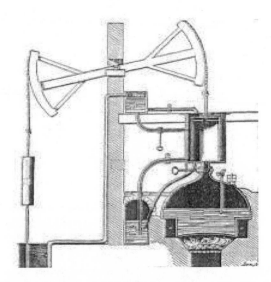

Figure 10.3 Thomas Newcomen's 'ENGINE for raising water (with a power made) by FIRE' (of 1717). http://www.newcomen.com/thomas.htm.

other end of the beam was connected to the apparatus used to pump water from the mine. The beam could swing up and down like a seesaw. In its equilibrium position, the weight of the pump caused the piston to rise towards the top of the cylinder (the pump apparatus sinking on the other side). To start the engine, steam, supplied by a boiler at the bottom of the apparatus, was introduced into the cylinder below the piston pushing out air that would otherwise have occupied this space. Injection of cold water into the steam subsequently served to create a partial vacuum below the piston. The pressure of the atmosphere above then caused the piston to move down initiating the first element of the engine stroke, lifting water from the mine on the other end through the coupled action of the interconnected chain–beam–chain combination. Reinjection of steam below the piston offset the net downward pressure of the atmosphere initiating the second (upward) component of the engine stroke (second). As before, the piston was allowed to rise responding to the combined weight of the pump and now the water on the other end, allowing the pump to sink once again into the mine water ready to lift an additional burden on the next stroke of the engine. An instructive animated illustration of the operation is presented in[1].

The work carried out by Newcomen's machine was defined by the product of the mass of water lifted in a particular stroke, the distance through which it was lifted (the stroke length), and the acceleration of gravity. All of this work was performed during the phase of the stroke in which the piston was driven downward in response to the pressure of the atmosphere acting on the near vacuum of the space below the piston. In this sense, Newcomen's machine was not strictly a steam engine but rather an atmospheric engine. Atmospheric pressure did the work. The primary purpose of the steam was to form the low-pressure regime on the underside of the piston (induced by injection of the cold water) without which the machine would have been incapable of accomplishing any work.

The operational mechanism of the pump is illustrated in Fig. 10.4. As the piston moves up, the pump sinks under its own weight into the mine water. Water is squeezed round the edges of the pump and is confined subsequently

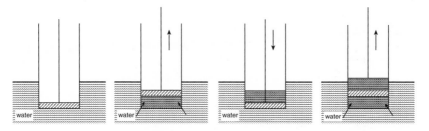

Figure 10.4 The operational mechanism of the pump. Water raised previously by the pump or to be raised in the next pump stroke is indicated by the darker shaded areas.

by the jacket of the pump as the pump element rises in response to the downward stroke of the piston. Additional mine water is raised on the next stroke and so on until the water that has been pumped reaches the top of the shaft. There it can be disposed of, or employed productively for some other purpose—as a source of potential energy for example for a water wheel capable of carrying out additional work (assuming of course that the mine water is lifted to some distance above ground level).

Newcomen's machine was simple in principle. To make it work in practice, however, required a great deal of additional innovation on Newcomen's part. Marsden (2002) summarizes the situation as follows: "There had to be a water supply, a cistern (or tank) for the injection water, a small pump to get the water up into the cistern, and pipes to spray it into the cylinder. There was a valve to regulate this spray of water, and another, called the 'snifting valve,' to get rid of the surplus air that clogged up the cylinder. . . . To make the cylinder air-tight there was leather packing inside and a pool of water over the top of the piston ('water packing'). There was a boiler, looking like a brewer's copper, and that had to have a safety valve, even though steam was rarely ever used at pressures more than a few pounds per square inch above that of the atmosphere (approximately fourteen pounds per square inch). The beam had its own (hopefully) robust gearing. . . . Newcomen had done his best to automate the action of the engine but there were people, too, in attendance. Engineers and stokers were forever on call to feed these temperamental mechanical beasts or to coax them into life."

The primary problem with Newcomen's machine was that it was energy inefficient. The temperature of the walls of the cylinder had to be raised before the cylinder could retain the steam admitted in advance of the expansion phase. This required an additional source of steam with a consequent unproductive expenditure of energy. Conversely, the walls had to be cooled before the steam could be eliminated to facilitate the downward stroke of the engine. Watt's primary contribution was to eliminate the need to cool the walls of the cylinder by introducing a small separate cylinder that would serve to extract the steam from the primary cylinder. Condensation would take place in this separate cylinder. This allowed the walls of the primary cylinder to remain at their initial high temperature eliminating the need for additional expenditure of energy to reheat these walls.

An estimate of the maximum efficiency of Newcomen's engine as modified by Watt is presented in Box 10.3. The computation indicates that a maximum of 6.5% of the energy expended could be converted to useful work, employed to raise water from the mine. The actual efficiency would have been less than this, by at least a factor of 2. The computation ignores for example the unavoidable loss of heat due to conduction through the walls of the cylinder, in addition to the loss of steam and heat through the top of the cylinder—through and around the edges of the piston—and energy consumed as result of the friction of the various moving parts. If we assume an efficiency of 3% for the ideal, Watt-modified, Newcomen engine, the efficiency of the original Newcomen

---

Box 10.3

Estimate the maximum possible efficiency of the Newcomen engine as modified using the separate condenser by Watt.

Solution:

Consider a cylinder of length L, cross-section area S. Suppose that the cylinder must be filled with steam to a pressure of 1 atmosphere (atm). First calculate the mass of H_2O corresponding to this steam.

Using the Perfect Gas Law: $P = NRT$

$$\rightarrow N = P/RT = 1 \text{ atm}/(8.31 \text{ J mol K}^{-1} \times 373\text{K})$$
$$= 1.013 \times 105 \text{ Pa}/(8.31 \text{ J mol K}^{-1} \times 373\text{K})$$
$$= 32.68 \text{ mol m}^{-3}$$

The mass of $H_2O = N \times L \times S \times 18 \text{ g mol}^{-1} = 588.23 \ LS \text{ (g)}$

Calculate then the heat required to produce this steam:

Heat (Q) required = heat to raise water from 293 K to 373 K + latent heat of evaporation.
$$Q = (558.23 \ LS)\text{g} \times 4.184 \text{ J K}^{-1}\text{g}^{-1} \times (373–293)\text{K} + (588.23 \ LS)\text{g} \times 2257 \text{ J g}^{-1}$$
$$= 1.51 \times 10^6 \ LS \text{ (J)}$$

The work (W) done in one stroke of the engine is given by:

$$W = 1 \text{ atm} \times S \times L = 1.013 \times 10^5 \ LS \text{ (J)}$$

It follows that the efficiency (E) is given by:

$$E = 1.013 \times 10^5 \ LS/(1.51 \times 10^6 \ LS) = 6.7\%.$$

---

engine could not have exceeded 0.5%. Energy equivalent to 200 units of coal would have been required to supply 1 unit of work. Where coal was readily available and cheap—in using the machine to pump water from a coal mine for example—this might not have posed a problem. But when the machine was employed to pump water from tin mines in Cornwall—where coal was not locally available and had to be shipped from elsewhere—it was clear that Watt's modification would enjoy an important competitive advantage. An estimate of the power required to raise 100 gallons of water per minute through 100 ft is given in Box 10.4.

Andrew Carnegie began his biography of Watt (Carnegie, 1905) as follows: "James Watt, born in Greenock, January 19, 1736, had the advantage, so highly prized in Scotland, of being of good kith and kin." Watt's father, James, was a merchant who supplied nautical goods and mathematical instruments. He was known for his skill in fabricating delicate instruments, a skill inherited by his son. His grandfather, Thomas, identified as a "Professor of Mathematicks" (Marsden, 2002), taught navigation and surveying. Watt's father fell on hard times in 1753—a ship in which he had invested was lost at sea—and the young

---

Box 10.4

Estimate the power required to raise 100 gallons of water per minute through 100 ft.

Solution:

First calculate the energy required to raise 100 gallons of water by 100 ft.

$$E = mgh$$
$$m = \text{density} \times \text{volume}$$
$$= 1 \text{ g/cm}^3 \times 100 \text{ gallons} \times 3785 \text{ cm}^3\text{/gallon}$$
$$= 3.785 \times 10^5 \text{ g} = 3.785 \times 10^2 \text{ kg}$$
$$h = 100 \text{ ft} = 100 \text{ ft} \times 0.305 \text{ m/ft} = 30.5 \text{ m}$$
$$E = 3.785 \times 10^2 \text{ kg} \times 9.8 \text{ m/s}^2 \times 30.5 \text{ m}$$
$$= 1.13 \times 10^5 \text{ J}$$

Now calculate the power required to raise 100 gallons of water by 100 ft per minute.

$$E = mgh = 1.13 \times 10^5 \text{ J}$$
$$\text{Power} = 1.13 \times 10^5 \text{ J / min}$$
$$= 1.13 \times 10^5 \text{ J / 60 s}$$
$$= 1.89 \times 10^3 \text{ W}$$
$$= 2.53 \text{ hp}$$

(1 hp = 745 W)

Watt was obliged to strike out on his own. He moved to Glasgow to take up apprenticeship as an instrument maker. There, family connections (his cousin, George Muirhead, was Professor of Latin Language and Literature) provided him with an introduction to Robert Dick, Professor of Natural Philosophy at Glasgow College. Dick would play an important role in the young Watt's subsequent career.

Dick arranged for Watt to move to London where he continued his education as a skilled instrument maker, working with John Morgan, "one of only five or six makers with the mix of knowledge that Watt demanded" according to Marsden (2002). When Watt returned to Scotland in 1756, Dick was again influential, arranging a room in the college where Watt could work and where he would be established eventually (Marsden, 2002) as "Mathematical Instrument Maker to the University," a position which, while enjoying no specific status and guaranteeing no specific income, provided Watt with an even more valuable resource—access to the considerable intellectual resources of the college.

Watt's interest in steam was sparked initially, reputably, by a conversation he had with John Robison, then an undergraduate at Glasgow College, later Professor of Natural Philosophy at the University of Edinburgh, subsequently a lifelong friend. Robison raised the possibility (in 1759) of harnessing steam to drive a wheeled carriage. Watt played around with the idea for some time, without success. But, as Carnegie (1905) put it, "the demon Steam continued to haunt him." As it happened, the college had in its possession a model of a Newcomen fire engine. Watt was engaged to try to make it work. This would turn out to be essentially impossible. The steam content of the cylinder of a Newcomen engine may be expected to vary in proportion to its volume. Heat loss from the cylinder is determined on the other hand by the area of its surface. The smaller the cylinder, the greater the area of its surface relative to its volume: consequently the more serious the loss of heat relative to requirements for its supply. The cylinder of the scale model engine available to Watt measured 6 in. high with a diameter of 2 in. In contrast the full-scale version of the Newcomen cylinder was almost 40 times larger. It had a diameter of several feet and stood close to 20 ft high (see the depiction in Fig. 10.3). Little wonder then that Watt was unable to make the miniature engine function like its larger prototype.

Struggling to make the miniature engine work, Watt embarked on a series of experiments to help him better understand the properties of steam. His inspiration and role model in this endeavor was Joseph Black who had joined the faculty of Glasgow College in 1756 as Professor of Anatomy and Botany and was appointed a year later to the Chair of Medicine. Black provided Watt with access to the library of the college and Watt set out to learn everything that was known (or at least written) about the properties of steam. From Black he learned about heat capacity and latent heat and proceeded to interpret his experiments using these concepts. Watt's ambition was to design what he described as "a perfect engine," one that could function with a single cylinder of steam, capable on the opposite stroke of achieving an essentially perfect vacuum. He focused soon on the critical obstacle to realizing this ambition, describing it as follows (Marsden, 2002): "To make a perfect steam-engine, it was necessary that the cylinder should always be as hot as the steam which entered it, and the steam should be cooled down below 100.00°C in order to exert its full powers. The gain by such construction should be double: first, no steam would be condensed on entering the cylinder; and secondly, the power exerted would be greater as the steam was more cooled." Watt's eureka moment came in 1765, the inspiration to install a separate condensing cylinder in the Newcomen engine eliminating thus the need for the repetitive heating and cooling of the main cylinder as required previously.

Having the idea was one thing: putting it into practice posed an entirely different, even more demanding, challenge. To demonstrate the advantages of his brainchild, Watt had to build a working model—not a miniature, for that could not work for the reasons discussed earlier. What was needed was a version somewhat akin to a full scale working Newcomen machine. That would require access to a major engineering works and capital—a lot of the latter. Black arranged for Watt to meet with one of Black's former students, John Roebuck, who not only had money but also owned Scotland's first heavy engineering works, the foundry at Carron. Watt set to work to construct his engine.

There were many pitfalls along the way but by mid-1768 sufficient progress had been demonstrated to persuade Roebuck to invest serious capital. The first step would be to obtain a patent. Roebuck agreed to pay for the patent in return for "two-thirds of the property of the investments." The patent was finally granted on January 5, 1769. A few weeks later, Watt would celebrate his 33rd birthday.

Birmingham by this time had emerged as a major center of engineering in England. Watt had visited Birmingham on a number of occasions prior to his application for the patent. He had made the acquaintance of Dr. William Small who shared Watt's fascination with steam engines. Small had been introduced to Matthew Boulton, a key figure in the Birmingham industrial establishment, by none other than Benjamin Franklin, the American scientist/statesman, and had taken a position as Boulton's personal physician. Small introduced Watt to Boulton in 1767. Boulton gave Watt a personal tour of his huge Soho Works, a facility unique for its time, a magnet drawing what Marsden (2002) describes as "technological tourists from all over Europe" who "marveled at its fine commodities, from trinkets to clocks, produced with state-of-the-art equipment in an ostentatiously progressive environment, and on the latest principles of factory organization." A few years later, in 1773, Boulton would acquire Roebuck's two-thirds ownership of the Watt patent, settlement of a debt as Roebuck fell on hard financial times. The following year Watt moved south taking up residence in Birmingham beginning what would develop as a lifelong association with Boulton.

Boulton had already had an important influence on Watt. As Watt prepared to apply for his patent, Boulton and Small encouraged him to be as expansive as possible. When granted, the patent basically locked up all possible applications of steam using a separate condenser. Notably, it envisaged also conversion of the up–down piston motion of his engine to drive rotary motion. As Marsden (2002) points out, "wonderfully simple in design, [this] proved impossible to build." But it gave Watt a foot in the door, one he would exploit with great skill later in his partnership with Boulton, formalized in 1775.

An early problem was to find a means to extend the patent, already 6 years old in 1775 with only 8 years to go and still no satisfactory working prototype. Boulton had the skills, and contacts, to solve the problem. Parliament was petitioned in February 1775 to consider an extension. In May 1775, the patent was extended to 1800 under an Act sanctioned by Royal Assent. The application for an extension of the patent described a scheme to use steam to both push and pull on the piston—a double acting engine—an innovation patented later in 1782, capable of doubling the power of the original, single-acting engine. A further innovation, patented also in 1782, allowed expansion of steam to carry out additional work without the need for an additional supply (achieved by allowing the steam to cool as it expanded).

With the patent secured for a further 25 years, the challenge for Boulton and Watt was to develop a market for their intellectual property. First they needed to produce a working model and to demonstrate its superiority over the competition—primarily the standard Newcomen design. A critical challenge was to produce a cylinder that would eliminate leakage of steam. To do so, the cylinder would have to be fabricated to an exacting standard. To meet this objective, the partners turned to John Wilkinson, a friend of Boulton, who had built (in 1776) an innovative boring mill designed to manufacture cannon for the Royal Navy at his plant in North Wales. Wilkinson succeeded in producing a 72-in. cylinder, the dimensions of which deviated over its length by less than one part per thousand, allowing the piston to slide back and forth with minimal resistance and with minimal loss of steam. For his contribution, Wilkinson would receive one of the first working Boulton–Watt engines, capable of driving a 700-lb hammer lifted 300 times a minute to a height of 2 ft. For close to 20 years Wilkinson would enjoy a virtual monopoly on production of cylinders for Boulton and Watt. The key market for the new engines, as it had been almost three quarters of a century earlier, however, would be to improve the efficiency of the machines used to pump water from the tin and copper mines in Cornwall.

Boulton and Watt took orders for two engines for Cornwall in 1776. They struck an arrangement by which the mine operators would pay for these engines based on a royalty tied to the savings in coal that would be realized using the new engines as compared to quantities of coal consumed by the existing Newcomen engines. One-third of this savings would accrue to Boulton and Watt. By 1800, 55 Watt engines had been installed in Cornwall and the Newcomen engine had been relegated to the status of a footnote in history.

In the early days of their partnership, Boulton and Watt lacked the facilities to assemble engines from scratch. They constructed their engines for the most part using components purchased from designated suppliers. Only the critical precision valves came from the Boulton–Watt factory. It was not until 1796 (with only 4 years to go on their patent) that they developed the ability to manufacture complete engines in house, in a dedicated foundry constructed at Soho. In the interim, most of their income was derived from royalties earned on the basis of the patents they had acquired rather than from construction and sale of complete engines.

While the early market could be satisfied by engines capable of driving reciprocal (up/down) motion (all that was required to pump water from mines), it was clear from the outset that a much greater payoff could be realized if the reciprocating Watt engine could be modified to accommodate rotational motion. In principle, this was not expected to pose too serious a problem. The obvious, well-established, approach was to couple the reciprocal motion of the piston to a crank that could convert the up/down motion of the piston to rotational motion of a crankshaft (for an animated depiction of the operation of a piston/crank combination see http://en.wikipedia.org/wiki/Crankshaft, read June 20,2007). Watt built an experimental model to do just that in 1779 but before he could patent his system he was beaten to the punch by James Pickard. Pickard's patent application envisaged the coupling of a crank to a Newcomen engine (the partners had declined to grant him a license to work with the Watt design). It was nonetheless a serious set back for the Watt/Boulton team. They could elect to pay a royalty to Pickard or take steps to develop an alternative approach to generate rotational motion—one that did not rely on a crank. Watt opted for the second option. In quick order he developed and patented (in 1781) five possible means to accomplish the objective, one of which, the so-called sun-and-planet system (Marsden, 2002), would eventually become the centerpiece of the Boulton/Watt rotative machines.

The sun–planet system consisted of a pair of toothed cogwheels, one of which, the sun, was fixed at the end of an axle that would constitute the driveshaft for the rotative engine. The second, smaller, cogwheel, the planet, was constrained to stay fixed to the first but was free to move in response to the up/down motion of the piston to which it was attached. As the piston moved up and down, the planetary cogwheel forced the larger (solar) cogwheel to turn and with it the driveshaft to which it was attached. Watt and Boulton were free now to develop their own rotative machines without need to compensate competitors.

The first Boulton/Watt rotative machine was completed in 1783. A working model was delivered to a Nottinghamshire cotton mill in 1785. The partners had hoped that publicity generated by this new facility would prompt additional orders but they were frustrated when the Nottinghamshire mill owners refused to allow outsiders to inspect their new machine. Boulton and Watt opened their own facility a year later, a flourmill in the heart of London. As Marsden described the event: "it was the talk of the town—and a hugely successful publicity stunt . . . grabbing attention from the great and the good at a splendid opening ceremony and a more considered, but still positive response from those less fashionable visitors that continued to flock in." By the end of the century (1800) more than 300 rotary Boulton/Watt engines had been installed in a variety of locations with a variety of applications. Glasgow alone had 18 cotton mills in which Boulton/Watt supplied machines were used both to spin and weave cotton thread (Marsden, 2002). The social structure of Britain was transformed as coal-fired steam engines replaced labor as the preferred means for production of industrial goods.

Workers losing their jobs did not always go quietly. Alarmed at their loss of livelihood, workers destroyed the showcase Boulton/Watt flourmill in London in 1791. The new industrial order triggered a flight from the countryside to cities that sprung up around the country often inadequately equipped to accommodate their expanded populations. Air quality suffered. Standards of hygiene deteriorated and a host of serious social problems ensued, as recorded for example by authors such as Charles Dickens. The political system was challenged for more than a century to develop an adequate response. And the revolution was not confined to Britain: it spread quickly to much of continental Europe—notably to France, Germany, Holland, and Russia—and took hold not much later in the New World. There would be no going back. Harnessing the power of steam revolutionized industrial life in Britain triggering a series of subsequent developments that would truly revolutionize the world.

By the time of Watt's death (in 1819) condensing engines persisted for the most part in use of steam at pressures as little as 5 psi above the pressure of the atmosphere (14.7 psi). Watt considered machines operating at higher pressure to be dangerous. His machines required between 5 and 8 lb of coal per unit of horse-power delivered per hour. By the end of the nineteenth century, steam engines were working at pressures as great as 250 psi driving ships and trains, thanks largely to the availability of improved materials, and the power delivered per unit of fuel had been enhanced by as much as a factor of five (Marsden, 2002). But the demand for goods had also increased and with it the demand for coal, setting the stage for the series of environmental and health problems discussed in Chapter 5.

10.4 STEAM AND DEVELOPMENT OF THE RAILROAD

As noted above, Watt's initial interest in steam, stimulated by his conversation with Robison, was directed at the question of whether steam could be used to drive a wheeled carriage. He abandoned this quest early, choosing to focus for the duration of his career on stationary sources of steam power. In this, he was influenced presumably by his skepticism as to the safety of high-pressure steam. Mobile applications of steam would require a high ratio of power delivered to the weight of the engine selected to deliver the power. This in turn would require necessarily a high-temperature, high-pressure, source of steam (the efficiency of a high temperature steam engine is intrinsically greater than the efficiency of an engine operating at lower temperature). It was left to Richard Trevithick (1771–1833), a Cornish mining engineer, and to Oliver Evans (1755–1819), an American inventor, to explore the potential of steam as a source of power for application to mobile systems.

Trevithick built a locomotive, propelled by high-pressure steam and designed to run on ordinary roads, in 1801. His locomotive functioned satis-factorily. It was not, however, a commercial success. It was too heavy for most of the roads existing at the time. He built a second locomotive in 1804, targeted in this case to run on a railway serving a coal mine in South Wales. Again, the locomotive worked but for a second time the weight proved excessive: it was too heavy to be supported by the light cast iron rails installed at the mine. Trevithick spent the rest of his career building pumping engines for Cornish mines, in which pursuit he was eminently successful (Cameron, 1997). It was left to others, notably George Stephenson (1781–1848), to show that steam could be successfully applied to run a railroad.

Evans was also a pioneer in the early applications of high-pressure steam. He won a U.S. patent for a high-pressure engine in 1804. He constructed an amphibious vehicle, drove it using steam at a pressure of 30 psi (pounds per square inch) from his workshop in Philadelphia in 1805 into the Schuylkill River, steaming downriver for a time before leaving the river and returning to the starting point. The machine was used successfully for several years to dredge the docks at Philadelphia under contract to the Philadelphia Board of Health (Evans et al., 2004).

Stephenson's father was a fireman for a coal mine near Newcastle: that is to say his job was to stoke coal into a Newcomen engine used to pump water from the mine. At age fourteen, Stevenson took a position as assistant fireman to his father. His mechanical skills were so impressive, however, that by the time he had reached age seventeen, he was given responsibility for his own fire engine. In 1812, at age 31, he was placed in charge of all of the machines owned by a cartel that controlled a major fraction of England's total coal reserves, the Grand Allies (Freese, 2003). At that point, he set out to develop an improved method to move coal from the mine mouth. The preferred approach at the time employed horses to haul coal along wooden tracks (yet another indication of the importance of access to wood). Stevenson embarked on a means to use coal-fired steam rather than horsepower to accomplish this task. In this he was eminently successful.

A 26-mile rail line was constructed linking the coal town of Darlington to the port of Stockton in 1825. Freese (2003) described the opening of this railroad thus: "the opening was attended by thousands who watched the procession of thirty-four wagons, carrying not just coal but six hundred passengers as well. It used locomotives on the flat sections of track and moved at a speed so slow that the procession was led by a man on a walking horse; where the track was steep, the cars had to be pulled uphill with ropes and stationary steam engines."

The construction of a rail line linking Liverpool and Manchester in 1830 signaled what might be considered the forerunner of the modern rail industry. Stephenson engineered the track and a locomotive of his design, the Rocket, drove the first train. The event, however, proved less than a public relations success. Again quoting Freese (2003): "As many as 400,000 people lined the route on that rainy September day in 1830. The Duke of Wellington, who had defeated Napoleon at Waterloo fifteen years before and who was now Prime Minister, was the guest of honor." One of the dignitaries participating in the opening was William Huskisson, Member of Parliament for Liverpool, a noted early champion of the railroad. Seven trains participated in the ceremonial opening. At the midpoint of the trip, the Duke's train stopped to take on water. Huskisson alighted from his train to greet the Duke. He did so while straddling a parallel track only to be knocked down and have his leg crushed by a train that had elected not to stop to allow its passengers to socialize with the Duke and Huskisson. Stephenson, who was driving the Duke's train, promptly unhitched his locomotive, loaded Huskisson on a wagon, and rushed him to the nearest town setting a speed record of 36 miles per hour along the way (Freese, 2003). Regrettably, it was too late. Huskisson died soon after reaching hospital, the first casualty of the railroad age.

The travails of the ceremonial inauguration of the Liverpool–Manchester rail link would not end with the untimely demise of Mr. Huskisson. An unruly crowd, including many mill workers who had lost their jobs courtesy of the new steam-based order, assembled to greet the entourage of dignitaries who arrived in Manchester several hours behind schedule. They proceeded to stone

the Duke's train forcing it to beat a hasty retreat, confirming the Duke's conviction, as summarized by Freese (2003), that "the lower orders were rotten to the
core and that revolution was imminent." The Duke's government would fall a
few months later, a response to the social unrest generated by the new industrial order. As an interesting footnote to his experience with the rail opening,
the Duke, who was initially lukewarm as to the prospects for the new railroad
technology, would subsequently change his opinion and go on to "make a great
deal of money speculating in railway stocks." The Liverpool-Manchester rail
link, despite its early problems, would prove a success and would play an
important role in the large-scale development of railroads that would rapidly
ensue not only in Britain but elsewhere.

Asa Whitney, an American merchant, traveled to England in 1830 to purchase items for his fancy goods business (Evans et al., 2004). While there, he
took a ride on the new Liverpool–Manchester railroad. He returned to the
United States captivated with the potential future of the railroad. Having made
his fortune, and after his wife died, Whitney elected to devote the rest of his
life in service to mankind. Visiting China, he returned to America in 1844
concerned with the plight of Chinese immigrants in the U.S. He saw the railroad, specifically one that would link the east and west coasts of America, as a
potential means to improve the lot of these people. As Evans et al. (2004) put
it, this could save "the immigrant millions from the tempting vices of our cities
[while] spreading America's civilizing influence among the starving, ignorant
and oppressed millions of Asia." He envisaged further "that the railway would
pay for itself by the sale of land adjacent to the right of way if only Congress
would set aside a 60 mile strip for the track."

It was left to Theodore Dehone Judah (1826–1863), a surveyor and committed railroad enthusiast, and his California partners Collis Potter Huntington
(1821–1900), Charles Crocker (1822–1888), Mark Hopkins (1802–1887), and
Leland Stanford (1824–1893) to see Whitney's vision to reality. The first transcontinental railroad was completed in 1869. Regrettably, Judah did not live to
experience the success of the project to which he had dedicated his life. He set
sail from San Francisco to Panama in 1863 on his way to New York to raise
funds for construction of the new railroad (the preferred route from the west
to east coast at that time was by ship to Panama, by land from the Pacific to the
Atlantic coast of Panama, and hence by boat to New York). Whitney contacted
yellow fever in Panama, dying 8 days later. His partners, on the other hand, all
made fortunes. Their names are perpetuated in a number of distinguished contemporary Californian institutions (The Huntington Library in Pasadena, The
Crocker National Bank before it was incorporated into Wells Fargo in 1986,
The Intercontinental Mark Hopkins Hotel in San Francisco, and Stanford
University).

By 1840, 2390 km of track had been laid in Britain, an extent that would
increase by close to a factor of ten over the following 30 years. The pace of
development was even more rapid in the United States: 4510 km of track in
1840, 84,675 km in 1870 following completion of the transcontinental rail link,

with a further extension to 410,475 km by 1914 (Cameron, 1997). While British trains were driven by steam generated by burning coal, it is interesting to note that trains in the U.S. were fueled initially by wood (another indication of the critical importance of wood as an energy source). Again quoting Freese (2003), "all along the tracks, people made money cutting wood to sell to the railroads— just one more way that trains would help turn the United States from a forested nation into an agricultural and industrial one."

The efficiency of steam locomotion increased steadily over the fifty years following Stephenson's successful demonstration of the Rocket, capitalizing for the most part on principles developed more than a half century earlier by Watt, incorporating not only single but also double, triple, and even quadruple expansion engines. Ever improving materials and increasingly more elaborate valves contributed to a persistent upgrading of both the performance and reliability of the early railways (Derry and Williams, 1960). So it would be until 1884, when a new technology, the steam turbine developed and patented by C. A. Parsons[2] exploiting principles developed earlier by James B. Francis for the water turbine (cf. Chapter 8), would usher in an era of even more efficient rail transportation.

Rail transportation continues to play an important role in all major world economies. It provides the least cost, most energy efficient, means for long-distance land-based transportation of goods. It provides the dominant mode for transport of low-sulfur western coal in the present day U.S. feeding the voracious power stations of the east. And it is similarly important in China where the bulk of that nation's coal is moved by rail from mine to market with coal accounting for more than half of the total rail freight. For close to a hundred years the railroads of the world were powered by steam generated by burning coal. Diesel oil, however, has now replaced coal as the dominant energy source and the characteristic puff-puff of the steam powered train is today but a distant memory.

10.5 STEAM POWERED SHIPPING

It is interesting to note that the important applications of steam to land-based transportation, specifically the railroad, did not occur until after the expiration of the Boulton/Watt patent in 1800. Applications of steam to power shipping began a little earlier. The key developments in this case took place in the U.S. The initial pioneer was an unlikely, jack-of-all-trades, a frontiers man named John Fitch (1743–1798).

Fitch's interest in steam began in 1785. He had heard about a steam engine that was used to pump water from a mine in New Jersey. Only later did he learn about the refinements of steam power achieved by Boulton and Watt. The British had banned the export of all manufacturing technology to the United States. Fitch set out to develop his own steam engine, one that could be used to power boats to move against the current of America's rich bounty of rivers and streams. The first American patent law was instituted in 1790. It authorized inventors to register their claims. It did not, however, require proof of originality allowing several individuals to hold patents simultaneously for the same idea (Evans et al., 2004). Partnering with a German born clockmaker, Henry Voight, his drinking buddy, Fitch demonstrated his working model of a steam-powered boat at the Constitutional Convention in Philadelphia in 1787. The demonstration was received as a curiosity by the delegates at the Convention but failed to attract funding for further development.

Fitch and Voight received licenses to operate powered boats on the rivers of Delaware, New York, and Pennsylvania in 1787. A larger, 60-ft boat with an 8-ft beam, driven by an improved engine designed by Voight, propelled by paddleboards installed at the stern, was launched on the Delaware River in 1790. Powered by steam generated by burning wood, with paddleboards turning at a rate of 76 strokes a minute (Evans et al., 2004), the boat went into commercial service in the summer of 1790, ferrying passengers between Philadelphia, Trenton, Bordentown, Bristol, and Burlington. The venture was a technical success but a commercial failure. To compete with competition from the stagecoach, Fitch and Voight were obliged to operate their service at a loss. They attempted to make up the revenue shortfall by serving "beer, sausages and rum in a pretty little cabin" (Evans et al., 2004) but the venture was doomed to failure from the start.

Fitch committed suicide in 1798, ingesting an overdose of opium pills prescribed for treatment of insomnia. His license to operate powered boats on the Hudson River was transferred in 1798 by the New York legislature to Robert Livingstone, an influential citizen of New York who had served at various times as Judge, Congressman, and Diplomat (he administered the oath of office installing George Washington as the first President of the United States in 1789). Livingstone had no particular skill as an engineer and his initial efforts to develop an operational steamboat on the Hudson River met with failure. He would achieve success, however, when he joined up with Robert Fulton (1765–1815), whom he met first at a dinner party in Paris in 1802, where Livingstone had recently arrived as the official representative of the U.S. government to France.

Fulton was born in Lancaster, Pennsylvania, in 1765. He moved to London in 1786, where he accepted the patronage of Viscount William Courtney, later the Earl of Devon, a notorious transvestite member of the British aristocracy. There, Fulton showed promise as an enterprising inventor. He developed a mechanical saw that markedly improved the efficiency of quarrying marble on Courtney's estate. He pioneered in construction of canals, building small boats with wheels that could be pulled up inclined planes using counterweights, eliminating the need for construction of expensive locks. In 1794, he wrote to Boulton and Watt enquiring as to the price of a small rotary engine that might be used to power a boat, signaling an early interest in powered shipping. He moved to Paris in 1797 convinced that revolutionary France could provide a more supportive environment in which to develop the multiple products of his fertile imagination.

With funding from a Dutch patron, he developed in France a submarine and a mine he described as a torpedo that could be used as a powerful new weapon of naval war. He was at somewhat of a loose end when he met up with Livingstone at the party hosted in Paris by an American couple, the Barlows, with whom Fulton had taken up residence (he was romantically involved at the time with Ruth Barlow, the wife of Joel, a wealthy American entrepreneur). Fulton's commitment to join with Livingstone to build a steam ship was sealed with a handshake at that party. Livingstone was assigned the task of finding "a way around the British export laws to get a Boulton and Watt engine to Fulton's specifications (Evans et al., 2004)." Fulton's role would be to design the boat and eventually to assemble the parts from which it would be constructed. As it turned out, it was Fulton who eventually arranged for the license to export the Boulton/Watt engine. Disillusioned with the French, Fulton returned to England in 1804 where he contracted to apply his skills to blow up French rather than British warships. Before doing so, however, he would have successfully demonstrated, in 1803, operation of a steam-powered boat on the Seine using an engine and boiler leased from a supplier in Paris. For his contributions to the British war effort, Fulton received a payment of 12,000 pounds and the permission he had requested to acquire and export a Boulton/Watt engine. Fulton would follow the engine to New York, arriving there in December of 1806.

Fulton set out to build his ship at a shipyard on the East River in New York. The ship, named the Clermont after the location of Livingstone's expansive estate on the Hudson, was finally ready for its maiden voyage on August 17, 1807. The Clermont was 146 ft long, 12 ft wide, with a 15-ft smokestack and was fueled by coal rather than wood. It set sail for Albany with plans to break the voyage at Livingstone's estate on the way. It made the 110-mile journey to Clermont in 24 h, continuing next day the further 40 miles to Albany, averaging over the entire journey a speed of close to 5 miles per hour. The return trip from Albany to New York, greeted by cheering crowds along the riverbanks, was accomplished in 30 h. A few weeks later, Fulton advertised to take paying customers initiating the first steam-powered New York–Albany ferry service.

Livingstone and Fulton elected to plow the profits from their ferry operation back into their steamship company. Fulton built new and improved boats and by 1812 was successfully operating ferry services with 21 boats not only on the Hudson River but also on the Delaware, Potomac, and James Rivers and on Chesapeake Bay. The ultimate goal, however, was to build a steamship that could traverse the entire length of the rivers from the industrial center at Pittsburg to New Orleans. Partnering with Nicholas Roosevelt (1767–1854), a New York born engineer credited with invention of the vertical steamboat paddle wheel, they fulfilled this ambition when the 310-ton New Orleans, captained by Roosevelt, successfully completed the 2000-mile journey from Pittsburg to New Orleans. The ship set off from Pittsburg in October 1811, docking in New Orleans in January 1812, surviving along the way an earthquake and an attempted raid by an Indian war party. They accomplished the trip at an average steaming speed of 8 miles per hour. It was clear though that the New Orleans would not have the power to make the trip in the return direction. It spent the rest of its working life ferrying passengers and goods between the Gulf of Mexico and Natchez, Mississippi, the purpose for which it was originally envisaged before finally foundering near Baton Rouge, Louisiana, in 1814.

Livingstone died in 1813. Fulton had lost a valuable confidant and his last days were unhappy. He was forced to go to court in Trenton, New Jersey, in 1815 to defend his precedence on the development of side wheels for his ships. Doing so, he produced a letter he claimed to have written in 1793 outlining his ideas. The opposing counsel held the paper to the light, however, showing that the letter was in fact postmarked in 1796. Thomas Addis Emmet, elder brother of the eminent Irish patriot Robert Emmet, defended Fulton. Thanks to a brilliant speech by Emmet extolling Fulton's contributions to mankind, Fulton survived this experience. Returning by boat from New Jersey to New York, however, Emmet attempted to reach the shore by walking across the ice that bordered the shore only to fall into the freezing waters of the river. He was rescued by Fulton but at a cost. Fulton fell victim to pneumonia from which he would succumb a month later on February 23, 1815 (Evans et al., 2004).

The steam age was by that time, however, well and truly launched. Twenty-three years later, the Steam Ship (SS) Great Western completed the first scheduled transatlantic crossing (in 1838). On October 6, 1848, the SS San Francisco steamed from New York, rounding Cape Horn, arriving in San Francisco on February 28, 1849 completing a voyage that lasted 4 months and 21 days launching the first east–west scheduled U.S. ship service.[3] Man would no longer have to depend on wind and sail in his quest to explore the oceans and exploit the resources of his planet.

10.6 CONCLUDING REMARKS

As discussed here, the steam age was established over an interval of approximately 150 years beginning with the patent for a fire engine granted to Thomas Savery in 1698. Patents were awarded first to glassmakers in Venice in the 1420s. King Henry VI granted a 20-year monopoly to John of Utyam to make stained glass in England in 1449. Responding to abuses in the patent system, the Parliament in England promulgated the Statute on Monopolies in 1624, which established time limits on award of patents (14 years or less) decreeing further that while the rights of inventors should be respected, there was also an obligation to consider the public interest. Federal authority over patents was legislated first in the U.S. with establishment of The Patent Commission of the U.S. in 1790. There are arguments both pro and con the patent system as a means to stimulate invention.

Newcomen was obliged to work with Savery to develop his fire engine. Watt, when he applied for his initial patent, granted in 1769, was advised by Roebuck and Boulton to make the patent application as broad as possible. It was thanks to the considerable political influence of Boulton that he was able to achieve an extension of the patent in 1775. Watt had the advantage of early access to the intellectual resources at Glasgow College. The key figure in his later career was Boulton who provided him not only with financial support but also with the contacts, such as Wilkinson, that would play a critical role in his ultimate business success. Watt proved adroit in exploiting the advantages of the patent system, not only for his initial refinement of the Newcomen engine but for the series of further inventions that broadened the scope of that invention and that ensured in particular his priority with respect to the applications of steam power to rotary motion. Again, Boulton was the indispensable partner.

It could be argued that the development of steam powered machines was stifled by the existence of the critical Watt/Boulton patents, at least up to the time that the original patent expired in 1800. It was only after 1800 that use of steam expanded to include applications at pressures much higher than the pressure of the atmosphere. High-pressure steam was essential not only for the development of the railroad but also for the development of steam powered shipping.

Stephenson rightfully deserves to be acknowledged as the father of the railroad, just as Judah should be recognized for his critical contributions to construction of the first transcontinental rail link in the U.S. But it was others who got rich as a result of the ingenuity of Stephenson and Judah, individuals such as the Duke of Wellington in England and the quartet of Huntington, Crocker, Hopkins and Stanford in the U.S. A similar pattern applied to the development of steam-powered shipping. Fitch built the first steam-powered boats in the U.S. He achieved a higher speed than the later boats developed by Fulton and Livingstone. Fitch died poor. Fulton, on the other hand, made a fortune, due in no small measure to the political acumen and influence of Livingstone. Livingstone was to Fulton what Boulton was to Watt.

A chronology of the major developments in the evolution of the steam age is presented in Table 10.1.

Table 10.1 Chronology of the major developments of the steam age

1698: Savery receives a patent for his "fire engine" identifying a variety of potential uses including but not restricted to pumping water from mines. Patent granted for an initial period of 14 years.

1699: Savery's patent extended to 1733.

1712: Newcomen, obliged under Savery's patent to collaborate with Savery, delivers his first fire engine to the Earl of Dudley's colliery. Atmospheric pressure, rather than steam, is the agent responsible for the work carried out by this machine.

1759: Robison excites Watt's initial interest in steam, raising the possibility that steam could be used to drive a wheeled carriage.

Table 10.1 continued

1765: After working with a miniature model of a Newcomen engine at Glasgow College and failing to make it work, Watt comes up with the idea for a separate condenser that could markedly improve the efficiency of the Newcomen steam engine.

1768: Watt initiates his partnership with Roebuck who provides funds and physical resources for development of Watt's engine design in return for two-thirds ownership in any eventual patent.

1769: Patent for Watt's steam engine.

1773: Boulton acquires Roebuck's share of the patent.

1774: Watt moves to Birmingham beginning his life-long partnership with Boulton.

1775: Boulton's influence responsible for an extension of the patent to 1800.

1781: Patents awarded Boulton and Watt for a variety of means to convert the reciprocal motion of steam driven pistons to rotary motion.

1782: Patent awarded to Boulton and Watt for use of steam to both push and pull a piston. Further patent awarded for use of the expansion of steam to carry out additional work significantly improving the efficiency of the Boulton/Watt steam engine.

1785: First Boulton/Watt rotary steam engine delivered to a cotton mill in Nottinghamshire initiating the use of steam in the textile industry.

1786: Boulton and Watt open their state-of-the-art steam-powered flourmill in the heart of London.

1787: Fitch and Voight demonstrate a steam-powered boat at the Constitutional Convention in Philadelphia. Receive licenses to operate steam-powered boats on the rivers of Delaware, New York, and Pennsylvania.

1790: A 60-ft boat powered by steam produced by burning wood launched by Fitch and Voight on the Delaware River initiating a commercial ferry service. Service proves to be a commercial failure.

1798: Livingstone acquires Fitch/Voight license to operate a steam-powered ferry service on the Hudson River.

1801: Trevithick builds the first railroad locomotive powered by high-pressure steam.

1803: Partnership between Livingstone and Fulton launched with a handshake at a dinner party in Paris.

1807: First Fulton/Livingstone steam-powered ship makes the 150-mile journey from New York to Albany at an average speed of close to 5 mph. Beginning of financially successful steam-powered ferry service between New York and Albany. By 1812, steam-powered ferry services are operating on the Delaware, Potomac, and James Rivers and on Chesapeake Bay.

1811–1812: Steam ship New Orleans makes the first successful steam-powered trip from Pittsburg to New Orleans.

1825: Stephenson operates his steam-powered train on a 26-mile rail link between Darlington and Stockton.

1830: Opening of Stephenson's rail link between Liverpool and Manchester.

1838: Steamship Great Western makes first scheduled transatlantic crossing.

1849: Steamship San Francisco completes voyage from New York to San Francisco sailing around Cape Horn.

1869: Completion of the first transcontinental rail link in the U.S.

1884: Steam turbine developed and patented by C.A. Parsons.

NOTES

1 http://www.keveney.com/newcommen.html (viewed May 13, 2007).

2 Charles Algernon Parsons, born at Hyde Park London in 1854, was the youngest son of an Anglo-Irish peer, the Earl of Rosse. His father was a distinguished amateur astronomer. Parsons received his early university education at Trinity College Dublin and later at St. John's College Cambridge where he graduated in 1877 with a First Class Honors Degree in Mathematics. He would go on to found C.A. Parsons and Company (now part of the German engineering conglomerate Siemens) dedicated to production of the steam turbines he had invented and later the Parsons Marine Steam Turbine Company. The Parsons building at Trinity College, housing the Department of Mechanical and Manufacturing Engineering, is named in his honor. The Charles Parsons Awards, instituted by the Irish government in 2006, provides funding for groups conducting energy research in Ireland.

3 (http://en.wikipedia.org/wiki/Steamboat, read July 6, 2007).

REFERENCES

Cameron, R.(1997). *A Concise Economic History of the World. From Paleolithic Times to the Present*, Third Edition. Oxford: Oxford University Press.

Carnegie, A. (1905). *James Watt*. Doubleday, Page and Company, New York. Available at http://www.history.rochester.edu/steam/carnegie/. Read April 1, 2007.

Derry, T.K. and Williams, T.I. (1960). *A Short History of Technology. From the Earliest Times to A.D 1900*. New York: Dover Publications.

Evans, H., Buckland, G., and Lefer, D. (2004). *They Made America: From the Steam Engine to the Search Engine: Two Centuries of Innovators*. New York: Little Brown and Company.

Freese, B.(2003). *Coal. A Human History*. Cambridge, MA: Perseus Publishing.

IPCC (2007). *Climate Change 2007: The Physical Science Basis. Contribution of Working Group 1 to the Fourth Assessment Report of the Intergovernmental Panel on Climate Change* (eds. S. Solomon, D. Qin, M. Manning, Z. Chen, M. Marquis, K.B. Averyt, M.Tignor, and H.L. Miller). Cambridge: Cambridge University Press.

Marsden, B.(2002). *Watt's Perfect Engine. Steam and the Age of Invention*. New York: Columbia University Press.

Wikipedia (2007). Timeline of steam power. http://en.wikipedia.org/wiki/Timeline_of_steam_power. Read May 9, 2007.

11
Electricity

11.1 INTRODUCTION

As discussed in the preceding chapter, low-pressure steam provided the energy source that fueled the early stages of the industrial revolution in the eighteenth century. By the turn of the century, engineers had mastered the ability to work with steam at high pressure, an advance that allowed, as we saw, for the development of steam-powered ships and railroad locomotives. What would truly transform society, however, would be the development and application of electricity over the second half of the nineteenth century.

It would be difficult to imagine a world today without electricity. Consider what would happen if one of our modern cities were to be deprived of its source of electricity, even for a period as short as a few days or a week. The city would be immediately dark at night. There would be no subway, no television, no radio, no refrigerator, no freezer, no computer, and no labor-saving domestic machines. As existing reserves of food spoiled in the absence of functioning refrigerator systems, food would soon be in short supply. Elevators in high-rise buildings would grind to a halt. A few floors above ground level, taps would run dry, as pumps that would normally supply water to these taps would lose their source of electrical power. If the power failure occurred in winter, there would be no way to heat buildings: delivery of hot water to radiators or hot air to heating vents depends on the availability of electrically powered pumps or fans to distribute the necessary source of heat. Similarly, there would be no means to cool buildings in summer. Soon, even the traffic on the streets would grind to a halt as service stations lost their ability to pump gasoline from underground storage tanks.

Ten million people in the Canadian Province of Ontario and close to forty million in eight States of the U.S. Midwest and Northeast were deprived of electricity for an interval that lasted in some cases as long as 2 days beginning on the afternoon of August 14, 2003. The interruption of power was triggered in this case by the cascading effect of a relatively minor accident in Ohio (caused by the failure of a regional power company, First Energy, to trim trees in their service area). Financial losses were estimated conservatively at $6 billion. For an account of the impact of this power failure, the largest ever to have occurred in North America, see[1]. Imagine what would have happened if the power failure had persisted for even a few more days.

The nature of electricity is discussed in Section 11.2. Applications of electricity to the development of the telegraph and telephone are discussed in Section 11.3. The pioneering contributions of Thomas Edison are treated in Section 11.4 while the relative merits of transmission of electricity by direct as compared to alternating current are described in Section 11.5. Summary remarks are presented in Section 11.6.

11.2 ELECTRICITY BASICS

As discussed in Chapter 9, an atom consists of a number of negatively charged electrons orbiting a positively charged nucleus. The magnitude of the charge on the nucleus depends of the number of protons it contains. The charge carried by an electron is equal and opposite in sign to the charge associated with a proton. In its normal state, an atom as a whole is electrically neutral: the number of orbital electrons is exactly equal to the number of protons contained in its nucleus.

Charged particles of opposite sign are attracted by a force the magnitude of which falls off as the square of the distance by which the particles are separated. In contrast, particles of the same sign experience a force that serves to drive them apart. In this case also, the magnitude of the force decreases as the square of the distance separating the particles. We can think of the electric force defining the interaction of two charged particles as analogous to the gravitational field that determines the interaction between two masses. A particle of mass m_1 is responsible for a gravitational force field that acts to attract other masses in its vicinity. The magnitude of the attractive force experienced by a second body of mass m_2 due to the presence of mass m_1 is proportional to the product of the masses of the two bodies and inversely proportional to the distance by which they are separated. Similarly, a particle of charge q_1 exerts an electric force on any second particle, charge q_2, in its vicinity, the magnitude of which varies as the product q_1q_2. As indicated above, in contrast to the gravitational case for which the force field is always attractive, the direction of the electric force field depends on the sign of the interacting charged particles—attractive if the particles have opposite sign, repulsive otherwise.

The magnitude of the force on a particle of charge q_2 due to the presence of a particle of charge q_1 separated by a distance r is given by Coulomb's Law named in recognition of the contributions of the French physicist Charles-Augustin de Coulomb (1736–1806) who, in a number of important papers published in 1785, first identified the nature of the (electrostatic) force between charged particles (Coulomb is credited also with important earlier investigations of the principles governing the laws of friction). Expressed in mathematical form, Coulomb's law states that the magnitude of the force, F, is given by

$$F = kq_1q_2/r^2 \tag{11.1}$$

The unit of charge in the SI system is also identified with Coulomb. With charges in Equation (11.1) defined in units of coulombs (C), distances in meters (m), and with k equal to 8.99×10^9 Nm^2C^{-2}, the electric force in Equation (11.1) is expressed in newtons (N), the basic unit of force in the SI system. Box 11.1 offers an illustrative calculation of the magnitude of the electrostatic force with which the electron is attracted to the proton in a hydrogen atom with a comparison of this force with the magnitude of the mass that would experience a similar force when exposed to the Earth's gravitational field.

The (electrostatic) force experienced by a particle of charge q is expressed as $q\underline{E}$. Here the symbol underlining $\underline{E}$ emphasizes that we are dealing with a vector quantity (see Chapter 3): to completely specify $\underline{E}$ it is necessary to define both its magnitude and its direction. Defined in this manner, $\underline{E}$ is referred to as the electric field. The electric field is analogous to the gravitational acceleration introduced in Chapter 3—the force exerted by the gravitational field on a particle of unit mass. Expressed in SI units, $\underline{E}$, defines the force experienced by unit charge and has dimensions of newtons per coulomb (NC^{-1}).

Box 11.1

Calculate the magnitude of the force with which the electron in the hydrogen atom is attracted to the proton of the hydrogen nucleus. Calculate the mass that would experience a force of similar magnitude in the gravitational field of the Earth.

Solution:

The force binding the electron to the proton in a hydrogen atom (mean separation 5.29×10^{-11} m) has magnitude 8.2×10^{-8} N:

$$F = (8.99 \times 10^9 \, Nm^2C^{-2})(1.602 \times 10^{-19} \, C)^2 / (5.29 \times 10^{-11} \, m)^2 = 8.2 \times 10^{-8} \, N$$

It is comparable to the force with which a particle of mass 8.4×10^{-9} kg would be attracted to the Earth by gravitation:

$$F = mg = (8.4 \times 10^{-9} \, kg)(9.8 \, m \, s^{-2}) = 8.2 \times 10^{-8} \, N$$

To put this in context, the mass of the electron is equal to 9.109×10^{-31} kg.

In the presence of an electric field, an electrically charged particle will be forced to move. Positively charged particles will take off in the direction of the field with negatively charged particles moving in the opposite direction. The field is said to do work on charged particles. The situation is analogous to what happens if a mass (a steel girder lifted by a crane at a construction site for example) suspended at some height in the atmosphere is released and allowed to fall. The gravitational field will do work on this mass, it will pick up speed (its kinetic energy will increase) and it will drop, accelerating to the surface. In a similar manner, charged particles can gain energy when exposed to the influence of an electric field.

A unit mass at a height h_1 in the atmosphere is said to possess potential energy in an amount equal to gh_1, where g defines the acceleration of gravity. Moving to a lower height, h_2, its potential energy is reduced and it gains kinetic energy as a consequence of the work done on it by the gravitational field. The work done by an electric field operating on a particle of unit charge can also be defined in terms of a change in potential (the electric potential in this case). A unit charge, forced to move by an imposed electric field from position a to position b will gain energy equal to the difference between the values of the potential at a and b, V_a and V_b. Positively charged particles will gain energy if $V_a > V_b$. For negatively charged particles to gain energy, the potential at position a must be less than the potential at position b: $V_a < V_b$. The gravitational potential energy corresponding to unit mass is expressed in the SI system in units of joules per kilogram ($J\,kg^{-1}$). The electric potential, with dimensions of energy per unit charge, is given in units of joules per coulomb, JC^{-1}, in the SI system. The unit of potential in the SI system (1 joule per coulomb) is referred to as the volt (V), named (in 1881) in honor of the Italian physicist, Alessandro Volta (1745–1827), credited in 1800 with invention of the first operational battery, an achievement for which he was awarded with the title of Count by Napoleon in 1810.

A material with the property that it can maintain a net flow of charge in the presence of an electric field is known as a conductor. A copper wire fits this bill (so also would silver, gold, and aluminum and a variety of other metals). The copper atoms that compose the wire readily surrender electrons. The wire is composed then of a matrix of positively charged copper ions with an approximately equal number of relatively mobile electrons. The electrons are accelerated in the presence of an electric field, moving as indicated above in the direction from low to high voltage (opposite to the sense of the imposed electric field). Their motion is interrupted by collisions with the positive ions that make up the bulk of the mass of the wire. In contrast, the much heavier positive ions are relatively immobile. If we were to take a snapshot of the electrons in the wire, we would see that they were moving at high speed (a speed determined by the temperature of the wire) but in essentially random directions (Fig. 11.1). Superimposed on this random motion would be a small drift towards the region of higher electric potential. Box 11.2 provides an estimate of the average speed of electrons in a wire at a temperature of 300 K (27°C).

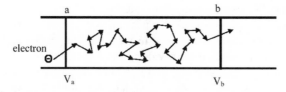

Figure 11.1 The number of electrons per unit time entering the region of the conductor between a and b at a is equal to the number of electrons exiting in the region at b, i.e. the current at a is equal to the current at b (charge is not allowed to accumulate between a and b). The figure illustrates the random walk of a typical electron moving from a to b depicting the effect of the interaction of the electron with the positive ions of the conductor. The voltages at a and b are taken as V_a and V_b respectively. The voltage difference between a and b is given then by $V = V_b - V_a$. As indicated in the figure, the electrons move from left to right, from low to high voltage. It follows that for this case $V_b > V_a$. As discussed in the text, it is conventional to think of the current as flowing from high to low voltage, as though the current was actually carried by positively charged particles. Current therefore for the present situation should be thought as directed from b to a. We are assuming here that the charge passing any position in the conductor in a time interval Δt is given by Δq. Hence $I = \Delta q/\Delta t$. There is a net loss of electrical energy due to the interaction of the electrons with the positive ions. The difference in voltage between a and b (and the associated force) accounts for the energy input required to offset the loss. The rate at which energy is supplied (or lost) is given by IV.

Box 11.2

Calculate the average speed of electrons in a wire at a temperature of 300 K.

Solution:

The kinetic energy of the average electron is given by $\frac{3}{2}kT$, where k is Boltzmann's constant. It follows that

$$\frac{1}{2}mv^2 = \frac{3}{2}kT$$

where m denotes the mass of the electron and v is its speed. Thus

$$v = \sqrt{\frac{3kT}{m}}$$

With the mass of the electron equal to 9.109×10^{-31} kg,

$$v = \sqrt{\frac{3(1.381 \times 10^{-23}\,\text{J/K})\,(300\text{K})}{9.109 \times 10^{-31}\,\text{kg}}}$$

$$= 1.17 \times 10^5\,\text{ms}^{-1}$$

Materials lacking the capacity to conduct electricity are referred to as electrical insulators. Examples of insulators include wood, glass, plastic, and cotton. What distinguishes conductors from insulators is the facility with which the atoms in the former are able to provide a source of free electrons.

Volta's battery was composed of two plates, one made of zinc the other of copper, immersed in a bath of sulfuric acid. In this configuration, a portion of the zinc dissolves in the sulfuric solution establishing an equilibrium described by

$$Zn \leftrightarrow Zn^{2+} + 2e \qquad (11.2)$$

As a result, a concentration of electrons, and consequently negative charge, builds up on the zinc plate. The zinc ions react with sulfate ions in solution to form zinc sulfate:

$$Zn^{2+} + SO_4^{2-} \rightarrow ZnSO_4 \qquad (11.3)$$

If a conducting wire is used to connect the zinc plate to the copper, a current of electrons will flow from the zinc to the copper plate driven by the excess negative charge on the zinc. The electron source, the zinc plate in this case, is referred to as the anode. The copper plate, which receives the electrons flowing through the connecting wire, is known as the cathode. The sulfuric acid solution in which the plates are immersed is referred to as the electrolyte. Electrons reaching the cathode combine with protons in the electrolyte to produce hydrogen gas:

$$2e + 2H^+ \rightarrow H_2 \qquad (11.4)$$

The build-up of electrons on the anode is sufficient to establish a voltage difference between the anode and cathode of a little more than 1 V. The voltage can be increased by stacking the zinc and copper plates and connecting anodes and cathodes in series; the more units, proportionally the higher the net resulting voltage. The electricity generated in this composite battery can be used to light a flashlight or to drive any number of simple electrical devices (more on this later).

A typical car battery incorporates an anode coated with spongy lead supported on a strip of metallic lead, a cathode the surface of which includes a coat of lead dioxide supported on a strip of metallic lead, and an electrolyte composed of sulfuric acid. The anode and cathode are separated using a strip of insulating material. When current is extracted from the battery (to start the car for example), electrons are drawn from the lead anode. The net reaction is represented in this case by

$$Pb \rightarrow Pb^{2+} + 2e \qquad (11.5)$$

The lead ions released in reaction (11.5) react with sulfate ions in the electrolyte to form lead sulfate ($PbSO_4$) that is deposited on the cathode:

$$Pb^{2+} + SO_4^{2-} \rightarrow PbSO_4 \tag{11.6}$$

The reaction that takes place when electrons from the anode reach the cathode may be summarized by:

$$2e + 4H^+ + PbO_2 \rightarrow Pb^{2+} + 2H_2O \tag{11.7}$$

The energy source that starts the car is derived from the sum of reactions (11.5), (11.6), and (11.7):

$$Pb + 2H^+ + H_2SO_4 + PbO_2 \rightarrow Pb^{2+} + PbSO_4 + 2H_2O \tag{11.8}$$

Reaction (11.8) is exothermic (it releases energy). A battery is simply a device that provides a mechanism for conversion of chemical to electrical energy.

The lead/acid car battery, in contrast to Volta's battery or the Duracell or Energizer batteries in common everyday use today (which typically incorporate zinc and manganese oxide electrodes immersed in an alkaline electrolyte such as potassium hydroxide), has the important property that it can be recharged. The chemicals depleted in reaction (11.8) can be restored by attaching the battery to a source of electrical power and running reactions (11.5), (11.6), and (11.7) in reverse. Most of us have had the experience at one time or another of having the battery in our car discharge: we try to start the car and the engine refuses to turn over. We solve the problem by finding a car with a healthy battery and attaching the batteries in an appropriate manner. The dead battery is conveniently brought back to life. A primer on how this works is presented in Box 11.3.

The charge passing by a point P on a conducting wire in unit time defines what is known as the current, I:

$$I = n_e q_e A v_d \tag{11.9}$$

where n_e denotes the number of electrons per unit volume (the number density of electrons), q_e is the charge per electron, A defines the cross-sectional area of the wire, and v_d is the translational or drift velocity associated with the bulk motion of the electrons. In the SI system, with n_e expressed in units of m^{-3}, q_e in C, A in m^2, and v_d in m s^{-1}, current has dimensions of coulombs per second (C s^{-1}). The unit of current in the SI system is known as the ampere (A) honoring the French physicist Andre-Marie Ampere (1775–1836). Ampere is credited with pioneering contributions establishing the connections between electricity and magnetism documented in a paper published in 1820 providing, at that time, the most comprehensive explanation of how an electric current can result in deflection of a magnetic needle. A current of 1 A is equivalent to a

---

Box 11.3

What are the steps involved in recharging a dead car battery?

Solution:

Find a good set of booster cables. Attach one end of one of the booster cables to the positive terminal of the discharged battery. Attach the other end of this cable to the positive terminal of the good battery. In the U.S., the portion of the jumper cables used to connect to the positive terminals is usually colored red. Now attach the other wire of the jumper cables to the negative terminal of the good battery. Attach the other end of this cable to an unpainted bolt or bracket of the car with the discharged battery, as far from the discharged battery as conveniently possible. This serves to provide a reliable ground for the connection. Electrons will flow now from the negative to the positive terminal of the functioning battery and from there through the booster cable to the positive terminal of the discharged battery. This source of electrons will drive reaction (11.5) to operate not from left to right but from right to left. The anode of the discharged battery will have its coating of lead replaced. Further, as the lead ion content of the electrolytic solution is depleted, the direction of reaction (11.6) will also reverse. Sulfuric acid will be restored to the electrolyte and the release of lead ions will trigger deposition of lead dioxide on the cathode returning the battery to at least some approximation of its prior functional, charged, condition.

---

flow of 1 C of charge per second past any given position of a conducting wire. An estimate of the drift speed associated with a current of 15 A flowing through a 14-guage copper wire is presented in Box 11.4.

A useful concept is to think of the flow of current in a conducting wire as analogous to the flow of water in a pipe. A pressure difference is required to maintain the flow of water in the pipe. Similarly, a change in voltage is required to drive the flow of charge in the wire. As electrons move from a region of low to one of high voltage, they pick up energy as a consequence of the work done on them by the electric field associated with the change in voltage. The interaction of the electrons with the positive ions that provide the bulk of the mass of the wire provides, however, a resistance to the flow of the electrons. Ohm's Law, named in honor of the Bavarian mathematician/physicist Georg Simon Ohm (1789–1854), defines the magnitude of the current flowing in a wire as a function of the change in voltage and the resistance. Ohm presented his Law in a book published in 1827 (The Galvanic Circuit Investigated Mathematically). The Law states that the change in voltage (V) is proportional to the magnitude

---

Box 11.4

The current available through an individual circuit of a typical home in the U.S. is about 15 A, sufficient to power sixteen 100 W bulbs. Assume that this current is supplied through a copper wire of radius $r = 0.01814$ cm (14 guage). Assume further that the density of free electrons in the wire is equal to the density of copper atoms, 8.2×10^{22} electrons cm^{-3}. Calculate the drift speed of the electrons constituting this current.

Solution:

The drift speed is given by

$$v_d = \frac{I}{n_e q_e A} = \frac{I}{n_e q_e \pi r^2}$$

Converting n_e and the radius of the wire to appropriate SI units, we have

$$n_e = 8.2 \times 10^{22} \frac{\text{electrons}}{\text{cm}^3} \cdot \left(\frac{100 \text{ cm}}{1 \text{ m}}\right)^3 = 8.2 \times 10^{28} \frac{\text{electrons}}{\text{m}^3}$$

and

$$r = 0.01814 \text{ cm} \cdot \frac{1 \text{ m}}{100 \text{ cm}} = 1.814 \times 10^{-4} \text{ m}$$

Hence

$$v_d = \frac{15 \text{A}}{\left(8.2 \times 10^{28} \dfrac{\text{electrons}}{\text{m}^3}\right)\left(-1.602 \times 10^{-19} \dfrac{c}{\text{electrons}}\right)\pi\left(1.814 \times 10^{-4} \text{ m}\right)^2}$$

$$= -0.011 \text{ ms}^{-1} \text{ or } 0.11 \text{ ms}^{-1}$$

(By convention, current is defined as positive in the direction towards which positive charges move.)

A comparison of the results obtained in Boxes 11.2 and 11.4 verifies that the drift speed of the electrons carrying the current is very much smaller than the average speed associated with their random motion.

---

of the current (I) flowing in the wire. The constant of proportionality is identified as the resistance (R):

$$V = RI \tag{11.10}$$

The resistance, a property of the composition of the wire, varies directly in proportion to the length of the wire and inversely as the magnitude of its cross-sectional area. It is easy to understand qualitatively the dependence of R on both the length and cross-sectional area: the longer the wire, the greater the number of positive ions the electrons must encounter in traversing the wire; the larger the cross-sectional area of the wire, the greater the current that can be maintained at a given voltage (similar to the situation with the flow of water in a pipe where the greater the cross-sectional area of the pipe the greater the flow of water that can be accommodated in response to a given drop in pressure). Resistance in the SI system is expressed in units of volts per amp (or $J\,C^{-2}s$). The unit of resistance in the SI system is referred to as the ohm (Ω).

The power, P, delivered by a current I across a voltage change V is given by IV or alternatively (using reaction (11.6)) by RI^2 or by V^2/R:

$$P = IV = RI^2 = V^2/R \qquad (11.11)$$

With I, V, and R expressed in SI units (amps, volts, and ohms respectively), power in Equation (11.11) is given in units of watts. Box 11.5 offers an estimate

---

Box 11.5

The voltage associated with a typical wall socket in the U.S. is equal to 120 V: that is to say, the increase in voltage from the hot wire supplying electrons to the return wire is equal to 120 V. Estimate the current required to supply the energy demand of a 120 W light bulb and the corresponding resistance (load) imposed by the filament. Compare results with the case of a 60 W light bulb and with 1200 W electric space heater. Estimate also the costs associated with operating these devices.

Solution:

Using Equation (11.10), it follows that

$$I = P/V = 120\ \text{W}/120\ \text{V} = 1\ \text{A}$$

The resistance of the filament is given by

$$R = V/I = 120\ \text{V}/1\ \text{A} = 120\ \Omega$$

For the 60 W light bulb, the current is reduced to 0.5 A while the resistance is increased to 240 Ω. We have ignored in this the resistance associated with the wires connecting the light bulbs to the wall socket.

The current required to supply the space heater is equal to 10 A and the resistance in this case is equal to 12 Ω.

Running the 120 W light bulb for an hour, the energy consumed is equal to 0.12 kWh (kilowatt hour), the unit in which the electric company typically charges for electricity. At the cost for electricity delivered to my home in Cambridge in January 2006 (19 cents per kwh), it cost 2.28 cents to operate the light bulb. The cost to operate the space heater for a comparable interval (an hour) would have amounted to 22.8 cents. Leaving the heater on for a 24-h period, the bill would come to $5.47.

of the current drawn from a typical wall electrical socket in the U.S. to supply a 120 W light bulb with an evaluation of the corresponding resistance of the filament of the bulb. The current required to supply a 60 W light bulb is equal to half that for the 120 W bulb. The resistance, however, is twice as high. The resistance imposed by an electrical device is commonly referred to as the load associated with the device. The box also includes an estimate of the current required to supply a space heater with an estimate of the corresponding resistance. Also included are estimates of the costs associated with operating the light bulbs and space heater on the basis of retail electricity rates prevailing in Cambridge Ma. during January 2006.

Electricity in the modern world is generated typically at large, widely separated, power stations. It is distributed over large distances by overhead wires suspended from steel towers. The challenge is to minimize the loss of energy associated with this long-distance transmission. With resistance fixed by the selection of material for the conducting medium (typically aluminum, less commonly more expensive copper) and by the length and cross-sectional area of the wire, we have a choice in transmitting electricity over a fixed distance. For any particular selection of conducting material, we can either increase the cross-sectional thickness of the wire (thus decreasing the value of R) or we can reduce the current that it carries compensating by increasing the voltage (cf. Equation 11.10). There are limits to what can be achieved with the first of these options—constraints imposed both by the cost of materials and by the need to support a heavier carrier. The preferred choice is to reduce the current while increasing the voltage.

As indicated in Equation (11.11), the power transmitted through the distribution system depends on the product of current and voltage. Suppose that a current of 1000 A at 220 V is produced at a central power station corresponding to a power output of 220 kW. If the current is passed through a transformer (more on this later) increasing the voltage to 220,000 V, the power can be transmitted using a current of 1 A (the value of the product IV is the same in both cases). Energy lost in the transmission system, calculated using Equation (11.11) would be reduced accordingly, by a factor of 10^6 if R were held fixed. In practice, companies involved in long-distance transmission of electricity take advantage of the increase in voltage to reduce the thickness of the carrier wires, capturing in the process significant savings in terms of requirements for materials (both the wires themselves and for the structures needed to support them).

To maintain a steady current in a conductor subject to some finite resistance requires a continuous input of energy. The device that supplies this source of electrical energy is called a seat of electromotive force, expressed in shorthand more commonly as a seat of emf. The electrical energy may be obtained at the expense of chemical energy as in the case of the batteries discussed above, or as mechanical energy in the case of a generator. The seat of emf serves to maintain the voltage differential required to sustain the current flowing in the circuit in face of the inevitable losses associated with the resistance of the circuit. As we have seen, electrons in the conductor are driven from regions of low to high voltage. For electrons, this is equivalent to gravitational masses rolling down hill. The seat of emf serves to reverse the process, to push the electrons back up the hill, to force them in this case to move from a region of high to a region of low voltage, permitting them once again to traverse the circuit. To accomplish this task requires a continuing input of work or energy.

To understand the mechanisms by which mechanical energy can be converted to electrical energy, we need to expand our treatment of electricity to include a discussion of magnetism. At one time or another, most of us have played around with magnets. We know that a magnet attracts certain materials, notably iron filings. We know that if we bring two magnets into close proximity, they will interact. A magnet has two poles; we refer to them as the north and south poles. When the magnets attach, the north pole of one will link up with the south pole of the other. The force field of a magnet is directed from the south pole of the magnet to the north. As with the electrostatic force, opposite poles attract, like poles repel. In contrast to the case of electricity, however, for which the force field can be associated separately with either a positive or a negative charge, magnetic poles always occur in pairs: if you split a permanent magnet in two, you will have two permanent magnets.

The Earth itself is a giant magnet. The poles of the Earth magnet are displaced somewhat with respect to the geographic poles (the magnetic pole in the northern hemisphere is located currently at 78 N, 104 W in the Elizabeth Islands of northern Canada; the pole in the southern hemispheric is at 66 S, 139 E, on the Adelie Coast of Antarctica). The pole in northern Canada is actually the south magnetic pole of the Earth while the pole in Antarctica corresponds to the Earth's north pole in terms of the configuration of the magnetic field. As illustrated in Fig. 11.2 , the lines of magnetic force for the Earth originate at the north magnetic pole in Antarctica, terminating at the south magnetic pole in northern Canada.

The magnetized needle of a compass points to the magnetic north. The north pole of the compass is attracted to the south magnetic pole of the Earth. If you are far removed from the geographic pole, the distinction between magnetic north and geographic north is relatively inconsequential. If you want a more precise definition of direction, or if you are too close to the pole, you need to refer to tables to find the appropriate correction factors.

We now know that electric currents are ultimately responsible for the force of magnetism. Ampere was the first to provide a quantitative description of this phenomenon. The magnetic field of the Earth originates in the Earth's iron-rich conducting core (near the center of the planet). The physical rotation of the Earth plays a critical role in generating the currents responsible for the magnetic field, accounting for the approximate alignment of the planet's geographic and magnetic poles. There is no particular reason why the north magnetic pole of the Earth should be located necessarily in the geographic southern hemisphere as it is today. Indeed we know that from time to time (approximately every million years) the planetary magnetic field reverses direction.

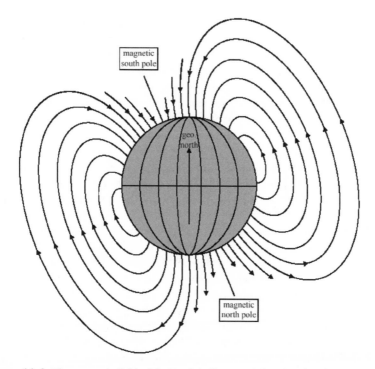

Figure 11.2 The magnetic field of the Earth is illustrated showing that the direction
of the field is oriented generally from south to north. The field lines are
more or less vertical at high latitudes, horizontal at low latitudes. As
indicated, the magnetic poles are displaced relative to the geographic
poles. This is the case for the magnetic equator, where the field lines
are horizontal: the magnetic equator is shifted relative the geographic
equator.

The north magnetic field was located in the geographic north about eight hundred thousand years ago. It flipped many times in the past and will surely do so
again, though probably not in our lifetimes (although the strength of the field
has been decreasing significantly over the past 150 years giving rise to specula-
tions that the next flip may be imminent).

We discussed earlier how charged particles are accelerated in the presence of
an electric field. They are subject also to a force associated with a magnetic
field. The force in this case is a little more complicated. Its magnitude is proportional both to the strength of the magnetic field and to the speed of the
charged particle. And it operates in a direction perpendicular both to the mag-
netic field vector, represented by $\underline{\mathbf{B}}$, and to the velocity vector, $\underline{\mathbf{v}}$. Expressed
mathematically, the force, $\underline{\mathbf{F}}$, on a particle of charge q is given by

$$\underline{\mathbf{F}} = q\,\underline{\mathbf{v}} \times \underline{\mathbf{B}} \qquad\qquad (11.12)$$

With $\underline{F}$ expressed in N, q in C, and v in ms^{-1}, $\underline{B}$ has dimensions of N s C^{-1} m^{-1} or N A^{-1} m^1. $\underline{B}$ is referred to as the magnetic induction, or simply as the magnetic field. The magnitude of the force experienced by a charge q is proportional to the magnitude of the charge on which it acts, the strength of the magnitude field, and the sine of the angle between the direction of motion of the charge (the orientation of the current) and the direction of the magnetic field. The notation $\underline{v} \times \underline{B}$ in Equation (11.12) denotes a vector aligned perpendicular to the directions of both the vectors $\underline{v}$ and $\underline{B}$ (the plane defined by the vectors $\underline{v}$ and $\underline{B}$). The direction of $\underline{F}$ is determined by what is known in vector notation as the right hand rule. Guess the direction of $\underline{F}$: place the thumb of your right hand along this direction; if the curl of your fingers moves $\underline{v}$ into the direction of $\underline{B}$ you have made the right choice; otherwise the direction of $\underline{F}$ is opposite to your initial choice.

The unit corresponding to the strength of magnetic induction or magnetic field in the SI system is known as the tesla (T) named for the Serbian-American inventor/engineer Nikola Tesla (1856–1943), credited over his lifetime with over 700 patents including patents relating to the use of alternating currents for the long-range distribution of electric power (see below). A magnetic field of 1 T will impart a force of 1 N to a charge of 1 C moving at a speed of 1 m s^{-1} in a direction perpendicular to the direction of the magnetic field. The strength of the Earth's magnetic field at mid-latitudes is equal to about 7×10^{-5} T.

Hans Christian Orsted (1777–1851), a Danish physicist, is credited as the first to recognize (in 1820) that an electric current can produce a magnetic field. His discovery was to some extent serendipitous. Working with a current loop connected to a battery, he observed that when the current was turned on it resulted in deflection of the needle of a nearby compass. Subsequent work by French physicists Ampere (introduced earlier), Jean-Baptiste Biot (1774–1862), and Felix Savard (1791–1841) provided a quantitative description of the strength and direction of the magnetic field resulting from any given configuration and intensity of current. The intensity of the magnetic field contributed for example by a current flowing in a long straight wire is proportional to the strength of the current and falls off as the square of the distance from the wire. Lines of constant magnetic field strength circle the wire as illustrated in Fig. 11.3. If the wire carrying the current is looped around an iron bar, as illustrated in Fig. 11.4, the bar becomes magnetized, with the magnetic field radiating from the bar assuming a configuration in this case similar to the configuration of the Earth's magnetic field (one end of the bar acts as the north magnetic pole, the other as the south magnetic pole). The greater the strength of the current carried by the wire and the larger the number of loops of wire around the bar, the greater the intensity of the resulting magnetic field. The latter arrangement defines what is referred to as an electromagnet and was exploited with great success in some of the early commercial applications of electromagnetism as discussed in Section 11.3.

William Sturgeon (1783–1850), a retired English artillery officer, was the first to demonstrate the power of even a simple electromagnet. He succeeded,

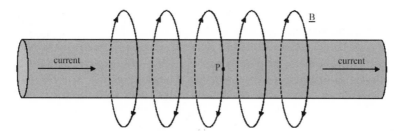

Figure 11.3 The figure illustrates the configuration of the magnetic field produced by a current flowing in a long straight wire. The strength of the magnetic filed decreases with radial distance, R, from the wire. The direction of the magnetic field at any point P is perpendicular to the plane defined by the wire and P as indicated by the circular lines of constant force strength depicted in the figure.

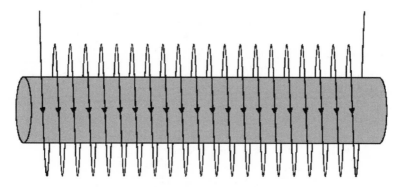

Figure 11.4 Illustration of the configuration of a solenoid with a conducting wire wrapped in a helical configuration around a central core.

in 1828, in lifting a nine-pound weight using a seven-ounce piece of iron wrapped with a wire supplied with current from a single battery.[2] Learning about Sturgeon's achievement, Joseph Henry (1797–1878), employed at that time as a teacher in a small school in Albany, New York, proceeded to carry out a series of even more extensive experiments. Henry would succeed eventually in wrapping coils of copper wire so tightly (separated with insulating strips obtained by tearing up his new wife's petticoats) that he was able to lift a weight of more than 1500 lb (Bodanis, 2005). He proceeded then to demonstrate the operational principles of the world's first telegraph. He placed what Bodanis describes as "something resembling a small clickable castanet, like a metal tongue" next to his electromagnet. With the current on, the castanet would be drawn to the magnet, responding with an audible click. Turn off the current, and the castanet would return to its original position with an additional click.

Henry realized that by varying the pattern of clicks—by altering the timing with which the current was turned on and off—he could send a message over a significant distance, using different arrangements of clicks to correspond to different letters of the alphabet. It would be left to another, Samuel Morse, to patent and exploit Henry's invention (more on this below). Henry would go on, however, to become one of the first great American-born scientists (his parents were Scottish immigrants), teaching at the College of New Jersey (renamed subsequently as Princeton University) and was appointed subsequently, in 1846, as the first Secretary of the Smithsonian Institution, a position he would hold until his death at age 81 in 1878.

A summary of the concepts introduced here is presented in Table 11.1.

Table 11.1 Summary of important concepts in electromagnetism

Concept	Definition	SI unit
Charge (q)	Property responsible for the electrostatic force by which particles of opposite charge are attracted while those of same charge are repelled	Coulomb (C): charge on the electron = 1.602×10^{-19} C
Electric field (E)	Force experienced by a particle of unit charge	Newtons per Coulomb (NC^{-1})
Voltage (V)	Analog of gravitational potential energy for charged particles. Positively charged particles gain energy moving from regions of high to low potential. Negatively charged particles gain energy when moving in opposite directions.	Volt (V): equivalent to $1 \, J \, C^{-1}$
Conductor	Medium than can maintain a concentration of free electrons and is consequently able to conduct electricity.	
Insulator	Medium that cannot provide a source of free electrons and is consequently unable to conduct electricity.	
Anode	The plate providing the source of electrons in a battery, the negative electrode	
Cathode	The plate in the battery to which the electrons are driven to flow from the cathode in a connecting wire, the positive electrode.	
Electrolyte	Substance which in solution can dissociate into positive and negative ions. Critical component of a battery.	
Current (I)	Quantity of charge passing a position P on a conducting wire in unit time.	Amp (A): equivalent to $1 \, C \, s^{-1}$

Table 11.1 continued

Resistance (R)	Measure of the resistance to flow of current driven by a specific voltage difference: $V = RI$ or $R = \dfrac{V}{I}$	Ohm (Ω): equivalent to $V\,A^{-1}$ or $J\,C^{-2}s$
Seat of electromotive force (emf)	Energy input required to produce the voltage difference required to maintain the flow of electrons in a circuit.	V
Magnetic field or Magnetic induction(**B**)	Analog of the electric field. Responsible for an additional force on a moving charged particle. Force oriented perpendicular both to the field and the direction of motion of the particle.	Tesla (T): 1 T provides a force of 1N to a particle of charge 1C moving at 1 M s^{-1}.
Electromagnet	Device by which an electric circuit can be employed to produce a magnetic field.	
Magnetic flux (Φ_m)	The magnetic flux through the area enclosed by an electric circuit is defined by the integral of the component of the magnetic field normal to the area over the area.	T m^2
Faraday's Law	A changing magnetic flux through a circuit will induce a current to flow in the circuit. The induced emf is equal to the negative of the time rate of change of the magnetic flux.	
Alternating current	Refers to a circuit in which the voltage and current oscillate in time at any given position in the circuit. The changing voltage causes electrons to move part of the time in one direction; in the opposite direction for the remainder of the time	
Transformer	Device that can be applied either to increase or decrease the voltage of a primary circuit. Permits a low voltage–high current source to be converted to a high voltage–low current source and vice versa.	

11.3 THE TELEGRAPH AND TELEPHONE

Samuel Morse (1791–1872), obsessed with the slowness of the mails, first conceived of the idea for the telegraph while on route to New York from Le Havre on board the French ship Sully in 1832. Morse had no particular understanding of electromagnetism when he embarked on his quest, the success of which would eventually revolutionize the technology for long-distance communications. He was totally unaware of the advances in electromagnetism achieved earlier by scientists such as Henry and Ampere. Ampere, for example, several years previously had suggested that an electromagnet could be used to move a needle to the position corresponding to either a particular letter or number but had declined to carry his idea further (Evans et al., 2004). And Henry, perfecting his electromagnet, had succeeded in transmitting electric signals over distances as great as a mile and had employed the magnetic field produced by his tightly wound electromagnet to ring a bell. Undaunted, Morse set out to build a working model of his telegraph when he arrived in New York at the end of his month-long voyage from France.

Morse's initial prototype included a pencil that could be forced to move by the electromagnet he connected to a simple battery. The pencil was expected to trace out horizontal lines on a piece of paper, the length of the lines corresponding to specific numbers. Morse went to the pain of associating every word in the dictionary with a particular number. The prototype was a failure. Even after adjustment, it was incapable of operating over a distance of more than 20 ft (Evans et al., 2004).

Morse consulted then with Leonard Gale, Professor of Geology and Mineralogy at New York University, an institution at which Morse had earlier accepted an unpaid position as Professor of Literature of the Arts and Design (in his early life Morse was a well-regarded portrait painter). Gale was familiar with the recent advances in electromagnetism and he quickly appreciated that the fundamental problem with Morse's prototype was that he was attempting to drive the current using too low a voltage. As discussed above, the higher the voltage, the easier it is to transmit current over a long distance. Gale made important improvements to Morse's machine. Morse and Gale formed a partnership that was joined later by Alfred Vail, a recent graduate of New York University, in return for an investment of $2000 (Morse had minimal personal means).

In the meantime, Morse abandoned the idea of using the electrical signals transmitted by the current to his electromagnet to program movements of a pen that could be interpreted in terms of specific words. He invented what became known as the Morse code, a series of dots and dashes communicated by either short or long pulses of electricity. Each letter of the alphabet was identified with a particular combination of dots and dashes, a space between the electrical transmissions signaling the end of a particular word. Morse code would become the standard for telegraph communication. It would continue in wide use for efficient communication even in the radio age to the middle of the twentieth century (when signals were sent not through a wire but through the air to bounce off the thin layer of charged particles present several hundred kilometers above the Earth in the region of the atmosphere known as the ionosphere). The international distress signal SOS, for example, was communicated by a series of 3 dots (S) followed by 3 dashes (O) followed by 3 dots (S).

Morse successfully demonstrated transmission of his code over two 5-mile segments of wire at his university in New York in 1838. He paid a visit later to

Henry, by that time installed in his laboratory at Princeton. Henry was generous with advice and provided Morse with a letter stating that Morse's invention was superior to telegraph systems patented earlier in England by Charles Wheatstone and William Fothergill. Morse was successful in his application for a patent in 1840, the application buttressed no doubt by the positive support he received from Henry.

With patent protection achieved, Morse's next challenge was to acquire the considerable funds required to implement his system. With support from Francis O. J. Smith, a devious Congressman from Maine, Chairman of the influential House Committee of Commerce, a sum of $30,000 was awarded by Congress in 1843 to build a trial line that would stretch 40 miles from Washington to Baltimore (Smith had earlier secretively negotiated for himself a 25% interest in the patent). Morse was allocated a salary of $2500 a year to oversee the project with Vail to be paid at a rate of $3 a day (Evans et al., 2004). Smith assumed control of the contract for himself. He elected to install the cable underground and spent close to $20,000 laying the first few miles. He neglected, however, to insulate it and predictably the cable proved incapable of transmitting any useful current. The contractor, Ezra Cornell, who was immediately aware of the problem, urged Morse to abandon the underground approach and instead to use overhead wires strung on 24-ft chestnut poles, separated by 200 ft, with glass drawer knobs employed as insulation to eliminate loss of current at the poles. The lines were strung along the Baltimore and Ohio (B and O) rail line. Cornell completed the installation remarkably in only a little more than a month and by May 1843 the system was ready to meet its first test.

Morse telegraphed Vail from the Supreme Court Chamber in the Capital in Washington. Vail responded from the B and O depot in Baltimore. His message, suggested by the daughter of a friend, the Commissioner of Patents, was famously: "what hath God wrought (Evans et al., 2004)." "Within five years, there were 12,000 miles of telegraph lines in America run by 29 different companies" (Evans et al., 2004) and Morse and his partners, including Cornell, were rich. Cornell, who was born in Westchester County, New York, used his new wealth to endow a public library for the citizens of Ithaca, New York, and would be influential later, in 1865, in the creation of the university named in his honor with funds secured under the 1862 Morrill Land-Grant Colleges Act by which Congress provided resources for universities that "would teach practical subjects as opposed to the classics as favored by more traditional institutions."[3] Morse would spend much of the rest of his life defending his priority with respect to the invention of the telegraph, downplaying in particular his debt to Henry. Like many of the successful industrialists of his time, however, such as Rockefeller and Carnegie, he would give away much of his money, notably as a patron of the arts, as a benefactor to his alma mater, Yale University, and, with his friend the brewer Matthew Vassar, as cofounder of Vassar College in Poughkeepsie, New York.

The first transatlantic cable went into operation in 1866 after several earlier abortive attempts foiled in part by insufficient understanding of the nature of electricity—failure to adequately insulate the conducting wire (Bodanis, 2005). The cable extended undersea from Newfoundland, Canada, to Ireland. The entrepreneur who made this happen was Cyrus West Field (1819–1892), a wealthy New York businessman. The scientist who advised on the project was a young Scot, William Thomson (no relation of J.J. Thomson credited more than 30 years later, in 1897, with the discovery of the electron, the basic carrier of electricity). Now finally, it was possible to send messages rapidly from the Old to the New World. No longer would it be possible to repeat what happened at the Battle of New Orleans in 1815 when the British and American forces engaged in the battle learned only weeks later that the war was over: that their fight was unnecessary.

Alexander Graham Bell, credited as the inventor of the telephone, was born in Edinburgh, Scotland in 1847. His father, Melville Bell, was a distinguished student of the physiology of speech. His mother was deaf. The young Alec, as he was known in his family, assisted his father in his work at University College London from 1868–1870. It was there that he developed his lifelong interest in the study of sounds and the mechanics of speech, inspired no doubt by the challenge to find an improved means to communicate with his mother.

The family emigrated to Canada in 1870, settling in Ontario. A few years later, in 1871, Bell relocated to the United States where he accepted a position on the staff of the School for the Deaf in Boston. It was there that he met the woman, one of his pupils, who would later become his wife, Mabel Hubbard, daughter of a wealthy Boston lawyer, Gardiner Hubbard. Mabel Hubbard had contracted scarlet fever as a child. The infection spread to her ears and she lost her hearing as a result (Bodanis, 2005). Bell opened his own school in 1872 dedicated to training teachers of the deaf. In 1873 he was appointed as Professor of Vocal Physiology at Boston University. By that time, though, he had developed an interest in exploring how electrical signals could be used to communicate sounds rather than mere clicks over an electrically conducting wire.

The Hubbards were initially cool to the courtship of their daughter by Alec. But by 1874, it was clear that the relationship was serious and Gardiner Hubbard elected not to persist in opposition but to invest in the fortunes of his future son-in-law. With funds provided by Hubbard, Thomas Watson, a talented machinist working at a local Boston electrical shop, was engaged to assist Bell in his research. Hubbard's objective for this work was to develop an improved means to transmit telegraphic signals over a standard telegraphic line, a capability to transmit multiple messages at the same time thus eliminating, or at least postponing, the need for installation of additional telegraphic wires. In this, Hubbard was inspired by his personal antipathy to the monopoly exercised by the Western Union Telegraph Company. As discussed above, the standard approach to transmission of a telegraphic signal involved operating with current switching intermittently between on and off. Bell's idea was to use what he described as "electrical undulations on the wire" to transmit information. Information in this case could be transmitted continuously, improving consequently the duty cycle of the telegraph. Bell's driving ambition, however, was to find a means to use the same approach to transmit voice, to allow individuals at two ends of a wire to engage in conversation.

From his earlier work with his father, Bell was well aware that sound was communicated by pressure waves transmitted through the atmosphere. With Watson, he rigged up a membrane and observed that voice patterns triggered characteristic vibrations of the membrane. The intensity of the vibrations was proportional to the amplitude (volume) of the signal. The frequency of vibrations responded to the pitch of the voice transmission. The challenge would be to convert the pattern of vibrations of a membrane into an electrical response that could be transmitted over a wire together with a means to convert this associated electrical signal into vibrations of a membrane at the other end of the wire that could be used to reproduce the fluctuations in atmospheric pressure responsible for the original voice message. Bell built a primitive prototype. Aware that his understanding of electricity was inadequate, he consulted with Henry, by then 78 years old, in 1875. Henry's response to Bell's confession of his insufficient understanding of electromagnetism was, famously: "Get it."

Bell worked diligently to perfect his brainchild. He applied for a patent on February 14, 1876 just a few hours before Elisha Gray's lawyer arrived at the patent office to file a competitive claim, actually in his case not an application for a patent but rather for what was known as a caveat, a provisional filing including drawings that allowed the inventor to delay filing a formal patent application while still registering a measure of priority. There is a dispute as to what actually took place. Evenson (2000) argues that Gray's caveat filing was taken to the Patent Office by his lawyer a few hours earlier than the filing of Bell's patent application. But the filing fee for Gray "was entered on the cash blotter hours after Bell's filing fee" according to Evenson's (2000) reconstruction of the sequence of events. In any event, Gray was persuaded by his lawyer to withdraw his caveat, effectively conceding Bell's priority. Bell, on March 7, 1876, was awarded U.S. patent 174,465 for the telephone.

Gray would later (in 1877) have a change of heart. He would file an application for the invention he had described under his earlier caveat challenging Bell's priority. Following 2 years of litigation, he would lose his attempt to rewrite history, a tribute to the skills of the Boston patent lawyer, Anthony Pollok, engaged by Hubbard to act on Bell's behalf. The essential elements of Bell's patent, abstracted from the actual language of the filing, are summarized in Table 11.2.

Morse's development of the telegraph made use of the fact that an electric current could produce a magnetic field. By the time Bell embarked on his quest to improve the telegraph and invent the telephone, understanding of electromagnetic theory had advanced to the point where it was well known, as Bell alluded to in his patent, that a time varying magnetic field could induce a current in a conducting circuit. The physics underlying this phenomenon is encapsulated in the law attributed to the English physicist/chemist Michael Faraday (1791–1841). According to this law, a change in magnetic flux through the area enclosed by a circuit is responsible for an induced electromotive force (emf)

Table 11.2 Basic elements abstracted from the patent titled "Improvement in Telegraphy" filed by his lawyer on February 11,1876 on behalf of Alexander Graham Bell. Patent approved March 7, 1876 as U.S. patent number 174, 465.

My present invention consists in the employment of a vibratory or undulating current in contradistinction to a merely intermittent or pulsating current, and by a method of, and apparatus for, producing electrical undulations upon the line-wire. . . . The advantages I claim to derive from the use of an undulating current. . . . are, first, that a very much larger number of signals can be transmitted simultaneously on the same current; second, that a closed circuit and single main battery can be used; third, that communications in both directions may be established without the necessity of special induction coils. . . . It has long been known that when a permanent magnet is caused to approach the pole of an electromagnet a current of electricity is induced in the coils of the latter, and that when it is made to recede a current of opposite polarity to the first appears in the wire. When a wire, through which a continuous current of electricity is passing, is caused to vibrate in the neighborhood of another wire, an undulating current is induced in the latter. Undulations are caused in a continuous voltaic current by the vibration or motion of bodies capable of inductive action; or by the vibration of the conducting wire in the neighborhood of such bodies. Electrical undulations may also be caused by alternately increasing or diminishing the resistance of the circuit, or alternately by increasing or decreasing the power of the battery. . . . The external resistance may also be varied. For instance, let mercury or some other liquid form part of a voltaic current, then the more deeply the conducting wire is immersed in the mercury or other liquid, the less resistance does the liquid offer to the passage of the current.

and thus for a change in the resulting current. The magnetic flux through the area enclosed by the circuit is given by integrating the magnitude of the component of the magnetic field strength perpendicular to the area over the area. If the field is aligned perpendicular to the area and if the strength of the field is constant over the area, the magnetic flux, Φ_m, is given simply by

$$\Phi_m = BA \tag{11.13}$$

where B defines the magnitude of the (perpendicular) magnetic field strength and A is the area enclosed by the circuit. Faraday's Law states that the emf, ε, induced in the circuit is given by the negative of the time rate of change of the magnetic flux:

$$\varepsilon = -d\Phi_m/dt \tag{11.14}$$

Faraday and Henry, in a series of experiments conducted in the 1830s, elucidated independently the details of the manner in which a time varying magnetic field could induce a current in a circuit.

It is useful at this point to check the compatibility of the units for the quantities on the two sides of Equation (11.14). In the SI system, ε is expressed in volts, equivalent to J C^{-1}. The right hand side has dimensions of $Tm^2 s^{-1}$ in the SI system. But T is equivalent to $NA^{-1}m^{-1}$ or $N C^{-1} s m^{-1}$ since A (the current measured in amps, not the area enclosed by the circuit) is expressed in $C s^{-1}$. It follows that the right hand side of Equation (11.14) has dimensions of $N C^{-1} m$, consequently the same as ε since Nm is equivalent to J.

The patent approved, Bell, Watson, Hubbard, and Sanders (father of another of Bell's deaf students and an early supporter of Bell's research) moved to form the Bell Telephone Company in 1877. Bell and Mabel Hubbard were married soon thereafter. The first telephone exchange went into operation under license from the Bell Company in New Haven in 1878. In 1880, the Bell Telephone Company merged with a number of newly formed private telephone companies to form the American Bell Company. The American Telephone and Telegraph Company (ATT) was incorporated in 1885 as a wholly owned subsidiary of American Bell charged with building out the long-distance telephone network. In 1899, ATT acquired the assets of the American Bell Company in 1899. ATT would continue as the dominant presence in the American telephone industry until the company was broken up in 1984 into eight separate corporate units as a result of an antitrust suit filed by the U.S. Government in 1974.

Bell would go on to participate in establishing Science Magazine and the National Geographic Society. He carried out pioneering research on air conditioning, helped develop the first iron lung, an improved strain of sheep, a sonar system for detection of icebergs, and set a world record for water-speed with a hydrofoil he designed in 1918.[4] He became a naturalized U.S. citizen in 1882, 11 years after moving to Boston from Canada. He died on August 2, 1922, at his summer retreat on Cape Breton Island, Nova Scotia, Canada, true to the end to his Scottish heritage.

11.4 EDISON, THE LIGHT BULB, AND THE DEVELOPMENT OF DISTRIBUTED ELECTRIC POWER

Thomas Alva Edison was the most prolific inventor of his time, arguably the greatest ever. Over his life, he accumulated no fewer than 1093 U.S. patents in addition to the significant number of patents he was awarded in Britain, France, and Germany. He is credited often with the invention of the light bulb. He did not invent the light bulb but he improved significantly on prior work, producing a light bulb that was sufficiently long lasting to be commercially successful. His major achievement, though, was to develop the infrastructure that allowed for the introduction of the first distributed electric power system, including the means to monitor and charge for power delivered, and his vision in creating the first integrated commercial research laboratory.

Edison was born in Milan, Ohio, in 1847. That made him exactly the same age as Bell. It is interesting to note that the two great inventive talents of the early telecommunications era both suffered from hearing deficiencies: Edison personally; Bell in his immediate family—his mother and his wife. As a boy,

Edison was an indifferent student, a result no doubt of the frequent illnesses from which he suffered, compounded by the problem with his hearing. He left school at age eight and received the remainder of his formal education at home from his mother who had been a village schoolteacher before she retired to bear her eight children (Thomas, or little Al as he was known in his family, was the youngest).

Edison had an early interest in electricity, one shared by many young boys of his age inspired by the newfangled technology of the telegraph. At age nine, his mother introduced him to A School Compendium of Natural and Experimental Philosophy by R.G. Parker (Evans et al., 2004), a book that included descriptions of a variety of experiments in chemistry and electricity that could be carried out by an enterprising young student. Edison tried them all, demonstrating remarkable early skill in mastering concepts and applying them in practice, traits that would serve him well in his future career as a polymath inventor. He demonstrated also early skills as an entrepreneur. His first job (at age 13) involved selling copies of the Detroit Free Press newspaper to passengers on the early morning train from Port Huron to Detroit. He hired two boys and set them to work selling foodstuffs he persuaded the train conductor to allow him to take on board. And he began writing, typesetting, copying, and selling his own digest of the weekly news. On the long daily layover in Detroit, Edison spent much of his time in the Detroit Free Library where among other ambitious initiatives he tried to study Newton's Principia, a task that would have been formidable even for a mathematically educated university student. Reputably, Edison came away from the experience with an abiding "distaste for mathematics (Evans et al., 2004)."

Edison's next formal job was as a telegraph operator for Western Union. It was there that he exhibited for the first time his creative talents. Messages came in over the telegraph at a rate of about 50 words a minute. It was difficult for operators to keep up with receipt of messages at this pace and transcriptions were plagued notoriously with errors. Edison found an original Morse paper tape machine and recorded incoming messages on tape. He then read the tape into a second telegraph transferring the data at a much slower speed. His transcription of messages was essentially flawless as a consequence.

Edison moved to Boston in 1868 (at age 21) where he was employed as a nighttime operator by Western Union. It was there that he launched his career as a serious inventor, taking advantage of funds provided by a backer to rent space for his first workshop. He improved on the existing stock ticker and with partners went into business selling a stock and gold quotation service to local banks and traders (Evans et al., 2004). He resigned from Western Union in 1869 announcing his intention to devote his life fulltime to the business of invention. He moved to New York in 1869 where he viewed opportunities for entrepreneurial activity as more promising. He met soon with success, selling an improved stock ticker to Western Union for $30,000, setting up as an independent manufacturer, contracting, together with a new partner, William Unger, a Newark, New Jersey machinist, to supply a variety of electrical equipment to Western Union including 1200 high-speed stock tickers. By dint of long hours and hard work he successfully met the terms of this contract. When Unger sought to withdraw from their business relationship, Edison bought him out. Soon, with borrowed capital, he was employing 50 or more pieceworkers at his plant in New Jersey and had set up a laboratory on the top floor of the facility to pursue his inventive activities (Evans et al., 2004).

Taking advantage of his new found resources, Edison proceeded to employ three individuals who would play a critical role in his subsequent career as a professional inventor: Charles Batchelor, an English machinist whose role would be to turn Edison's rough drawings into more precise renderings; John Kruesi, a Swiss watchmaker whose job it would be to construct the models that would be submitted in connection with patent applications; and Edward Johnson, an engineer who would assume important administrative responsibilities including oversight of applications for patents by the new organization. A particular initial target was to develop a more efficient means for transmission of telegraphic messages. To aid in this work, Edison negotiated an arrangement whereby he had access to the experimental facilities of Western Union, in turn for which he agreed that Western Union should have first claim on any viable inventions that should arise as a result of this access. Edison hit pay dirt with the invention of the quadruplex, which allowed for simultaneous transmission of four distinct messages, two in each direction along a single telegraph wire, an invention that was worth at least $20 million to Western Union according to Josephson (1959).

Strapped for money, Edison offered rights to his invention to Western Union for $25,000 payable in two installments with $17,000 per year to be paid over the 17-year life of the patent. When Western Union demurred, Edison made an alternate arrangement with Jay Gould, owner of the Atlantic and Pacific Telegraph Company, a rival of Western Union. For Edison's half share of the patent, Gould paid $30,000 in cash with a commitment for shares in the Atlantic and Pacific Company valued at the time at $75,000 (Evans et al., 2004). By 1875, these shares were worth $250,000. Gould, unscrupulous robber baron that he was, reneged on the share deal. Edison was happy though with the cash infusion, particularly so since Western Union had agreed in the interim to Edison's original proposal for the rights to the other half of the patent. Edison put his new found wealth to good effect setting up the research laboratories (and his new primary residence) at Menlo Park, New Jersey, that would provide the site for much of his work over the rest of his life.

Edison was driven to succeed. He was accustomed to working 16-h days and expected no less from his associates. When he married his first wife, Mary Stillwell, he did so on Christmas Day, 1871. Even on this holiday wedding day, Edison elected to return to his laboratory after dinner where he remained until close to midnight (Evans et al., 2004). Edison worked diligently to assimilate the scientific advances of the time that could influence his ambition to invent. He purchased and studied assiduously Michael Faraday's three-volume treatise on Experimental Researches in Electricity. Recognizing the importance of chemistry in his work, he took courses in chemistry at Cooper Union, complementing his classroom work by reading papers in the original literature. He was instinctively tuned to the advice Bell received from Henry: know what you need to know, and if you don't know it, "get it."

Edison was also an excellent judge of people. He hired talented individuals and created a work environment where they could flourish. He understood the world of business and had a keen appreciation for advances for which there would be an immediate market. He was also an excellent promoter when it came to his inventions with a flair for the dramatic. And he knew how to protect his intellectual property by aggressive use of the patent system. The Menlo Park team turned out inventions earning patents at a prodigious rate—no fewer than 75 in 1882 alone. Many of these inventions represented merely incremental improvements on the current state-of-the-art. But some were truly revolutionary, notably the invention of the first phonograph in 1877, which earned for Edison popular acclaim and the title by which he was known subsequently in the media as the "Wizard of Menlo Park." But his most lasting achievements were undoubtedly associated with the work he initiated in 1878, the twin goals to develop a long-lasting light bulb and with it the infrastructure that would allow the benefits of electricity to be available not just to specialist organizations such as the telegraph and telephone companies but to as wide a segment of the general public as possible.

The principles underlying the incandescent light bulb were well known when Edison began his quest. Connect an electrical current to a filament and if the resistance of the filament is great enough and the supply of electrical power sufficient, the filament could be induced to glow emitting visible light. Edison announced that: "with a process I have just discovered, I can produce a thousand—aye, ten thousand lamps from one machine. Indeed, the number may be said to be infinite. The same power that brings light to you will also bring power and heat . . . with the same power you can run an elevator, a sewing machine, or any other mechanical contrivance, and by the means of the heat you may cook your food (Evans et al., 2004)." This was a grand claim since Edison had not yet embarked on the research required to support the conviction that he could deliver what he promised. But his reputation was such that New York capitalists, notably Cornelius Vanderbilt and John Pierpoint Morgan, were eager to invest. A sum of $300,000 was raised to form the Edison Electric Light Company with $50,000 in cash for Edison in addition to 2500 of the 3000 shares issued in return for allocation to the company of any patents or improvements that might ensue as a result of this investment over a 5-year period (Evans et al., 2004). Edison now had the resources to set to work.

There were two problems that had to be resolved to develop a long-lasting bulb. The first was to find material for a filament that would not simply melt or react with oxygen in the air and decay in face of the exceptionally high temperatures required for the filament to emit significant quantities of light in the visible. Recall that the temperature of the region of the sun responsible for the visible light that illuminates the Earth during daytime is close to 6000°C

(Chapter 3). For the filament to be useful as a source of light it would need to be able to maintain temperatures of at least 2500°C.[5] The second challenge was to find a material for the filament whose resistance would be great enough so that the bulb could be supplied using a relatively modest current of electricity (see Equation (11.11) and Box 11.4). Edison settled on coiled threads of carbon with a resistance of a little more than 100 Ω for the filament and thanks to the talents of a German glassblower he engaged, Ludwig Boehm, he succeeded in placing the filament in a glass bulb from which all but 1% of the atmosphere had been excluded thus prolonging the time that would elapse before the filament would burn. Edison received a patent for a high resistance lamp in 1879.

Development of a long-lasting light bulb was but the first step in Edison's ambition to provide lighting and power to a wide range of commercial customers. Evans et al. (2004) succinctly summarized the challenge: "he had to build a central power station; design and manufacture his own dynamos economically to convert steam power into electrical energy; ensure an even flow of current; connect a 14-mile network of underground wiring; insulate the wiring against damaging moisture and the accidental discharge of electrical charges; install safety devices against fire; design commercially efficient motors to use electricity in daylight hours for elevators, printing presses, lathes, fans and the like; design and install meters to measure individual consumption of power, and invent and manufacture a plethora of switches, sockets, fuses, distributing boxes and lamp holders." The capital costs alone would be enormous and would be supplied primarily by Edison himself who fortunately by that time was rich as a result primarily of the income he continued to derive from his telegraph and telephone patents (his personal wealth was estimated at about $500,000). Political skills would be required also to persuade the politicians to provide permission to tear up the streets to lay cable. All of this came together on September 3, 1882. The dynamos were installed in a power station based in a number of dilapidated warehouses Edison acquired at 255–257 Pearl Street in Lower Manhattan. The demonstration was staged in the offices of Drexel, Morgan and Company on Broad Street, and in the offices of the New York Times. When the engineer at Pearl Street threw the switch at 3 pm, 106 lamps wired in parallel[6] turned on in the offices of Drexel, Morgan, with a further 52 at the New York Times. Edison had successfully engineered the beginning of the modern electrical age. There was a battle, however, yet to be waged: the question of whether the power for this new electrical world should be supplied by direct current, Edison's choice, or by the alternative, alternating current, favored by George Westinghouse and Nikola Tesla.

11.5 ARGUMENTS FOR DIRECT VERSUS ALTERNATING CURRENT

With direct current, the flow of charge past any given position in a conduction wire is constant in time. With alternating current, not only does the current vary in time, changing sign or direction, but so also does the voltage. As an

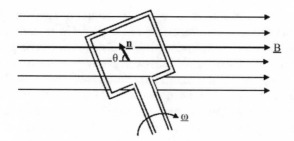

Figure 11.5 The figure illustrates a conduction coil rotating at angular velocity ω in a constant magnetic field. The angle θ defines the direction between the magnetic field and a vector oriented perpendicular to the plane defined by the coil.

indication of the properties of an alternating current and how such a current might be generated, consider a conducting coil undergoing uniform (constant) rotation in a constant, fixed direction, magnetic field as illustrated in Fig. 11.5.

The magnetic flux through the coil (see Equation (11.13) for a definition of magnetic flux) is a maximum when the coil is oriented at right angles to the magnetic field. It drops to zero when the coil is aligned with the magnetic field. Let θ denote the angle between the magnetic field and the direction perpendicular to the plane of the coil: θ = 0 when the plane of the coil is perpendicular to the field: θ = 90 when the plane of the coil is aligned with the field. Assume at time zero that the coil is aligned perpendicular to the field, θ = 0. Let ω denote the (constant) rotation rate of the coil expressed in units of radians per second. In this case, θ = ωt. For simplicity, assume that the strength of the field is constant equal to B. The magnetic flux corresponding to orientation θ is given then by BAcosωt where A denotes the area included by the coil.

The change of magnetic flux with time may be obtained by differentiating this quantity with respect to time:

$$\frac{d\Phi_m}{dt} = \frac{d}{dt}[BA\cos(\omega t)] = -BA\,\omega\sin(\omega t) \tag{11.15}$$

It follows, according to Faraday's Law (11.14), that the emf at time t, $\omega(t)$, is given by

$$\varepsilon(t) = BA\,\omega\sin(\omega t) \tag{11.16}$$

Suppose that the source of emf is connected to a circuit with resistance R. According to Ohm's Law, the voltage associated with the emf, ε, is related to the current and resistance according to the relation $\varepsilon = IR$. It follows that

$$I(t) = BA\,\omega/R\,\sin(\omega t) \tag{11.17}$$

The magnitude of the induced emf and the strength of the associated current are zero initially. They increase with time reaching a maximum value of $BA\omega$ at $\omega t = \frac{\pi}{2}$ in the case of the emf, $BA\,\omega/R$ in the case of the corresponding current. They decrease subsequently, changing sign at $\omega t = \pi$, falling to minimum (negative) values at $\omega t = \frac{3\pi}{2}$, returning subsequently to the initial conditions when the coil has described a complete loop around the magnetic field. The induced emf and the associated current oscillate repetitively in time. The emf changes sign and the current switches direction on a time interval given by $\Delta t = \frac{\pi}{\omega}$. The overall pattern repeats on a cycle specified by $\omega t = 2\pi$ corresponding to a time interval of $\Delta t = \frac{2\pi}{\omega}$. The current generated by this time varying emf is referred to as an alternating current. The frequency of the alternating current (expressed in cycles per second) is given by $\nu = \frac{\omega}{2\pi}$.

The power dissipated by the induced current (equal to the power produced by the time varying magnetic flux in the generator) is given by ϵI or by $I^2 R$. The instantaneous rate of power generation is defined thus by the relation

$$P(t) = I^2 (t)R = B^2 A^2 \omega^2/R \sin^2(\omega t) \qquad (11.18)$$

Noting that the time-averaged value of $\sin^2(\omega t)$ is $1/2$, it follows that the time-averaged value of the power produced in the alternating generator is given by

$$P_{av} = 1/2\ B^2 A^2 \omega^2/R \qquad (11.19)$$

The necessary power must be supplied mechanically to maintain the required rotation of the coil. We postulated here that the coil intersecting the magnetic field had a single loop. If the coil had incorporated N loops, the resulting emf would have been greater by a factor of N and the power required (or delivered) would be enhanced by a factor of N^2.

We assumed in the preceding analysis that the configuration of the magnetic field was fixed and that the variation of the magnetic flux through the coil (including its various loops) was produced by rotation of the coil. An entirely equivalent result would have been obtained had we assumed that the coil was fixed and that the rotational motion was imparted to the field rather than to the coil. The power delivered to the electrical circuit depends on the relative velocity of the coil/field system rather than on the absolute velocity of either.

The advantage of alternating current over direct current is that there is a ready means to either step-up (increase) or step-down (decrease) voltage in the case of the former that is lacking for the latter. The device that allows this to take place is known as a transformer. The operation of a transformer is illustrated schematically in Fig. 11.6. It envisages a primary circuit driven by a generator providing a source of alternating emf represented by $\varepsilon(t)$. The primary circuit includes a solenoid with N_1 loops wound around a core composed of soft iron. A secondary circuit with N_2 loops is attached to the same core. The presence of the soft iron core ensures that the change in magnetic flux through a loop of the primary circuit is essentially the same as the change in magnetic

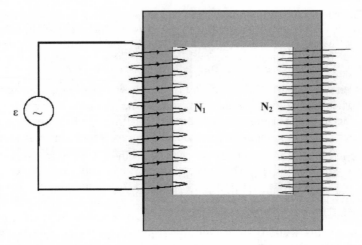

Figure 11.6 Transformer with N_1 loops in the primary and N_2 loops in the secondary.

flux through a loop of the secondary circuit (assuming that the area encompassed by a loop of the primary circuit is the same as a loop of the secondary). If the voltage induced across the solenoid of the primary circuit is given by V_1 and the voltage across the solenoid of the secondary circuit is given by V_2, the voltage of the secondary circuit will differ from that of the primary circuit by a factor equal to the ratio of the number of loops in the secondary circuit to the number of loops in the primary circuit:

$$V_2 = (N_2/N_1)\, V_1 \qquad\qquad (11.20)$$

To increase the voltage of the secondary circuit with respect to the primary, the necessary condition is that the secondary circuit incorporate a greater number of loops than the primary, $N_2 > N_1$. Conversely, to decrease the voltage of the secondary circuit with respect to the primary the required condition is that the number of loops included in the secondary circuit should be less than the number incorporated in the primary, $N_2 < N_1$. A transformer employed to increase voltage is known as a step-up transformer. Used to decrease voltage, it is referred to as a step-down transformer.

The presence of an alternating current in the primary circuit is critical to the ability of a transformer to either increase or decrease voltage in a secondary circuit. If the primary circuit were energized by a constant emf, the magnetic field induced in the solenoid would also be constant. There would be no resulting change in magnetic flux with time and consequently no induced current in either the primary or secondary circuit, thus no opportunity to adjust voltage. The ability to increase voltage is critical for efficient long-distance transfer of electrical energy. As indicated earlier, the power carried over a wire is given by VI, where V defines the voltage change and I denotes the magnitude of the current. The higher the value of V, the lower the value of I required to deliver a given power source. As indicated by Ohm's Law, Equation (11.11), loss of energy in the wire is given by I^2R, where R measures the resistance of the wire. The lower the value of I required to deliver a given power source, the higher the tolerable value of R. Since R varies in proportion to the length of the wire and inversely in proportion to its cross-sectional area, this means that the higher the voltage the thinner the wire required to transport power over a given distance and thus the lower the cost of the infrastructure required to transport the energy.

Edison's pioneering Pearl Street system was based on delivery of power in the form of direct current. From the beginning, Edison was categorically opposed to the use of high voltage delivery systems and consequently to the use of alternating current. The downside of his choice, though, was that Edison could only deliver useful power over a radius of about a mile from his power generating station. Not only that, but when different devices, power tools for example, required a supply at different voltages, the power had to be delivered on separate wires. The advantages of alternating current were championed early by George Westinghouse (1846–1914), joined later by Nikola Tesla (1856–1943), triggering what came to be known as the War of Currents.

Westinghouse, like Edison, was a talented inventor/engineer. His first major invention, patented in 1872, was a braking system for trains based on use of compressed air supplied from the locomotive and delivered to each car individually by a pipe running the length of the train. The system allowed brakes to be applied or released on all cars simultaneously, markedly improving the safety of rail transport. The Westinghouse Air Brake Company was formed to manufacture and market the invention. It achieved immediate success ensuring Westinghouse's status as a bonafide member of the distinguished class of great American entrepreneurs of the last quarter of the nineteenth century.

Westinghouse began experimenting with alternating current (AC) as a power source in 1885 (3 years after Edison's path breaking demonstration of the potential for distributed power in New York). He imported transformers designed by Lucien Gualard in France and John Gibbs in England, together with an AC generator developed by Siemens in Germany. He improved the function of both the transformers and the generators, developing efficient procedures for the manufacture of both. The first AC power distribution system was installed in Great Barrington, Massachusetts, by William Stanley and Westinghouse in 1886. Power was produced by a hydroelectric generator at 500 V AC, stepped up to 3000 V for transmission, stepped down subsequently to power lights at 100 V. The problems of dealing with high voltage electricity were highlighted when Westinghouse's assistant Franklin Pope was electrocuted by a malfunctioning AC converter in the basement of his home.[7] Undaunted, Westinghouse proceeded to form the Westinghouse Electric and Manufacturing Company in 1886, renaming it the Westinghouse Electric Corporation in 1889. He was joined by Nikola Tesla in 1888 who would play a major role not only in the science of alternating current but also, with Westinghouse, in its application.

Tesla was an ethic Serb, born in what is now Croatia. He emigrated to the U.S. in 1884. He worked initially for Edison but left within a year when Edison declined to pay him the sum Tesla felt he had been promised ($50,000) for improvements he had implemented in Edison's motors and generators. Tesla developed and successfully demonstrated, in 1888 (at the American Institute of Electrical Engineers, now the Institute of Electrical and Electronic Engineers, IEEE), the first polyphase AC electric motor.[8] Tesla carried out pioneering research in a variety of areas of electromagnetism over the course of his life. As noted earlier, his contributions were recognized in 1960 when the international scientific community elected to name the unit measuring the strength of the magnetic field in his honor. Given their early history, and the differing positions they held on the relative merits of direct and alternating current, it is not surprising that Edison and Tesla were bitter adversaries over most of their lives. Ironically, Tesla's contributions would be recognized by the Institute of Electrical Engineers with award of the Edison medal. Conferring the medal, Vice President Behrend of the Institute commented[9] that: "Were we to seize and eliminate from our industrial world the result of Mr. Tesla's work, the wheels of industry would cease to turn, our electric cars and trains would stop, and towns would go dark and our mills would be idle and dead." He concluded paraphrasing Alexander Pope's tribute to Isaac Newton: "Nature and nature's laws lay hid by night. God said 'Let Tesla be' and all was light."

Westinghouse and Tesla would eventually win the War of Currents but not before Edison would put up a stiff fight. In an effort to discredit AC, Edison presided over the use of high voltage currents to electrocute a number of animals, mainly stray cats and dogs, but famously on one occasion Topsy, an elephant that had run amok killing three men in a circus at Coney Island. The first electric chair, using AC current, was designed by Harold Brown, an employee of Edison. While Edison was personally opposed to capital punishment, it appears that his parochial interests would win out in resolving the conflict. The electric chair was used to execute William Kemmer in New York's Auburn Prison in 1890. It was a particularly brutal performance. A 17-s pulse of electricity was administered initially and served not to kill but to simply render Kemmer unconscious. In a follow-up, Kemmer was hit with 2000 V and his body caught fire. The entire operation lasted for close to 8 min.[10] Westinghouse commented: "they would have done better to have used an axe." Despite the adverse publicity, AC was the winner. The Westinghouse/Tesla design would become the standard for distributed power, based in the case of the U.S. on

3-phase AC delivered at a frequency of 60 cycles per second (60 hertz, written as 60 Hz). The frequency was selected to be high enough so that there would be no perceptible flicker in lamps illuminated using this current but not so high that significant amounts of power would be lost from power lines due to emission of electromagnetic waves.

The Edison Electric Light Company merged with a number of other electric companies in 1891 to form the Edison General Electric Company, a merger in which Edison had a 5% interest and for which he would receive $1.75 million through a combination of stock and cash. The Edison General Electric Company merged in 1892 with Thomson-Houston, an electric company based originally in Lynn, Massachusetts, in what was essentially a reverse takeover of the larger company (Edison General Electric) by the smaller (Thomson-Houston) orchestrated by J.P. Morgan and the Vanderbilts convinced that the smaller company had better management. The merger created the General Electric Company, which persists today as one of the world's largest corporate conglomerates. Edison would take no part in the management of the new company although he accepted initially a seat on the board. According to Evans et al. (2004), his reaction was: "I simply want to get as large dividends as possible from such stock as I own." When he died in 1931, Edison's estate was worth $12 million.

11.6 CONCLUDING REMARKS

We began this chapter with a conjecture as to what would happen today were we to be deprived of electricity for even a brief period of time. It would surely not be an exaggeration to conclude that electricity, more than any other technological innovation, has been responsible for shaping the nature of our modern world and the prosperity we enjoy (at least those of us who live in the so-called developed world). The seeds of this success were sown more than a hundred years ago. It is a remarkable fact that most of the important work was carried out before scientists (specifically J.J. Thomson in 1897) had identified the electron as the key carrier of the energy that fueled this revolution. We highlighted the contributions of distinguished scientists such as Coulomb, Volta, Ohm, Ampere, Orsted, Biot, Savard, Sturgeon, Henry, Faraday, and Tesla. To this list we should add the Scottish mathematician/physicist James Clark Maxwell (1831–1879) whose equations provided the definitive description of the interrelated properties of electricity and magnetism. Einstein is quoted (by Patrick Mc Fall in the April 23, 2006 issue of the Sunday Post) to have described Maxwell's contribution as "the most profound and the most fruitful that physics has experienced since the time of Newton." As important as the work of the scientists, however, were the contributions of the inventors and the financial sponsors—the venture capitalists of the day—who made the technical advances possible. A summary of the major technological milestones is summarized in Table 11.3.

Table 11.3 Chronology of important technological applications of electricity

Date	Innovation
1800	First operational battery demonstrated by Alessandro Volta.
1820	Hans Christian Orsted discovers that a magnetic field is produced by an electric current.
1828	William Sturgeon demonstrates the first electromagnet.
1831	Joseph Henry uses electromagnetism to remotely activate ringing of a bell setting stage for development of the telegraph.
1837	William Cooke and Charles Wheatstone awarded patent for first electric telegraph.
1838	Samuel Morse successfully demonstrates transmission of a Morse coded message over two 5-mile wires in New York.
1844	Samuel Morse installs and successfully demonstrates a telegraph linking Washington D.C. and Baltimore using Morse code to send messages.
1849	Samuel Morse receives U.S. patent for his telegraph.
1866	First successful transmission of messages over the transatlantic cable linking Ireland and Newfoundland.
1874	Thomas Edison awarded patent for the quadruplex permitting four messages to be transmitted on a single wire simultaneously, two in each direction.
1876	Alexander Graham Bell awarded patent for telephone.
1876	Thomas Edison invents an improved transmitter for the telephone making use of the changing resistance of carbon grains whose density responds to pressure charges associated with sound waves triggered by the human voice. Remained the standard for the telephone up to the 1960s.
1877	Thomas Edison patents first phonograph.
1879	Thomas Edison receives patent for his high-resistance lamp.
1882	First demonstration of distributed electrical power. Direct current transmitted by Thomas Edison over a 1-mile radius from Pearl Street in Lower Manhattan, New York.
1884	Charles Algeron Parsons receives patent for the steam turbine that would become the standard for generation of AC electric power.
1886	First AC power distribution system installed in Great Barrington, Massachusetts by William Stanley and George Westinghouse.
1888	Nikola Tesla demonstrates first polyphase electric motor.
1895	George Westinghouse delivers AC electric power from Niagara Falls to run street cars in Buffalo, New York.
1902	First modern electric air conditioning system invented by Willis Haviland Carrier; controlled both heat and humidity.
1923	First self-contained refrigerators introduced by Frigidaire.
1944	Mark I computer developed by Howard Aiken and Grace Hopper at Harvard University.
1947	Introduction of first top-loading automatic clothes – washing machine.
1947	Development of the transistor by John Bardeen, Walter Brattain, and William Shockley at Bell Laboratories.
1958	Development of the integrated circuit by Jack Kilby at Texas Instruments and Robert Noyes at Fairchild Semiconductor.
1969	Prototype of the modern internet developed by the U.S. Advanced Research Projects Agency (ARPA)
1981	Introduction of the first person computer (PC) by IBM

NOTES

1 (http://en.wikipedia.org/wiki/2003_North_America_blackout, read July 11, 2007).

2 (http://en.wikipedia.org/wiki/William_Sturgeon, read July 19, 2007).

3 (http://en.wikipedia.org/wiki/Ezra_Cornell, read July 20, 2007).

4 (http://www.bookrags.com/biography/alexander-graham-bell-woi/, read July 25, 2007).

5 Incandescent light bulbs in use today normally include filaments composed of tungsten, which can withstand exceptionally high temperatures without melting. The filament, in the form of a thin-coiled wire, is enclosed in a glass bulb the atmosphere of which is composed of inert gases such as nitrogen and argon. The temperature of the filament is typically about 2500°C.

6 Wiring devices in parallel means that the electrical current flows independently through the different devices in contrast to an arrangement where the devices are connected in series. The effective resistance of an arrangement in which devices are arranged in parallel is less than one where they are connected in series. Ohm's Law implies as a consequence that a greater current is required to supply the parallel arrangement as compared to the series connection. The disadvantage of the series arrangement is that if one device fails the entire connection may be broken. This is what happens with a string of light bulbs on a Christmas tree: one bulb blows and the tree goes dark. In contrast, with the parallel arrangement, failure of any of the components will not jeopardize the function of the rest. This feature was important for Edison when he staged his high-risk demonstration in New York.

7 (http://en.wikipedia.org/wiki/George_Westinghouse, read August 2, 2007).

8 Tesla's invention represents the prototype for most of the large electrically driven machines in operation today. The underlying principle is relatively straightforward. Imagine three separate sources of AC current connected to solenoids located at clock positions 12, 4, and 8. The currents differ in phase by 120°. As the current in the circuit at 12 o'clock peaks and begins to decline, the current at 4 o'clock picks up. Similarly, as the current at 4 o'clock crests, the current at 8 o'clock begins to make its presence felt. The result is production of a magnetic field that rotates around the clock face, completing revolutions tuned to the cyclic oscillations of the AC current sources. The rotating magnetic field provides the force field used to spin the operational elements of the motor.

9 (http://www.teslasociety.com/biography.htm, read August 2, 2007).

10 (http://en.wikipedia.org/wiki/Electric_chair, read August 2, 2007).

REFERENCES

Bodanis, D. (2005). *Electric Universe*. New York: Crown Publishing Group.

Evans, H., Buckland, G., and Lefer, D. (2004). *They Made America: From the Steam Engine to the Search Engine: Two Centuries of Innovators*. New York: Little Brown and Company.

Evenson, A. E. (2000). *The Telephone Patent Conspiracy of 1876: The Elisha Gray– Alexander Bell Controversy*. Jefferson, NC: McFarland.

Josephson, M. (1959). *Edison: A Biography*. New York: McGraw-Hill.

12

Automobiles, Trucks, and the Internal Combustion Engine

12.1 INTRODUCTION

If, as asserted at the conclusion of the preceding chapter, the development of the electric sector has had the most significant influence on modern life, arguably the impact of the internal combustion engine must rank a close second. In 2004, there were 199 million licensed drivers in the U.S. (out of a population of 293 million) with 237 million registered motor vehicles. That is to say, there were more vehicles registered than people authorized to drive them. The transportation and electricity sectors in combination were responsible for 70% of total U.S. emissions of CO_2 in 2004—32% from transportation, 38% from generation of electricity. If we are to arrest the growth, or preferably reverse the upward trend in emissions of CO_2 in the U.S., it is clear that both the electricity and transportation sectors must be targeted for action. In the case of electricity, it might be possible to continue our use of coal by implementing measures to capture and sequester the associated source of CO_2, a topic explored in Chapter 14. A more promising immediate approach would appear to be to increase the efficiency with which we use electrical power while ideally at the same time reducing demands on the fossil fuel component employed to generate it. For the transportation sector, capturing CO_2 would not appear to be an option. The obvious strategy in this case would be to implement measures to increase the efficiency with which fossil energy is used to drive cars and trucks, searching at the same time for alternative, non-fossil, sources of energy to reduce current reliance on oil-derived gasoline and diesel fuels (electrically propelled vehicles for example with the electricity produced from non-fossil sources).

Table 12.1 Data relating to driving practices in the United States

Year	Registered vehicles (millions)	Total vehicle miles (billions)	Miles per gallon (miles)	Fuel consumed (billion gallons)
1960	74.5	718.8	12.4	57.9
1965	91.7	887.8	12.5	71.1
1970	111.2	1109.70	12	92.3
1975	138	1327.70	12.2	109
1980	161.5	1527.30	13.3	115
1985	177.1	1774.80	14.6	121.3
1990	193	2144.30	16.4	130.8
1991	192.3	2172.20	16.9	128.6
1992	194.4	2247.20	16.9	132.9
1993	198	2296.40	16.7	137.3
1994	201.9	2357.60	16.7	140.8
1995	205.4	2422.70	16.8	143
1996	210.4	2485.80	16.9	147.4
1997	216.6	2561.70	17	150.4
1998	215.5	2631.50	16.9	155.4
1999	220.5	2691.10	16.7	161.4
2000	225.8	2746.90	16.9	162.6
2001	235.3	2797.30	17.1	163.5
2002	234.6	2855.50	16.9	168.7
2003	236.8	2890.50	17	170.1
2004	237.2	2964.80	16.6	178.5
2005	241.2	2989.80	16.7	179.1

A summary of data relating to driving practice in the U.S. is presented in Table 12.1.[1] The total number of miles traveled per vehicle per year has risen only modestly over the past 45 years, from 9700 miles in 1960 to 12,400 miles in 2005. The number of vehicles on the road has increased however over the same period from 74.5 million in 1960 to 241.2 million in 2005. The number of miles traveled has risen accordingly, from 718.8 billion miles in 1960 to 2989.8 billion miles in 2005. Viewed as miles traveled per person per year, allowing for the growth in U.S. population from 180 million in 1960 to 297 million in 2005, the number of miles traveled per person per year has more than doubled over this period from 4000 miles per person per year in 1960 to more than 10,000 miles per person per year in 2005. Fuel economy averaged 12.4 miles per gallon in 1960 and changed little over the decade of the 1960s and the early 1970s. The oil crises of the 1970s triggered an increased demand for fuel-efficient vehicles, a pattern that developed first in the late 1970s and continued through the 1980s with fuel economy measures per vehicle rising from 12.2 miles per gallon in 1975 to 16.9 miles per gallon in 1991. There has been little change in fuel efficiency standards however since 1991, a trend attributed to the increasing popularity of pick-up trucks and heavy-duty sports utility vehicles (SUVs). Increases in the fuel efficiency of passenger vehicles, to close to 30 miles per gallon today, has been offset in recent years by increasing numbers of fuel-inefficient pick-up trucks and heavy duty SUVs.

We begin in Section 12.2 with a description of the internal combustion engine. The early history of the auto industry is treated in Section 12.3 highlighting the role of Henry Ford in developing an automobile—the Model T—that would be accessible to a relatively large fraction of the general public and not just to the super rich. We continue in Section 12.4 with an account of developments in the post World War II period. More recent history, highlighting the introduction of hybrid electric vehicles capable of running on a combination of electricity (generated on board) and gasoline, with their promise for increased fuel efficiency, is treated in Section 12.5. Section 12.4 includes also an account of potential future automotive prospects, with a particular emphasis on hybrid vehicles that can satisfy their requirements for electricity not only by generation on board but by plugging into the external electric grid (plug-in hybrids). It includes a discussion also of the prospects for electric vehicles whose requirements for electrical energy could be met by generation on-board using hydrogen fuel cells. Summary remarks are presented in Section 12.6.

12.2 THE INTERNAL COMBUSTION ENGINE

The energy released in an internal combustion engine is coupled directly (internal to the engine) to the performance of mechanical work. The internal combustion engine differs thus from the steam engine in that the energy liberated by burning fuel in the latter is several steps removed from the harnessing of this energy to perform useful work. The chemical energy released by burning fuel in a steam engine is converted first to thermal energy or heat. This thermal energy is used then to convert water to steam. Work is performed subsequently by the resulting steam. In the case of the internal combustion engine, work is carried out directly by expansion of the gases produced by combustion of the fuel.

For most cars in operation today, the energy source applied ultimately to turn the wheels is provided by what is known as a 4-stroke engine. An illustration of the separate elements of the 4-stroke engine is presented in Fig. 12.1.

The engine consists of a cylinder containing a piston to which is attached a rod (the piston rod) by which the reciprocal (up/down or back and forth) motion of the piston is transferred to a crankshaft and converted thus to rotational motion, delivered ultimately through a system of gears in the transmission and used to turn the wheels of the vehicle. A mixture of air and vaporized fuel is admitted to the cylinder through an open valve and, responding to the pressure of the atmosphere, the mixture is employed to trigger the first stroke of the engine. The piston moves down and the piston rod initiates rotation of the crankshaft. When the piston reaches the lowest point of its excursion, the valve admitting the fuel–air mixture is closed, at which point the cylinder is sealed. The piston then begins to move upward initiating the second stroke of the engine. The crankshaft continues to rotate and the fuel–air mixture is compressed in response to the decrease in the volume of the cylinder above the piston. When the piston reaches the high point of its reciprocal excursion,

Four-stroke cycle

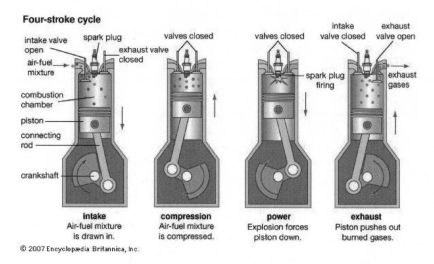

Figure 12.1   Illustration of the operation of a 4-stroke engine. From Encyclopedia Britannica, 2007. Reproduced from http://cache.eb.com/eb/ image?id=93572&rendTypeId=34, read August 23, 2007.

a spark provided by a high-voltage source of electricity is used to ignite the fuel. The pressure in the cylinder increases then in response to the high temperature gases produced by combustion of the fuel, triggering the third (power) stroke of the engine. The camshaft continues to rotate. When the piston reaches the low point of its trajectory, when the maximum work it can supply has been expended, a second, exhaust, valve is opened, allowing the products of combustion to exit the cylinder, to be emitted to the atmosphere through the tailpipe of the vehicle. The associated reduction in pressure forces the piston to ascend initiating the fourth stroke of the engine and then the process repeats. The camshaft undergoes two rotations in response to the double up/down (or twice back and forth) motion of the piston associated with the four strokes of the engine.

In cars manufactured prior to about 1985, the mixture of air and fuel delivered to the engine was supplied through what is known as a carburetor. The accelerator pedal of the car controlled the supply of air to the carburetor. Suppress the pedal, and the airflow was increased. The pressure in the carburetor was reduced accordingly as required by Bernoulli's Principle (Equation (8.1)): the lower the pressure, the greater the supply of fuel to the carburetor. In a well-tuned car, the admixture of air and fuel is just right to ensure complete combustion of the fuel when the air–fuel mixture is admitted to the engine cylinder. In cars manufactured more recently (post 1985), what is known as a fuel injection system is used to control the supply of air and fuel to the engine, dispensing with the need for a carburetor. Fuel in this case is pumped to the engine cylinder through a small nozzle. A computer regulated feedback loop employing an O_2 sensor is used to ensure the appropriate mix of air and fuel. Introduction of the fuel injection system, ensuring a more reliable appropriate supply of air and fuel to the engine, resulted in a significant improvement in the operation of automobiles not only in terms of fuel efficiency but also in terms of the quality of the exhaust gases emitted to the atmosphere.

The German inventor/engineer Nicolaus Otto (1832–1891), working with Gottlieb Daimler (1834–1900) and Wilhelm Maybeck (1846–1919), is usually credited as the inventor of the 4-stroke engine. In recognition of his contribution, the principle by which the 4-stroke engine operates is commonly referred to as the Otto cycle. The engine as originally introduced by Otto was fueled by gas (not gasoline) and was designed for stationary applications. It required a pilot light to ignite the gas and was consequently unsuitable for mobile applications. Otto was originally awarded a patent for his invention (in 1877) but the patent was subsequently ruled invalid by the German courts on the grounds that the concept for his engine had been anticipated earlier and described in a privately published pamphlet by the French inventor Alphonse Beau de Rochas (1815–1893). Otto would later introduce, in 1884, a liquid-fueled version of his engine, one in which the fuel was ignited by an electric spark setting the stage for the subsequent development of the modern gasoline-fueled automobile.

An alternative to the 4-stroke gasoline engine was developed and patented in Germany by Rudolf Diesel (1858–1913) in 1892. He was awarded a U.S. patent for his invention 6 years later, in 1898. Diesel was born in Paris of German parents. He received his early education in Munich and worked subsequently not only in Germany and France but also in Switzerland. His design was intended from the outset to improve the efficiency of the engine invented earlier by Otto. The diesel engine takes in air that is raised to high temperature by compression prior to admission of the fuel. The fuel is ignited directly by contact with the high-temperature compressed air without the need for a spark. The higher temperature of the fuel–air combustion mix in the diesel engine ensures that it can achieve a significantly higher efficiency than is possible given the lower compression ratio (and consequently lower operating temperature) of the conventional gasoline-powered 4-stroke engine. This advantage is achieved however at a cost in terms of a higher content of particulates and other pollutants in the exhaust. Admission of fuel before the temperature of the compressed air is high enough to ensure complete combustion results in production of relatively high levels of soot and uncombusted hydrocarbons. The higher temperatures of combustion in the diesel engine favor also emission of higher levels of nitrogen oxides, NO_x. These problems have largely been eliminated in diesel engines installed in more recent diesel-powered automobiles through a combination of more precisely controlled combustion, use of filters to trap particulates, and the introduction of catalytic converters to reduce emission of NO_x and other gaseous pollutants. Diesel-powered automobiles have achieved increasing popularity as a consequence, particularly in Europe, in response to the greater fuel efficiency offered by these vehicles, the higher cost of fuel in Europe as compared to the U.S., and the greater urgency accorded by European governments to the need to reduce emissions of CO_2.

Karl Benz (1844–1929) is usually credited with the invention of the first successful automobile, known as the Motorwagen. He received a patent for the Motorwagen in 1886, having earlier, in 1878, earned a patent for his invention of a 2-stroke version of the internal combustion engine. The Motorwagen had three wheels. It was steered by turning the single wheel located at the front with the two wheels at the rear used to support the weight of the passengers. The wheels were constructed of wood (yet another illustration of the importance of wood as late as the last quarter of the nineteenth century). The Motorwagen incorporated an accelerator pedal, a carburetor, and a battery-powered electrical ignition system and was fueled by gasoline. The cars designed and manufactured by Benz were marketed by Benz and Cie, the company Benz founded with partners in 1883. By the end of the century, Benz and Cie was established as the world's largest and most successful automobile company, producing no fewer than 572 cars in 1899.

The Daimler Motor Company, constituted in 1889 by Gottlieb Daimler (Otto's partner in the early development of the 4-stroke engine) to market the world's first four-wheeled automobile, would emerge eventually as the most significant competitor to Benz and Cie. Eric Jellinek, a rich Austrian businessman, was an important early customer for Daimler cars and an important intermediary in their sales outside of Germany. Jellinek adopted the name of his daughter, Mercedes, as a pseudonym when he participated using a Daimler automobile in one of the world's first automobile races in Nice, France, in 1899. Under pressure from Jellinek, the name Mercedes was adopted for all engines manufactured by Daimler after 1902. The Benz Company would merge eventually with Daimler Motors in 1926 to become the Daimler-Benz Company. Cars produced by the merged company were subsequently marketed as Mercedes-Benz. Daimler-Benz combined with the American automobile manufacturer, Chrysler, in 1998 to form the Daimler-Chrysler Company, a marriage that would be dissolved less than 10 years later.

12.3 EARLY HISTORY OF THE AUTOMOTIVE INDUSTRY IN THE U.S.

Automobiles received a rather tepid reception when they were first introduced in the U.S. in the last few years of the nineteenth century. They were considered initially a play toy for the rich, accentuating, as Evans et al. (2004) summarized it, "the gulf between the pretentious rich and the working poor." Woodrow Wilson, then President of Princeton University, later President of the United States, is quoted (Evans et al., 2004) in 1907 to the effect that "nothing has spread socialist feeling in the country more than the use of the automobile, a picture of the arrogance of wealth." Laws were passed to restrict the influence and spread of the automobile. The State of Vermont, for example, decreed that every motorist driving a car should be preceded by "a person of mature age" carrying a red flag. Of 4200 cars built in the U.S. in 1900, only one-quarter was propelled by internal combustion engines. Most were driven by steam.

The twin brothers Francis and Freeland Stanley, schoolteachers from Maine, developed the Stanley Steamer and for a number of years held the record for the fastest cars on the road. Their Stanley Rocket set records for speed on the racetrack in Daytona, Florida, in 1906 realizing speeds in excess of 125 mph.[2] The steam-powered car failed, however, to achieve commercial success. It was difficult to start. It had problems working under cold weather conditions (the water required to produce the steam was given to freezing), and an accomplished engineer was required to operate it. Within a few years, the steam-powered car was abandoned, as was the electric car that was an early competitor to the internal combustion-fueled automobile (the demise in this case related to the difficulty of finding a battery that could allow the car to travel for more than a few miles before the battery had to be recharged), and America would begin its long-lasting love affair with the gasoline-fueled automobile, 34,528 of which were sold in the U.S. in 1910 alone. The person who was most influential in bringing about this momentous change in attitude was Henry Ford (1863–1947), responsible for the development of what became known as the People's Car, the Ford Model T.

Henry Ford was born in Michigan a few miles from the city of Detroit that would emerge later as the center of the U.S. automotive industry. His father's family, poor tenant farmers, had emigrated from Cork, Ireland, in 1847 to escape the Irish famine. Unlike most of the poor people who fled Ireland at that time, the Ford family was Protestant, a circumstance that might well have contributed to the eventual success of the family in a country where the church one attended on Sunday was still regarded as a social discriminant. Ford, having served several unsuccessful apprenticeships as an engineer in Detroit, returned to the family farm in 1883. He took a job with Westinghouse fixing farmers' steam traction engines, taking a variety of classes in night school in Detroit—in business administration, accounting, and typing (Evans et al., 2004). At age 25, in 1888, he married Clara Jane Bryant and for a time it appeared that he might settle down to life as a farmer.

The conventional wisdom is that Ford was first inspired to develop an automobile in 1890 when he had the opportunity to inspect first-hand a 4-stroke, gasoline-powered, Otto engine installed at a soda plant in Detroit. He set to work immediately to construct his own gasoline-powered engine. Three years into their marriage, in 1891, the Ford family abandoned their farm in Dearborn and moved to rental quarters in Detroit where Henry accepted employment as the nighttime supervisor of electricity power transmission at the Edison Illuminating Company of Detroit. He rose quickly through the ranks of the company assuming eventually the position of chief engineer overseeing a staff of 50 with a salary of $1000 a year, handsome remuneration for the time. In his spare time, he rented space and with help from a number of his friends he continued to work on perfecting his engine and on building his first car. Success came in 1896, when Ford drove his first car out of the shed in which he had built it, knocking down the door of the shed in the process (he had failed to notice that the width of his car was greater than the width of the door to the shed). The car weighed 500 lb and was equipped with a two-cylinder, four-horsepower engine. For its inaugural voyage, Ford drove the car 8 miles to visit his family in Dearborn taking along his wife Clara with baby son Edsel on her lap. He averaged close to 20 mph on the trip.

Ford was introduced in 1896 to Thomas Edison, a person he had idolized from his youth. Edison listened to Ford's plans for the automobile. After hearing him out, Edison pounded his fist on the table and advised Ford (Evans et al., 2004) that: "Electric cars must keep near to power stations. The storage battery is too heavy. Steam cars won't do either, for they have a boiler and fire. Your car is self-contained, carries its own power plant, no fire, no boiler, no smoke and no steam. You have the thing. Keep at it." The reaction he received from the great man provided just the encouragement Ford needed to proceed with his vision. He set up a machine shop, investing his personal funds on machine tools. Two years later, in 1898, he produced his second car and a year later his third and the business community of Detroit, led by the mayor, William Maybury, a Ford family friend, was persuaded to invest funds to set him up to manufacture cars at the Detroit Automobile Company. Ford, accepting a small stake in the new enterprise, elected to terminate his comfortable employment at the Edison Company and to set out on his own. The new enterprise would prove, however, a failure. Ford would devote his efforts, not to building expensive cars that could be sold to the few that could afford them as his backers desired, but to building heavy delivery trucks and, with even greater enthusiasm, a race car. The Detroit Automobile Company was dissolved in 1901 but the directors of the Company permitted Ford to continue at the plant to complete construction of his racing car.

Ford would achieve unexpected fame as both racecar designer and driver when against all odds he would defeat Alexander Winton, considered the best racecar driver of his generation, in a highly publicized event staged around a newly constructed dirt racetrack at Grosse Point, Michigan, in 1901. The Henry Ford Company was incorporated a month later with the goal to develop a lightweight car that could sell for $1000. But Ford's interest in building racecars would win out once again over the objectives of his investors to make money. He proceeded to build a giant-sized racing car with four cylinders, delivering as much as 70 horsepower that would set a record for speed in October 1902. The investors lost patience with Ford, bringing in Henry M. Leland, an expert on precision machine tools, to run the Company. Ford resigned, or was fired. He was awarded severance pay of $900 but was accorded the right to use his name in any subsequent business venture. The Henry Ford Company was renamed the Cadillac Motor Company, honoring the founder of Detroit, Antoine de la Cadillac. The Ford Motor Company, with initial capital of $28,000, was formed by Ford together with eleven other investors in 1903.

A Canadian, James Couzens, was brought in to oversee the finances of the new company and would play a critical role in its future success. Historian Douglas Brinkley (2003) ranked the founding of the Ford Motor Company "among the most significant events in twentieth-century U.S. industrial history." From the outset, the objective was to build affordable cars for the mass market. A series of cars, named after the letters of the alphabet, were produced in quick order, the first of which, the Model A, was sold in July 1903, a little more than a month after the date when the Company was formally incorporated. It was sold to a Chicago dentist, Ernst Pfenning, for $850. By the end of its first year in business, the Ford Motor Company had sold 1000 cars and 125 people were employed at its assembly plant in Detroit. It was immediately profitable: investors recovered 100% of their initial investment in the form of dividends over the first 15 months of the Company's operation. But its greatest success would come later, in 1908, with the introduction of the Model T.

From the outset, Ford's ambition for the Model T was that it should be large enough to accommodate five passengers, light enough to be able to achieve moderately high speeds while minimizing wear and tear on the tires. It should be simple to operate, and reliable, in recognition of the variety of uses to which it would be applied and the uneven quality of the roads over which it would have to travel. Most cars at the time were manufactured using standard steel. To meet the requirement for strength and light weight, Ford focused on a vanadium–steel alloy that had been employed earlier in French luxury cars. To ensure an adequate supply of this material, he had to invest in his own dedicated steel plant. Other requirements for his model car would be satisfied by careful attention to detail in design. Cost containment would be an overriding goal. The car had to be inexpensive, available to as large a cross-section of the public as possible. Ford's objective was that people making more than $2500 a year should be able to afford a Model T. The first Model T introduced in 1908 sold for $850. By the end of 1924, Ford had succeeded in reducing the price to $290.[3] His success in lowering the cost of the Model T was largely attributed to the innovations he was able to introduce over time in its manufacture.

The most important innovation was the introduction of the assembly line to speed up and increase the efficiency of the manufacturing process. Initially, the idea was to specialize the tasks assigned to individual workers. The worker who inserted a bolt was not required to tighten the bolt. That task was left to another. A second worker would be charged with tightening the bolt. Ford engineers conducted detailed analyses of the entire manufacturing process measuring the time required for workers to complete individual tasks. A series of incremental improvements was implemented culminating eventually in the introduction of the conveyor belt. The unfinished chassis of the car moved steadily along the conveyor belt. As it passed individual workers they completed the specific task to which they were assigned. The time available to complete the task was regulated by the speed with which the chassis moved along the belt, a speed that was optimized to ensure maximum production with minimum expenditure of labor. The number of man-hours required to assemble the car decreased steadily from 12.5 h to 93 min and with it the cost for production of the car. Ford summarized the system thus (Evans et al., 2004): "save ten steps a day for each of 12,000 employees, and you will have saved fifty miles of wasted motion and misspent energy." Ford was not the first to introduce the concept of a conveyor belt to speed up the manufacturing process. He was first, however, to apply the concept to the automobile industry, an innovation for which he was richly rewarded. By 1927, when the last car rolled off the production line, no fewer than 15 million Model Ts had been sold worldwide with assembly lines operating in no fewer than 21 different countries (Evans et al., 2004).

Ford was not only an innovator in manufacture, he was also a trendsetter in terms of the business practices and the labor relations he adopted. Relying initially on outside suppliers for many of the parts for his cars, he switched eventually to a more vertically integrated structure through which he controlled the supply of most if not all of the intermediate components needed to produce his cars. He pioneered in the establishment of a network of dealers to market his cars. And in the face of a national labor crisis in 1914, he responded by reducing the work day for his employees from 9 to 8 h, rewarding them with a then unprecedented share of the profits of the Company, an increment in pay that would amount to no less than $5 a day for the lowest paid worker. Ford proved also to be a tough opponent when his interests were threatened. George Seiden, a Rochester lawyer, following a series of complex legal maneuvers, was awarded a U.S. patent for a road vehicle fueled by gasoline in 1895. He assigned the patent to a group of wealthy New York financiers in 1897. An organization known as the Association of Licensed Automobile Manufacturers (ALAM) was set up demanding licensing fees amounting to 1.25% of sales for all cars manufactured in the U.S. The majority of car manufacturers reluctantly went along with this demand. Ford, however, declined to join. He was taken to court by ALAM in 1909. The patent was ruled initially to be valid. Ford appealed the verdict and the decision was reversed in 1911, largely as the result of the ability of the Ford lawyer to demonstrate that an expert witness who had testified for the plaintiffs in the original trial had changed his mind in the interim, according credit belatedly but deservedly for the early development of the gasoline-powered automobile, not to Seiden, but to Otto and Benz.

Ford bought out the minority shareholders of the Ford Motor Company in 1919, taking the Company private. The Company continued to prosper but the essentially exclusive emphasis on the production of sturdy cheap cars by Ford up to 1927, when the last Model T was produced, provided an opportunity for competitors to make inroads by moving into niches left vacant in the middle and higher income segments of the market. The General Motors Company and the Chrysler Motor Company emerged as Ford's major competitors in the late 1920s and early 1930s as the Model T came to the end of its useful life.

The General Motors Company was founded in 1908 incorporating the Buick Motor Company founded by David Dunbar Buick in 1903. The Company grew rapidly drawing under its banner within 2 years the Olds Motor Vehicle Company (formed by Ransom E. Olds in 1897), the Cadillac Motor Company (successor to the original Henry Ford Company), and the Rapid Motor Company of Pontiac, Michigan (predecessor of GMC truck). The Chevrolet Motor Company of Michigan was formed in 1911 and was absorbed subsequently into General Motors in 1918. Following Edison's earlier example at Menlo Park, General Motors elected early to invest in a dedicated research facility. The General Motors Research Department, charged with providing engineering services to all constituent parts of the company, was established in 1911. Pierre S. du Pont, then Chairman of the Board, became President of General Motors in 1920 and was succeeded 3 years later in both roles by Alfred P. Sloan. Sloan would articulate his vision for General Motors in a message to shareholders in 1924: to produce "a car for every purse and purpose." General Motors would surpass Ford as the world's largest automobile company in 1927. A few years later, in 1933, Ford would drop to number three, supplanted by the Chrysler Motor Company formed by Walter P. Chrysler in 1925. Twenty-five years would elapse before Ford would recover the number two position in 1950.

Ford Motor Company, taken private by Henry Ford in 1919 as noted above, resumed its status as a public company in 1956. The Ford family continued, however, to exercise a controlling interest through their holdings of a special class of preferred stock. William Clay Ford Jr., great grandson of the founder Henry Ford and great grandson also of Harvey Firestone founder of the Firestone Tire and Rubber Company, was appointed Chairman of the Ford Motor Company on January 1, 1999. In 2001, he assumed the additional positions of President and Chief Operating Officer (CEO). Bill Ford, as he is known, is a committed environmentalist. In 2000, he announced the goal of improving the fuel efficiency of the Company's light trucks and SUVs by 25% by mid decade. The Company was obliged to back down from this objective, however, in 2003 when it admitted that it would be unable to meet the target given competitive market conditions and the public's continuing preference for powerful engines and heavy-duty trucks. Ford stepped down as President and CEO on September 5, 2006 naming former Boeing executive Alan Mulally as his successor. He continues to serve as Chairman of the Board's Executive Committee.

12.4 DEVELOPMENTS POST–WORLD WAR II

The automobile industry in the U.S. enjoyed an unprecedented boom in the period immediately following the end of World War II. Sales reached 6.7 million vehicles in 1950, climbing steadily to 9.3 million by 1965. Changes in the social structure of the nation contributed to this expansion in demand. Large numbers of new family units were formed and families increasingly migrated to the suburbs triggering the phenomenon of urban sprawl. New institutions—the drive-in movie theatre, the shopping mall, and fast food restaurants such as McDonalds for example—were introduced to cater to this changing pattern of lifestyle. The automobile became not simply a luxury but a necessity. Cheap prices for gasoline (see Fig. 6.4) contributed to the growth. Cars grew in size and new models, notably the station wagon and the convertible, were introduced to cater to the changing pattern of demand. Growth lasted until the early 1970s when the first of the two oil shocks that occurred in that decade brought growth to at least a temporary halt. The changing pattern of oil consumption displayed in Fig. 6.3 over the 1970s and early 1980s mirrors the changes in the behavior of the American driving public that set in over this period.

The immediate response to the oil shock was a demand for more fuel-efficient cars. Detroit was slow to respond to this abrupt change in market conditions and much of the new demand was met by imports. Smaller, more fuel-efficient cars increased their share of the U.S. market, climbing to 26% of the total domestic sales by 1980. Much of this increase in demand was satisfied by imports from Japan.

The automobile industry played a major role in the recovery of Japan following the devastation to its economy resulting from World War II. The Ministry of International Trade and Industry (MITI) saw the automobile industry as a potential catalyst to trigger the recovery of Japan's post war economy. The domestic car industry, led by Toyota and Nissan, succeeded in producing 32,000 vehicles by 1950. To put this in context, production of automobiles in Japan in 1950 amounted to no more than one and a half days of production in the same year in the U.S. The industry took off, however, following the end of the American occupation in 1952. Japan focused from the outset on the potential opportunities available to it as an exporter of high quality automobiles. By 1960, production of automobiles in Japan had risen to 482,000 vehicles, 8% of which were exported. A decade later, Japan produced 5.3 million cars, 1.1 million of which (21%) were sold abroad. By 1980, Japan had emerged as the world's largest automobile producer with 54% of its production sold outside Japan. And, as Landes (1998) points out, Japan was not competing simply on the basis of price but rather on the basis of quality.

Detroit had been spoiled by success. Henry Ford's initial accomplishment in forging a constructive relationship between management and labor as a result of the initiative he took in 1914 reducing the work day for his employees and rewarding them with a share of Company's profits, gave way to a more adversarial relationship. Workers did no more and no less than that to which they were obligated under union negotiated contracts. These legalistically binding but inflexible arrangements notably inhibited spontaneity in innovation on the production line. Japan learned from Henry Ford and simply took his early initiative several steps further. Success of the company was inculcated as the overriding objective, more important than the success of either labor or management viewed separately. What emerged was a much more flexible system of design and production. Japanese manufacturers developed a keen sense of what the market needed and the ability to respond rapidly to changing customer preference. Where it took 60 months to craft and test a new model in the U.S., Japanese manufacturers succeeded in doing it in 46 months. And they were able to ensure quality of production much faster than their American competitors, within 1.4 months of introducing a new model in Japan as compared to 11 months in the U.S. (Womack et al., 1990). They had the advantage also of more modern manufacturing plants since their industry was relatively new as compared to the more mature status of the industry in the U.S. And they exploited these advantages to the limit. In July 2007, for the first time ever, less than 50% of the cars sold in the U.S. were produced by the Big Three—General Motors, Ford, and Chrysler. Japanese companies, notably Toyota, were the primary beneficiaries of this changing of the guard. Now, to an increasing extent, these Japanese cars are being produced, not in Japan but in the U.S., by Japanese companies free of the pension and benefit rights to which U.S. manufacturers are obligated with respect to workers who are either retired or who were laid off involuntarily during the slowdown in the U.S. industry that began in the 1980s.

The mood of the American consumer changed in the 1990s as quickly as it did in the 1980s. The oil crisis was over. The price of gasoline declined and Americans resumed their love affair with large cars. To an increasing extent, light trucks and SUVs emerged as the motor vehicles of choice for American consumers and, as indicated at the outset, fleet average fuel economies dropped accordingly. It became a self-sustaining prophecy. People bought large cars or trucks because they viewed it as simply unsafe to drive a small light car when so many of the cars on the road were multi-ton behemoths. Gasoline prices, however, are again on the rise, with increasing concern about the vulnerability of the U.S. economy to supplies of oil from regions of the world viewed as politically unstable. There are worries also that the global climate system may be destabilized by increasing emissions of CO_2. The question is whether this most recent change of mood will prove durable or whether it will turn out to be as ephemeral as the change that took hold in the 1980s.

The changing configuration of automobiles marketed in the U.S. since the introduction of Ford's Model T is illustrated in Fig. 12.2. The small cars of the pre-World War II period (exemplified by the Model T) gave way to the giant vehicles of the 1950s and 1960s, to be replaced temporarily by more fuel-efficient imports in the 1980s, only to give way once again in the 1990s and early years of the 2000s to gas guzzling trucks and heavy duty SUVs exemplified for example by the tank-like construction of the General Motors Hummer. The critical question is whether the future lies with the Hummer and similar behemoths or with more fuel-efficient alternatives such as the electric hybrid (both plug-in and free-standing), a topic addressed in the section that follows.

12.5 RECENT DEVELOPMENTS AND FUTURE PROSPECTS

An important innovation in the automotive industry in recent years has been the introduction of what are known as hybrid electric vehicles (HEVs)—cars, trucks, and SUVs that can be driven either electrically or by a conventional gasoline-powered internal combustion engine. Japanese companies (notably

(a) The 1908 Ford's Model T. Reproduced from http://www.hfmgv.org/exhibits/showroom/1908/model.t.html, read August 28, 2007.

(b) The 1957 Ford. Reproduced from http://www.loti.com/57ford.jpg, read August 28, 2007.

(c) The 1986 Ford Taurus. Reproduced from http://www.hfmgv.org/exhibits/showroom/1986/taurbig.jpg, read August 28, 2007.

Figure 12.2 Illustrating the changing configuration of automobiles.

(d) The 2000 Hummer H1. Reproduced from http://www.edmunds.com/used/2001/hummer/h1/100000509/photos.html#, read August 28, 2007.

(e) The 2007 Toyota Prius. Reproduced from http://www.toyota.com/prius/exterior.html, read August 28, 2007.

Figure 12.1 *continued*

Toyota and Honda) pioneered in the early development of the hybrid. The Prius, produced by the Toyota Motor Company, continues to enjoy the status as the most popular hybrid accounting for a little more than 50% of all hybrid vehicles sold in the U.S. in the first half of 2007. On June 7, 2007 Toyota announced the sale of their one-millionth hybrid vehicle worldwide, more than half of which were sold in the U.S. and Jim Lentz, Executive Vice President of Toyota Motor Sales U.S., predicted that combined sales of Toyota and Lexus hybrids would surpass a quarter of a million vehicles in the U.S. during calendar year 2007.[4] J.D. Powers and Associates, the global marketing information company, projects that there will be as many as 65 different hybrid models on the market by 2010—28 cars and 37 light trucks—and that these vehicles will account then for 4.6% of the U.S. market.[5]

The Prius incorporates a conventional internal combustion engine working in consort with an electric motor supplied with energy from a nickel-metal-hydride (NiMH) storage battery. The battery is charged on board the vehicle as required using an electric generator run by the internal combustion engine. The overall energy efficiency of the vehicle is increased by capturing energy that would normally be dissipated as heat on braking the car, directing a portion of this energy to run the generator. When the car is forced to come to a halt as in stop-and-go city driving, the engine turns off, conserving fuel. When it is time to restart, the electric motor is used to provide the necessary start-up power. Torque is delivered instantaneously to the wheels by the electric motor minimizing fuel that would be expended otherwise in a conventional car as the internal combustion engine works to generate the torque required to provide the necessary acceleration. The hybrid enjoys its major advantage under city driving conditions where fuel can be conserved when the car is stationary and where a major fraction of the propulsion may be provided by the electrical power system. The U.S. Environmental Protection Agency (EPA) rates the fuel economy of the 2007 Toyota Prius 4-door Hatchback, which provides comfortable accommodation for five passengers, at 60 miles per gallon (mpg) for city driving, 51 mpg for driving on the highway. By way of comparison, the Toyota Highlander Hybrid SUV is rated at 31/27, the standard gasoline-powered Toyota Highlander at 20/25, the Ford Escape Hybrid SUV at 32/29, and the standard Ford Escape at 21/24 where the first of the numbers quoted here refers to the fuel efficiency (mph) for which the vehicle is rated for city driving, the second the rating for driving on the highway. For a complete list of fuel efficiencies for vehicles rated by the EPA, see[6].

The savings in CO_2 emissions that could be realized if the entire U.S. automotive fleet was converted to HEVs would amount to about 30% with additional benefits in terms of improvements in air quality (Evans, 2007). Even greater savings could be realized if HEVs were replaced by so-called plug-in hybrids. With HEVs, the energy used to drive the vehicles is provided primarily by an oil-derived fossil source, mainly gasoline at present to be complemented potentially in the future by increased use of more efficient diesel (complemented by the current use of ethanol as an additive to gasoline accounting typically for about 10% of fuel consumed in the U.S. but with minimum savings in terms of greenhouse gas emissions as discussed in Chapter 15). With plug-in hybrids, a portion of the electrical energy consumed by the vehicles would be derived from the grid and, depending on the energy source used to generate this electricity, the savings in terms of CO_2 emissions could be even more significant (more on this later in the wrap-up Chapter 16).

The Electric Power Research Institute(EPRI) reported in 2001 a study of the potential advantages that could be derived from the introduction of plug-in hybrids (EPRI, 2001). Using a simulation program (ADVISOR) developed by the National Renewable Laboratory of the U.S. Department of Energy, they studied four cases: a conventional vehicle (CV), a standard HEV without plug-in capability (HEV 0), a plug-in vehicle with sufficient battery capacity to provide for an all-electric range of 20 miles (HEV 20), and a plug-in option with an all-electric range of 60 miles (EV 60). It was assumed that all three hybrid vehicles would use state-of-the-art NiMH batteries and that the energy deployed in braking could be recovered and used to recharge the battery. Performance characteristics were taken as comparable for all four vehicles: gasoline storage on-board sufficient to provide for a range between refueling of 350 miles; a minimum top speed of 90 mph; and a time to accelerate from 0 to 60 mph of less than 9.5 s. Properties of the four vehicles defined as meeting these constraints by the ADVISOR model are summarized in Table 12.2 adopted from Evans (2007). The power required to be delivered by the gasoline engine decreases steadily from the conventional gasoline-powered vehicle (CV) to the 60-mile range plug-in hybrid option (HEV-60), from 127 kW to 38 kW, and the weight and complexity of the internal combustion engine required to meet the stated driving requirements are reduced accordingly. The HEV 0 vehicle requires battery capacity of 2.9 kWh as compared to 17.9 kWh for the HEV 60. The weight of the HEV 60 is approximately 6% greater than the weight of the conventional gasoline-powered option (CV), 1782 kg as compared to 1682 kg.

The EPRI analysis explored also the implications for emission of CO_2 of the four choices of vehicles considered in their study, accounting not only for emissions associated with consumption of the gasoline used in driving the cars but also for emissions associated with refining of the crude oil to produce the gasoline. For the plug-in options, they assumed that the electricity taken from the grid was generated by combined cycle power plants consuming natural gas as feedstock with an efficiency for conversion of chemical to electrical energy of 50%. The conclusion was that the reduction in emissions associated with the HEV 20 option as compared to the conventional gasoline-powered auto-mobile could amount to as much as 50% (as compared to 30% for the conventional HEV 0) and could increase to 60% in the case of the HEV 60. If the electricity supplied by the grid to power the batteries of the plug-in HEV was produced using non-fossil sources of energy, nuclear or wind power for exam-ple, it is clear that the savings could be much greater, approaching 100%

Table 12.2 Properties of vehicles studied by EPRI (2001).

Vehicle	CV	HEV 0	HEV 20	HEV 60
Gasoline engine power (kW)	127	67	61	38
Electric motor power (kW)	–	44	51	75
Battery capacity (kWh)	–	2.9	5.9	17.9
Vehicle mass (kg)	1682	1618	1664	1782

(in a more comprehensive analysis it would be important to consider the fossil energy consumed in producing the batteries and the extra energy expended in the manufacture of the weightier plug-in vehicle). An estimate of the relative cost of driving using gasoline in a conventional internal combustion engine as compared to driving using electricity taken from the grid is presented in Box 12.1.

The hydrogen fuel cell offers another potential future source of automotive power. The fuel cell provides a means for direct conversion of chemical energy to electricity. It is composed of two electrodes, an anode and a cathode, separated by an electrolyte. Hydogen is supplied to the anode which is coated typically with a noble metal catalyst, most commonly platinum in current designs. Interacting with the noble metal catayst, the hydrogen molecules are converted to protons and electrons:

$$H_2 \rightarrow 2H^+ + 2e \tag{12.1}$$

The electrons formed in reaction (12.1) pass through an external circuit where they are employed to drive an electric motor. The protons produced in reaction (12.1) travel through the electrolyte which is composed of a material that allows

---

Box 12.1

Estimate the relative costs for driving using gasoline in a conventional internal combustion engine as compared to driving using a battery charged with electrical energy taken from the grid.

Solution:

The energy content of a gallon of gasoline is equal to about 1.15×10^5 BTU equivalent to 33.7×10^1 kWh (see Table 3.1 for factors used to convert from BTU to kWh).

Approximately 80% of the energy content of gasoline consumed in a motor vehicle is wasted, converted to heat. Only 20% is employed ultimately to turn the wheels. It follows that the energy delivered to the wheels by a gallon of gasoline in a conventional internal combustion engine run motor vehicle is equal to about 6.7 kWh.

Storing energy in a battery and subsequently employing it to drive a motor vehicle is much more efficent. Losses in this case amount to only about 10%. It follows that to obtain the same driving power as that derived from a gallon of gasoline we would need about 7.4 kWh of electricity.

As indicated in Chapter 3, the cost for electricity delivered to my home in Cambridge in January 2006 amounted to 19 cents per kWh (17.8 cents per kWh in August 2007). The cost to obtain the same driving capacity with a battery charged with grid supplied electricity as that delivered by a gallon of gasoline would amount in this case to $1.41, approximately half of the retail price for a gallon of gasoline prevailing in Massachussets in August 2007. Clearly, at present prices for gasoline and electricity, there is a future for plug-in hybrids.

protons to pass unimpeded while inhibiting passage of electrons. The material most commonly employed for this purpose is a fluorocarbon-based polymer produced by the du Pont Company under the tradename Nafion. The protons and electrons arriving at the cathode combine then with oxygen molecules delivered in an air stream to the cathode, forming water, completing the composite reaction:

$$2H_2 + O_2 \rightarrow 2H_2O \qquad (12.2)$$

The chemical energy released in reaction (12.2) is converted to electricity with an efficiency of about 50%.

The challenge with the hydrogen fuel cell option is to identify a suitable source of hydrogen. In prototype hydrogen fuel cell cars currently under development by a number of auto companies, the hydrogen is produced on board the cars by reforming methane in natural gas. The efficiency for conversion of the hydrogen in CH_4 to H_2, however, is only about 50%. Combining the efficiencies for conversion of CH_4 to H_2 with the inherent efficiency of the fuel cell, the overall efficiency of the composite system is not much better than that of a modern diesel or gasoline-powered internal combustion engine and less than that of hybrids already in service. Further, if the hydrogen is supplied from methane, the savings in terms of CO_2 emissions would be minimal.

A number of writers, Rifkin (2002) for example, have trumpeted the merits of a future hydrogen-based energy economy. The hydrogen could be formed by hydrolysis, by conversion of water to H_2 and O_2 using electricity, with an efficiency potentially as high as 90% (the current state-of-the-art is somewhat less than that, between 70% and 80%). If the electricity was produced using non-fossil sources of energy, nuclear or wind for example, this could certainly contribute to a significant reduction in emission of greenhouse gases such as CO_2. There would be problems, however, in ensuring an adequate supply of H_2 to meet local and regional demand. It would require an entirely new distribution and delivery system. While the energy per unit mass available in H_2 is superior to that of gasoline by a factor of 2.6, the disadvantage is that H_2 at ambient temperatures and pressures is a gas. Consequently a much larger volume would be required to store the quantity of H_2 needed to replace an equivalent supply of energy in the form of gasoline. Even if the hydrogen were maintained at a pressure as high as 200 atmospheres, the volume discrepancy favoring gasoline would exceed a factor of 16 (Evans, 2007). To store the quantity of hydrogen on board a fuel cell powered vehicle required to accommodate an acceptable driving range would require high pressure cylinders the mass of which could add as much as 25% to the total weight of the car and would take up space equivalent to or even greater than the space currently allocated to the car's trunk (Evans, 2007). There would be an inevitable increase in the price of materials needed to construct the body of the car. If not compensated by an even greater reduction in the cost of the fuel cell drive system as compared to the drive system for the conventional gasoline-powered car or for the hybrid, it is difficult to envisage a rosy future for the hydrogen fuel celled option particularly in view of the major expense that would be required to ensure the infrastructure required to supply the necessary hydrogen.

12.6 CONCLUDING REMARKS

The automobile has evolved over the past century from a luxury to a necessity. The basic technology introduced more than a hundred years ago by the German engineers Otto, Daimler, Diesel, and Benz continues to retain its status as the technological backstop of the industry although advances in the interim have markedly enhanced the efficiency with which automobiles are produced, the ease with which they can be operated, their reliability, and the comfort level they afford to modern day consumers. We have come a long way from Henry Ford's model T but we are now at a crossroads.

We face two serious challenges. Cars have increased in size and numbers, not only in the U.S. but also worldwide. The automotive industry for example is one of the fastest growing sectors of the Chinese economy and the pattern promises to repeat in other parts of the developing world. With the increase in numbers and sizes of cars worldwide has come an increased demand for oil to supply the gasoline and diesel fuel these cars need to operate. The bulk of this demand must be met by sources derived from notably unstable regions of the world raising the risk of serious economic, and even social, dislocations that could be triggered by sudden interruptions in supply. The oil crises of the 1970s gave us a taste of what this could imply but our present vulnerability is now many times greater given the global reach the automobile has achieved in the interim.

The second challenge relates to the impact on global climate change of the carbon dioxide gases emitted by automobiles, trucks, and SUVs. The transportation sector is a major contributor to the contemporary increase in the concentration of atmospheric CO_2. Our ability to arrest growth in the use of fossil fuel sources of energy in the transportation sector, or indeed to reverse this growth, will have much to do with the success we might hope to achieve in the future in addressing the issue of climate change.

Much of the increase in consumption of gasoline in recent years can be attributed to the increased popularity of trucks and heavy-duty SUVs in preference to conventional automobiles. While the use of these vehicles for commercial purposes can be justified, it is clear that much of their use is not for this application but rather reflects consumer discretionary choice. A modest increase in gasoline prices, as has occurred in recent years (prior to more rapid rise over the more immediate past), is unlikely to reverse the trend. Even at 10 mpg, the cost of operating one of these vehicles for a year is unlikely to prove a determinative factor in their future market success. The expense for gasoline required to drive 10,000 miles a year at $3 a gallon is but a small fraction of the cost of operating these vehicles, less than 10% of the purchase price. A surcharge on the price for purchase of these vehicles indexed to fuel consumption, if sufficiently prohibitive, could be successful, however, in reversing the trend.

Revenue raised through such a surcharge could be used to offset other tax levies or to subsidize public investments that could promise a more productive return. Use of the vehicles for legitimate commercial purposes could be exempted from any such surcharge.

We discussed the advantages of the hybrid vehicle in terms of its fuel economy and related reduction in emission of pollutants, not only CO_2 but also NO_x and hydrocarbons that contribute to local and regional air quality (notably elevated levels of ozone). Particularly attractive, we judged, would be the introduction of a plug-in option, especially if this could be linked to the use of electricity generated using non-fossil sources of energy such as nuclear and wind. The higher price for these vehicles represents an impediment to their greater current market acceptance. Payback times in the U.S., compensating for their higher price allowing for their greater fuel economy, are estimated at 5 years or longer even with existing, historically elevated, prices for gasoline. Subsidies introduced a few years ago in the U.S. to encourage increased use of so-called alternative vehicles (mainly hybrids) were tied to the number of such vehicles that were sold subject to a sunset clause requiring that the subsidy lapse once sales surpassed a specified threshold. It would be constructive to eliminate this restriction. The subsidy for fuel-efficient vehicles could be tied to a corresponding surcharge on fuel-inefficient vehicles with minimal impact on net revenue expense.

We discussed also the merits of a potential future technology option in which vehicles would be driven electrically with electricity generated on board the vehicles by hydrogen fuel cells. If the hydrogen were produced using a fossil energy source such as natural gas, the current preferred option, we concluded that the advantages of the fuel cell technology would appear to be limited, that the plug-in hybrid might offer a better choice in terms of the savings in fossil fuel use that could be realized. If the hydrogen could be supplied by non-fossil sources such as wind or nuclear, the situation could be different. We noted though the significant challenges that would have to be met to ensure a reliable supply of hydrogen, not only in terms of its production and distribution but also in terms of the requirements that would have to be met to ensure an adequate supply of fuel stored on the vehicle, including the implications this would have for increased weight and consequently material costs for the vehicle. Ultimately, the fate of hydrogen fuel cell vehicles will depend on considerations of economics, costs to produce the vehicles, costs to establish the infrastructure needed to support them, and costs for competitive options such as the plug-in hybrid. It is our judgment that, at least at present, the advantage lies with the plug-in hybrid option though both options could contribute to important savings in terms of emissions of CO_2 with ancillary benefits in terms of improvements in air quality.

We did not include in this chapter any discussion of prospects for replacing fossil-derived fuels such as gasoline and diesel with alternatives such as ethanol or methanol produced from biological feedstocks. Treatment of this option is postponed to Chapter 15.

NOTES

1 (http://www.bts.gov/publications/national_transportation_statistics/, read August 8, 2007).

2 (http://www.autolife.umd.umich.edu/Environment/E_Overview/E_Overview, read August 13, 2007).

3 (http://www.princeton.edu/~hos/h398/readmach/modeltfr.html, read August 14, 2007).

4 (http://www.toyota.com/about/news/product/2007/06/07-1-hybridsales, read August 17, 2007).

5 (http://www.jdpower.com/corporate/news/releases/pressrelease.aspx?ID=2007127, read August 17, 2007).

6 http://www.fueleconomy.gov/feg/FEG2007_GasolineVehicles.pdf, read August 17, 2007.

REFERENCES

Brinkley, D. (2003). *Wheels for the World: Henry Ford, His Company, and a Century of Progress, 1903-2003.* New York: Viking Press.

EPRI (2001). Comparing the benefits and impacts of hybridelectric vehicle options. Electric Power Research Instiute Report 1000349.

Evans, R.L. (2007). *Fueling Our Future: An Introduction to Sustainable Energy.* Cambridge: Cambridge University Press.

Evans, H., Buckland, G., and Lefer, D. (2004). *They Made America: From the Steam Engine to the Search Engine: Two Centuries of Innovators.* New York: Little Brown and Company.

Landes, D.S. (1998). *The Wealth and Poverty of Nations: Why Some are so Rich and Some are so Poor.* New York: W.W. Norton.

Rifkin, J. (2002). *The Hydrogen Economy: The Creation of the Worldwide Energy Web and the Redistribution of Power on Earth, the Next Great Economic Revolution.* New York: Jeremy P. Tarcher/Penguin.

Womack, J.P., Daniel, T.J., and Daniel, R. (1990). *The Machine that Changed the World.* New York: Rawson Associates.

13

The Challenge of Global Climate Change

13.1 INTRODUCTION

The energetics of Earth's climate reflects a balance between energy absorbed from the sun and energy radiated to space, primarily in the infrared region of the spectrum. The flow of solar energy per second crossing an area equal to 1 m^2 oriented perpendicular to the direction at which the solar radiation is incident at the Earth's average distance from the sun is defined by what is known as the solar constant. Expressed in units of watts per square meter (W m^{-2}), the solar constant is equal to 1370 W m^{-2} (Chapter 3).

The Earth presents a target of area πR^2 to this incident solar radiation, where R defines the radius of the Earth. The total rate at which solar energy is intercepted by the Earth is equal therefore to 1.75×10^{17} W (Box 3.3). Approximately 30% of the incident sunlight is reflected back to space under current climate conditions, mainly from clouds and bright regions of the surface (snow- or ice-covered areas for example). The fraction of the incident light reflected back to space is specified by what is known as the Earth albedo (0.3 for the present Earth as discussed in Chapter 3). The balance (70%) is absorbed. On a global scale, the rate at which energy from the sun is absorbed by the Earth is equal to 1.23×10^{17} W (Box 3.3). Summed over the course of a year, Earth absorbs an amount of energy from the sun equal to 3.5 million Quad (see Chapter 3 for definition of various units used to specify energy: 1 Quad $= 10^{15}$ Btu $= 10^{10}$ therm $= 2.93 \times 10^{11}$ kWh $= 1.05 \times 10^{18}$ J).

It is instructive to compare the energy absorbed by the Earth from the sun over the course of a year with the total primary energy (fossil, nuclear and renewable) consumed by humans on a global scale. Details of this calculation are presented in Box 13.1. The energy absorbed by the Earth from the sun in

---

Box 13.1

As indicated, the rate at which the Earth absorbs energy from the sun is equal to 1.23×10^{17} W. In 2006, primary global consumption of energy by humans amounted to 435 Quad (BP, 2008). How does human consumption of energy compare with the amount of energy absorbed by the Earth from the Sun?

Solution:

The rate of absorption of solar energy, 1.23×10^{17} W, is equivalent to absorption of energy at a rate of 1.23×10^{17} J s^{-1}.

Multiplying by the number of seconds in a year, 3.15×10^7, the total energy absorbed by the Earth over the course of a year, T, is given by:

$$T = (1.23 \times 10^{17} \text{ J s}^{-1})(3.15 \times 10^7 \text{ s}) = 3.88 \times 10^{24} \text{ J}$$

Converting from J to Quad, noting that J $= 9.48 \times 10^{-19}$ Quad,

$$T = (3.9 \times 10^{24} \text{ J})(9.48 \times 10^{-19} \text{ Quad J}^{-1}) = 3.68 \times 10^6 \text{ Quad}$$

The ratio of solar energy absorbed to total human use, R, is given by:

$$R = \frac{3.68 \times 10^6 \text{ Quad}}{4.35 \times 10^2 \text{ Quad}} \approx 8500$$

2006 exceeded the total energy consumed by humans over the course of the same year by a factor of a little less than 8500.

In a stable or steady state climate, the energy absorbed from the sun should be equal to the energy returned to space in the form of infrared radiation. As indicated in Box 13.2, emission of radiation by the Earth at an average (black body) temperature of 255 K (-18°C) would be sufficient to establish a balance between energy in and energy out. The average temperature of the Earth's surface, however, is almost 35 K greater than this, about 290 K (17°C). How do we account for the apparent discrepancy? The answer involves the Greenhouse Effect. The Earth does not radiate to space to any significant extent from the surface but rather from a level in the atmosphere approximately 5 km above the surface. Infrared radiation emitted from the surface is absorbed by selected constituents of the atmosphere, notably H_2O, CO_2, CH_4, and N_2O. These species, referred to collectively as greenhouse gases, act as insulating agents (inhibiting release of Earth energy or heat directly from the surface to space) maintaining the surface at a temperature much warmer than would be the case in their absence.

Water vapor is the most important of the greenhouse gases. It absorbs radiation over an extensive portion of the infrared spectrum. In contrast, CO_2, CH_4, and N_2O absorb only over limited regions of the infrared spectrum. Notably, the major constituents of the atmosphere, O_2 and N_2, are ineffective

Box 13.2

Calculate the average temperature the Earth would need to attain in order to be able to radiate back to space the energy absorbed from the Sun. Assume that the Earth can radiate to space with equal efficiency from all positions on its surface (day and night, equator to pole).

Solution:

Denote the solar constant (W m^{-2}) by S. Assume that 70% of the incident solar energy is absorbed by the Earth. The total rate at which energy is absorbed, A, is given then by

$$A = (0.7) \, S \, (\pi \, R^2)$$

where πR^2 defines the effective target area offered by the Earth to intercept sunlight (R is the radius of the Earth).

Assume that the Earth radiates as a black body at temperature T (a blackbody refers to an idealization, an object capable of radiating and absorbing light at all wavelengths, the spectrum of which is determined solely by its temperature).

The energy emitted per unit time from unit surface area of a black body at temperature T is given by σT^4 where σ is a physical constant (the Stefan–Boltzmann constant). With T expressed in K and with the rate at which energy is emitted expressed in W m^{-2},

$$\sigma = 5.67 \times 10^{-8} \text{ W m}^{-2} \text{ K}^{-1}$$

The energy emitted from the entire surface of Earth (assumed to be approximated by a black body at temperature T) is given by

$$E = (4\pi R^2)(5.67 \times 10^{-8}) \, T^4$$

Equating A and E, implies

$$(0.7) \, S \, (\pi R^2) = 4\pi R^2 (5.67 \times 10^{-8}) \, T^4$$
$$\Rightarrow T^4 = \frac{(0.7) S}{4(5.67 \times 10^{-8})}$$

Taking $S = 1.37 \times 10^3$ W m^{-2} (see above),

$$T^4 = 4.228 \times 10^9 \text{ K}^4$$

Hence

$$T = 255 \text{ K } (-18°\text{C})$$

as greenhouse agents. If the atmosphere were composed solely of O_2 and N_2, infrared radiation emitted from the surface would pass freely through the atmosphere to space. If the albedo of the Earth were the same as today, the average temperature of the surface for this hypothetical Earth would be equal to 255 K (−18°C) as calculated in Box 13.2. The surface, however, would most likely be covered in snow or ice in this case. An even greater fraction of the incident sunlight would be reflected back to space (the albedo would be greater than 0.3) and the average temperature of the surface would be even lower than 255 K (−18°C).

The abundance of H_2O in the atmosphere depends on temperature. If the atmosphere were cold, its capacity to hold H_2O would be limited. In the absence of what the Intergovernmental Panel on Climate Change (IPCC, 2007) refers to as the long-lived greenhouse gases (LLGHGs), notably CO_2, CH_4, and N_2O, the Earth as indicated by the preceding analysis would be freezing cold. The abundance of H_2O in the atmosphere would be extremely low and the contribution of H_2O to the greenhouse impact would be minimal. Add LLGHGs to the atmosphere and the surface of the Earth would warm up. The abundance of H_2O in the atmosphere would increase leading to additional warming. More water vapor in the atmosphere would result most likely, however, in an increase in cloud cover. The reflectivity (albedo) of the Earth would increase in this case prompting a decrease in the fraction of the incident sunlight absorbed by the Earth. This provides an example of what is referred to as negative feedback, an offset to the original positive (increasing temperature) response.

To calculate the response of surface temperature to a specified increase in the abundance of greenhouse gases requires a model capable of assessing the combined impact of all relevant feedbacks, both positive (as in the case of the increase in water vapor) and negative (as in the case of the expected increase in cloud cover). Further complicating matters, an increase in cloud cover at high altitude (where temperatures are much lower than in the region occupied by low altitude clouds) would be expected to amplify the greenhouse effect, contributing an additional source of positive feedback (reflecting the fact that the interaction of the small ice crystals that compose these clouds with long wavelength infrared radiation is more significant than the interaction with the shorter wavelengths characteristic of visible solar radiation).

Human activity over the past several centuries has resulted in an unprecedented (at least over the past 650,000 years) rise in the concentration of LLGHGs as discussed in Section 13.2. An increase in the concentration of a specific greenhouse gas would imply that radiation emitted directly to space by this gas must originate from a higher level of the atmosphere (the probability that a photon emitted from a given level of the atmosphere can be transmitted directly to space is determined by the total overlying concentration of the gas available to absorb the photon). Since temperature generally decreases with altitude in the region of the atmosphere from which infrared photons communicate directly with space, it follows that the net emission of radiation to space must decrease as the proximate result of an increase in the concentration of individual greenhouse gases. The immediate impact of a reduction in emission of infrared radiation to space will be a tendency for the Earth to warm up (an imbalance between energy absorbed from the sun and energy emitted to space favoring the former). The magnitude of the energy imbalance triggered by a specific change in the concentration of greenhouse gases is defined by what is referred to as the radiative forcing associated with the indicated change in composition. The concept of radiative forcing is discussed in Section 13.3 which includes estimates for the magnitude of the radiative forcing (both positive and negative) associated with the variety of disturbances to the Earth's radiative state that have ensued over the past several centuries, attributable for the most part to diverse forms of human activity. The potential of individual gases to contribute to climate change, the global warming potential is discussed in Section 13.4.

Translating an estimate of radiative forcing to prediction of a specific change in climate requires a complex multidimensional model incorporating feedbacks not only on the atmosphere, but also on the biosphere (the living part of the Earth), the cryosphere (ice, both on land and on the ocean, snow cover, and frozen ground including permafrost), the hydrological cycle (rainfall, runoff, and evaporation), and the ocean. A simple metric involves assessment of the impact of a given radiative forcing on global average surface temperature. Changes in global average temperature over the past 150 years (the period for which modern thermometer-based measurements are available) are discussed in Section 13.5. The additional heat absorbed by the Earth as a consequence of past radiative forcing has been stored preferentially in the ocean, more than 80% of the total added to the Earth system over the past 40 years according to IPCC (2007). The implication of this is an important lag between radiative forcing and its ultimate impact on climate. Expected future changes in climate are discussed in Section 13.6. Concluding summary remarks are presented in Section 13.7.

13.2 THE CHANGING COMPOSITION OF THE ATMOSPHERE

Measurements of gas trapped in polar ice provide an important record of past changes in the composition of the atmosphere. The existing record extends back to 650,000 years before present. Changes observed in the concentrations of CO_2, CH_4, and N_2O are displayed in Fig. 13.1. The record for CO_2 and CH_4 is essentially continuous over this entire interval. Data for N_2O are more limited but nonetheless valuable. The figure includes also a record of measurements of the deuterium concentration of the ice from which the gas was extracted. These data provide a proxy for the temperature of the environment in which the ice was deposited. The figure includes also a record of measurements of the relative abundance of ^{18}O in the shells of benthic organisms preserved in marine sediments. These data provide a proxy for concurrent changes in the volume of land-based ice associated with waxing and waning of the continental ice sheets (water removed from the ocean to build the continental ice

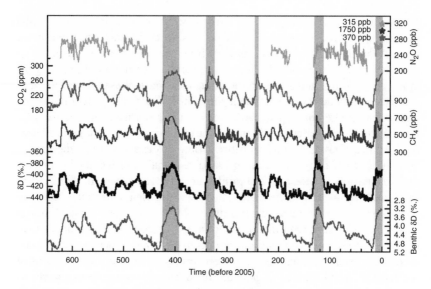

Figure 13.1 Changes in concentrations of the greenhouse gases CO_2, CH_4, and N_2O derived from measurements of air trapped in ice cores drilled in Antarctica. Variations of the relative abundance of deuterium in the ice (δD) provide a proxy for local temperature. The benthic ^{18}O data provide a proxy for changes in the volume of land-based ice. From IPCC (2007).

sheets is characteristically isotopically light: organisms growing in the ice age ocean build their shells consequently using enhanced concentrations of ^{18}O derived from isotopically heavy water). Climate over the past several million years was generally cold, punctuated by relatively brief episodes of warmth. Shaded areas in the figure indicate the timing of the present and four previous interglacial periods.

The concentrations of CO_2 displayed in Fig. 13.1 vary from a low of about 180 parts per million by volume (ppm) to a high of about 290 ppm. High values are associated generally with times when the global climate was warm (interglacial epochs). Low values coincide with periods when the climate was cold (the depths of ice ages).

Concentrations of CH_4 vary also, as indicated, in tune with changes in global climate, from peak values of about 700 parts per billion by volume (ppb) to low values of about 350 ppb although the variability exhibited by CH_4 is greater than that indicated by CO_2. The more limited data for N_2O indicate a range of concentrations between about 200 and 280 ppb.

A more detailed record of the changes in CO_2, CH_4, and N_2O over the most recent 20,000 years is presented in Fig. 13.2. Data displayed here reflect a combination of measurements obtained from analyses of gas recovered from ice cores taken both from Greenland and Antarctica (both from the ice itself and from firn, the unconsolidated surface layer of the ice) and, for the most recent several decades, direct measurements of the concentration of these gases in the atmosphere. The near vertical lines representing the trend in concentrations over the recent past clearly attest to the significance of the contemporary disturbance. Concentrations for all three gases are now significantly higher than the levels observed over the entire 650,000-year record summarized in Fig. 13.1. The pace of change observed over the past several centuries is also unprecedented.

Combustion of fossil fuels—coal, oil, and natural gas—is primarily responsible for the contemporary increase in the concentration of CO_2 with an additional contribution from the manufacture of cement. There is a contribution also from deforestation, mainly in the tropics, though the impact of this source is offset to a significant extent by regrowth of vegetation at northern mid-latitudes. Worldwide fossil fuel related emissions of CO_2 increased by 19.4% between 1984 and 1994 and by a further 24.6% between 1994 and 2004. Global emissions amounted to 7.9 billion tons of carbon in 2004 (multiply by 3.67 to convert to tons of CO_2) apportioned as follows: oil, 35.6%; coal, 35.9%; natural gas, 18.1%; cement, 3.8%.

The U.S. was responsible for 21% of 2004 global emission of CO_2: 43.9% from oil, 35.7% from coal with 19.5% from natural gas. China with 17.3% of the global total ranked number two (but has since moved to number one). Coal accounted for 98.7% of Chinese emissions in 1950 with minimal contributions from either oil or gas. Oil is now responsible for 17% of emissions from China with the contribution from coal dropping to 71.9% despite rapid growth in consumption of both fuels in the interim. Manufacture of cement accounted for 9.7% to Chinese emissions in 2004 reflecting the rapid recent pace of development of that country's infrastructure (China accounted for 44% of total global production of cement in 2004). Data quoted here are taken from Marland et al. (2007).[1]

The concentration of atmospheric CO_2 increased by 4.1% between 1984 and 1994 (from 344.4 ppm to 358.63 ppm) and by an additional 5.3% (from 358.63 ppm to 377.55 ppm) between 1994 and 2004.[2] It has risen more recently (by the end of 2007) to more than 382 ppm. Given any reasonable projection of future energy use, the concentration of atmospheric CO_2 is expected to climb to levels in excess of 500 ppm by mid-century. Approximately half of the CO_2 added to the atmosphere as a result of the combined influences of fossil fuel burning, cement manufacture, and changes in land use remains in the atmosphere with the balance absorbed by the ocean. An estimate of the contribution of emissions associated with use of fossil fuel to the airborne fraction, the fraction of CO_2 emitted as a result of fossil fuel burning and cement manufacture in recent years that is retained by the atmosphere, is presented in Box 13.3. For an account of the chemical processes regulating uptake of CO_2 by the ocean, see McElroy (2002).

Processing of organic matter by anaerobic bacteria (organisms active under low oxygen conditions) represents the primary source of atmospheric CH_4.

Box 13.3

A quantity of CO_2 equal to 5.92×10^{10} tons C was added to the atmosphere between 1984 and 1994 due to the combined influence of burning fossil fuels and manufacturing cement with an additional 6.87×10^{10} tons C added between 1994 and 2004. The abundance of atmospheric CO_2 increased by 14.23 ppm between 1984 and 1994 and by 18.92 ppm between 1994 and 2004. Calculate the airborne fractions that would appear to have applied over these intervals.

Solution:

An increase of 1ppm in the concentration of CO_2 in the atmosphere corresponds to an additional 2.12×10^9 tons C. The abundance of C in the atmosphere increased therefore by 3.02×10^{10} tons C ($2.12 \times 10^9 \times 14.23$) from the beginning of 1984 to the end of 1993 and by 4.01×10^{10} tons C from the beginning of 1994 to the end of 2003.

The implied airborne fractions (dividing the magnitude of the atmospheric increase by the amount added) are 51% for 1984–1994 and 58% for 1994–2004.

For the entire interval 1984–2004, the implied airborne fraction is equal to 55%.

Important natural sources include swamps and waterlogged soils, environments in which production of organic matter exceeds the rate at which oxygen can be supplied to oxidize it (the supply of O_2 in the case of swamps is inhibited by stagnant water overlying the organic-rich sediments of the swamps; in the case of organic-rich soils, the supply of O_2 may be limited by water clogging the pore spaces of the soil).

Cultivation of rice in flooded rice paddy fields contributes an important human-influenced or anthropogenic source of CH_4, a source that has increased steadily in recent years reflecting the growth of human populations in regions of the world where rice provides a critical food source. Ruminants—cattle, sheep, and goats—are responsible for additional production. The increase of CH_4 over the past several centuries is attributed generally to the combination of rice cultivation and to the growth of the population of ruminants (there are more than 1.3 billion cattle in the world at the present time, as many as a billion sheep and more than 700 million goats). Additional production is associated with biomass burning (clearance of land largely for purposes of agriculture) and with decomposition of organic matter deposited in landfills. The modern rise in CH_4 reflects thus primarily trends in the growth of the global human population, changes in land use, and related trends in human dietary preferences. The energy sector is currently a minor source. It is conceivable, however, that with increased dependence on fossil fuels the energy-related source of CH_4 (emitted as a by-product of coal mining and oil and gas production) may increase significantly in the future.

The abundance of CH_4 in the atmosphere is regulated by a balance between global production and loss. Methane is removed from the atmosphere by reaction with the hydroxyl radical OH produced through a complex sequence of chemical reactions initiated by absorption of ultraviolet sunlight by ozone (O_3). The abundance of CH_4 has increased in the atmosphere over the past 250 years from a pre-industrial value of 715 parts per billion (ppb) in 1750 to 1775 ppb in 2005 (IPCC, 2007). The rate of increase has slowed significantly however over the past 20 years, from a growth rate of more than 15 ppb per year in the 1970s to close to zero since 2000. The slow-down is attributed to a stabilization of emissions rather than to a change in the efficiency of the removal process (an increase in the abundance of OH for example).

Given the diversity and spatial heterogeneity of sources of CH_4 and the lack of quantitative understanding of underlying source mechanisms, it is not possible at this time to provide a reliable projection of the future course of the concentration of the gas in the atmosphere. Of particular concern is that changes in climate (increasing temperatures and changing patterns of rainfall) could result in a significant increase in production of CH_4 from sources such as soils and swamps that might otherwise be classified as natural. Soils at high latitude, for example, contain abundant sources of organic matter. These soils are frozen over for most of the year. As a consequence, their store of organic matter is essentially immune to attack by decomposing organisms. An increase in temperature accompanied by melting of the permafrost normally present in these soils could result in a major increase in production and release of CH_4 to the atmosphere prompting a further increase in temperature, an example of what is referred to as positive feedback (a disturbance, the consequence of which is responsible for further amplification of the original disturbance, as would be the case if release of CH_4 resulted in warming prompting additional release of CH_4 leading to further warming).

The concentration of N_2O has increased over the past several centuries from 270 ppb in 1750 to 319 ppb in 2005. Concentrations of N_2O were relatively constant for at least 2000 years prior to 1750. The increase over the past several decades has averaged a relatively constant 0.26% per year (IPCC, 2007). Similar to the case for CH_4, microbial processes dominate production of N_2O. Nitrous oxide is produced both by oxidation of ammonium (NH_4^+) and by reduction of nitrate (NO_3^-). The former process is referred to as nitrification, the latter as denitrification. Processing in both cases is effected by bacteria.

The yield of N_2O produced by nitrification is enhanced at low levels of O_2. Denitrification, which proceeds under anoxic or near anoxic conditions, can result in either a source or a sink for N_2O. As supplies of NO_3^- are depleted in the course of denitrification, the bacteria responsible for denitrification are capable of switching to exploit N_2O as an oxygen source. Observations of N_2O dissolved in low oxygen waters associated with a region of high biological productivity (the result of locally intense upwelling of nutrient rich waters) off the coast of Peru offer definitive evidence for the importance of this sequence. In the core of the low oxygen waters, where O_2 concentrations are essentially

Table 13.1 Budgets of N_2O for the pre-industrial and contemporary
environments

	Pre-industrial	1988	2005
Concentration (ppb)	280	307	319
Inventory (Mt N)	1327	1455	1512
Growth rate (ppb year^{-1})	0	0.7	0.8
Growth rate (Mt N year^{-1})	0	3.4	3.9
Loss rate (Mt N year^{-1})	11.4	12.5	13.0
Production, ocean (Mt N year^{-1})	4.0	4.0	4.0
Production, land natural (Mt N year^{-1})	7.4	7.4	7.4
Total production (Mt N year^{-1})	11.4	15.9	16.9
Anthropogenic production (Mt N year^{-1})	0	4.5	5.5

zero, N_2O levels are undetectably low. On the edge of the anoxic zone, concentrations of N_2O are high, significantly greater than levels that would apply if the waters were in equilibrium with the atmosphere. For a more comprehensive account of the global nitrogen cycle and the microbial processes responsible for production of N_2O see McElroy (2002).

A budget for N_2O is presented in Table 13.1. The data presented here include an estimate for the distribution of sources and sinks in the pre-industrial environment following a tentative budget developed by McElroy (2002). The loss rate for atmospheric N_2O is relatively well defined: it is determined by absorption of ultraviolet radiation resulting in decomposition of N_2O to N_2 and O, primarily in the stratosphere (more than 20 km above the Earth's surface). The sink for N_2O amounted to 11.4 million tons N/ N_2O per year (Mt N year^{-1}) in the pre-industrial era, assumed to be equal to the magnitude of the natural source at that time. Based on analysis of existing data, McElroy (2002) attributed 35% of the natural source to microbial processes in the ocean with the balance ascribed to microbial processes on land. By 1988, the sink, tracking the increase in the concentration of N_2O in the atmosphere, had risen to 12.5 Mt N year^{-1}. Combined with the observed increase in the burden of N_2O in the atmosphere, 3.4 Mt N year^{-1}, it follows that the source of N_2O had increased in the interim to 15.9 Mt N year^{-1}. Attributing the increase in production to anthropogenic activity on land, it follows that the anthropogenic contribution to production in 1988 amounted to close to 40% of the combined land–sea global natural source. By 2005, the anthropogenic component had grown to 5.5 Mt N year^{-1}, corresponding to 48% of the natural source and shows little sign of abating.

A major fraction of the Earth's nitrogen is present in the atmosphere as N_2. In this form, nitrogen is relatively inaccessible to biological organisms. Before it can be incorporated in living tissue, the triple bond binding the atoms of N in N_2 must be sundered: the N atoms must be incorporated in more accessible compounds such as NH_4^+ and NO_3^-. Nitrogen in biologically available form is said to be fixed. The natural source of fixed nitrogen is estimated at about 200 Mt N year^{-1}. Anthropogenic activity, however, associated notably with combustion of fossil fuels and with the manufacture of chemical fertilizer, is responsible for an additional, and increasing, source of fixed nitrogen, estimated to contribute at the present time as much as 50% of the natural source (McElroy, 2002). The increase in N_2O is attributed to acceleration of the global nitrogen cycle driven by an enhanced, human-induced, applications of fixed nitrogen (McElroy, 2002). Ultimately, as was concluded for the case of CH_4, the driving force for the increase in N_2O may be attributed to the increasing demand for food to feed a growing world population, especially the more affluent fraction of this population, combined with practices for disposal of increased burdens of human and animal waste (McElroy and Wang, 2005).

13.3 RADIATIVE FORCING

The concept of radiative forcing induced by a change in the concentration of a specific LLGHG involves an estimate of the resulting net change in energy absorbed by the Earth (energy absorbed from the sun minus energy radiated to space in the infrared). By convention, this change is referenced to the tropopause, the uppermost boundary of the lower atmosphere (the troposphere) where changes in climate are expressed most directly. In calculating the change in radiative flux (energy in minus energy out) at the tropopause, it is assumed that the temperature of the atmosphere above the tropopause, the region known as the stratosphere, responds radiatively to the assumed change in composition (that is to say the temperature of the stratosphere is adjusted so that the energy absorbed by the stratosphere is precisely equal to the energy emitted). In computing the radiative forcing associated with any particular change in composition, the temperature at the surface and of the atmosphere between the surface and the tropopause is held fixed. So defined, radiative forcing is intended to provide an indication of the potential for a given change in composition to alter climate. When forcing is positive, the energy absorbed by the Earth from the sun exceeds the rate at which energy is emitted to space. When forcing is negative, the energy balance is altered in the opposite sense: the energy emitted to space exceeds the energy absorbed from the sun. In the former case, the planet may be expected to warm up to restore energy equilibrium. In the latter case, it would be expected to cool.

The radiative forcing in 2005 attributed to the increase in the concentration of LLGHGs since 1750 is estimated at 2.63 W m^{-2} (IPCC, 2007). Of this, 1.66 W m^{-2} is associated with the post-industrial increase in the concentration of CO_2 (from 278 ppm to 379 ppm): 0.48 W m^{-2} to CH_4 (increase from 715 ppb to 1774 ppb); 0.16 W m^{-2} to N_2O (increase from 270 ppb to 319 ppb); with the balance due to SF_6 and a variety of anthropogenically associated halocarbons (including a number of chlorine- and bromine-containing halocarbons implicated not only in radiative forcing of the climate system but responsible also for reduction of the abundance of ozone in the stratosphere). In position number one in terms of its contribution to the change in radiative forcing since 1750 is CO_2: CH_4 ranks number two; position number three is occupied, however, not by N_2O but by CF_2Cl_2 (CFC-12). Emissions of CF_2Cl_2 are regulated under the Montreal Protocol, the agreement reached by the international community in September 1987 in Montreal, Canada, to phase out emissions of a number of gases, including CFC-12, that were implicated in depletion of stratospheric O_3. The lifetime in the atmosphere of CF_2Cl_2 is exceptionally long, 108 years (McElroy, 2002). As a consequence, even though emissions have been curtailed, the concentration CF_2Cl_2 remains high and will decrease only slowly (with an e-folding time constant of 108 years equal to the lifetime) in the decades ahead. Given the steady increase in the concentration of N_2O, though, we may expect N_2O to assume the position of number three in the radiative forcing hierarchy in the not too distant future (most likely within a decade or so).

Trends in radiative forcing associated with variations in the concentrations of CO_2, CH_4, and N_2O over the past 20,000 years are illustrated in Figs. 13.2a–c. Temporal rates of change of radiative forcing associated with changes in the concentrations of these gases are presented in Fig. 13.2d. Notable here is the unprecedented rapidity of the change in radiative forcing since 1750. The change in radiative forcing introduced by the increase in the concentrations of CO_2, CH_4, and N_2O since the end of the last ice age up to the beginning of the modern industrial era (taken as 1750 AD) amounts to about 2.6 W m^{-2}, approximately equal to the change that has developed subsequently.

Not all of the changes in radiative forcing over the modern industrial era have been positive. Positive forcing driven by the increase in the concentrations of LLGHGs and additional positive forcing induced by an increase in the abundance of O_3 in the troposphere (O_3 is also a greenhouse gas) have been offset to some extent by negative forcing caused by an increase in the abundance of aerosols in the atmosphere (particulate matter suspended in the air). The increase in the abundance of tropospheric O_3 is attributed to a combination of fossil fuel related emissions of nitrogen oxides and enhanced emissions of hydrocarbons from a variety of industrial sources (for a comprehensive review of ozone chemistry see McElroy, 2002). A variety of sources contribute to the increased burden of atmospheric aerosols. Combustion of sulfur-rich fossil fuels is responsible for important emissions of SO_2, which is oxidized in the atmosphere to form H_2SO_4 (sulfuric acid). The combination of H_2SO_4 and NH_3 (ammonia) results in production of highly reflective (high albedo) particulate matter composed of ammonium sulfate. Oxidation of nitrogen oxides provides a source of HNO_3 (nitric acid), which in combination with NH_3 can contribute an additional source of reflective aerosols (ammonium nitrate in this case). Soot, black carbon, is responsible for another source of particulate matter. Radiative forcing is positive in this case reflecting the low albedo of this black material.

Aerosols can have both a direct and an indirect effect on radiative forcing. The direct effect relates to the immediate impact of the aerosols on the transmission of visible solar radiation. Addition of a significant quantity of reflective aerosol (ammonium sulfate or ammonium nitrate for example) increases the overall albedo of the Earth: a greater fraction of the incident sunlight is reflected back to space contributing thus to negative radiative forcing. On the other hand, addition of black carbon can enhance the fraction of incident sunlight that is absorbed rather than reflected by the Earth: the forcing is positive in this case.

The indirect effect of aerosols relates to their impact on the properties of clouds. Aerosols such as ammonium sulfate or ammonium nitrate are hygroscopic (they readily absorb water) and can serve consequently as important sites or nuclei for condensation of water. Distribution of a given amount of condensable water over a larger number of condensation nuclei would imply that the resulting cloud would be composed of a larger number of cloud particles of smaller average size than would be the case with a smaller number of condensation centers. The reflectivity of the cloud with the larger number of smaller size scattering particles would be greater than the reflectivity of the cloud with the smaller number of scattering centers: increasing the number of aerosols that can serve as condensation nuclei may be expected thus to contribute to negative radiative forcing (a larger fraction of the incident sunlight would be reflected back to space).

A summary of the various contributions to radiative forcing as recommended by IPCC (2007) is presented in Fig. 13.3, defined for the present environment relative to conditions in 1750. The combination of LLGHGs and O_3, according to this analysis, is responsible for net radiative forcing of $+2.9\pm0.3$ W m^{-2}. Larger uncertainty is associated with the composite impact of aerosols, estimated according to the analyses summarized by IPCC (2007) to lie in the range -2.2 to -0.5 W m^{-2} with a median value of -1.3 W m^{-2}. The IPCC recommendation for the composite effect of all radiative forcing for the period 1750 to 2005 is $+1.6$ W m^{-2} with a 90% confidence range of between $+0.6$ and $+2.4$ W m^{-2}.

If we were to successfully eliminate emissions of black carbon, and there are good reasons to do so relating to their impact on public health, net radiative forcing could be reduced to about 0.7 W m^{-2}. Were we to also seriously cut back on emissions of the compounds responsible for the reflective aerosols, and there are good reasons also to do this (to reduce the incidence of acid rain for example), radiative forcing would rise back to the level set by the build-up of greenhouse gases, about 3 W m^{-2}.

Box 13.4 provides an estimate of how the energy gained by the Earth as a result of current radiative forcing compares with current global consumption of primary energy.

13.4 GLOBAL WARMING POTENTIALS

The potential for a particular gas emitted to the atmosphere to contribute to global warming is defined by convention in terms of the total radiative forcing for which the gas may be expected to be responsible over some selected time

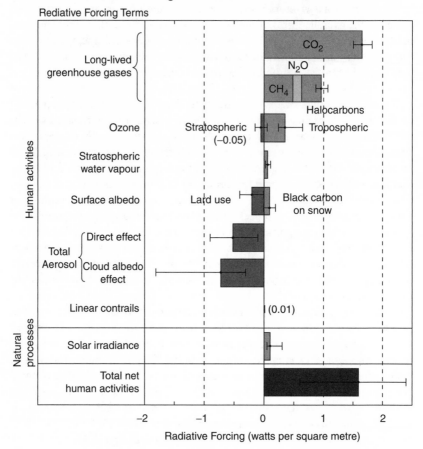

Radiative forcing of climate between 1750 and 2005

Figure 13.3 Summary of the principal components of the radiative forcing of
climate change. All these radiative forcings result from one or more
factors that affect climate and are associated with human activities
or natural processes as discussed in the text. The values represent the
forcings in 2005 relative to the start of the industrial era (about 1750).
The only increase in natural forcing of any significance between 1750
and 2005 occurred in solar irradiance. Positive forcings lead to warming
of climate and negative forcings lead to a cooling. The thin black line
attached to each shaded bar represents the range of uncertainty for the
respective value. From IPCC (2007).

Box 13.4

Comparison of current radiative forcing with current global commercial energy use

Solution:

Contemporary radiative forcing = 1.6 W m^{-2}

Integrated over the surface area of the Earth, forcing (F) is given by

$$F = (1.6 \text{ W m}^{-2})(5.1 \times 10^4 \text{ m}^2) = 8.16 \times 10^{14} \text{ W}$$

Total energy absorbed by the Earth over the course of a year due to the radiative imbalance (T) is given by

$$\begin{aligned} T &= (8.16 \times 10^{14} \text{ J s}^{-1}) \ (3.15 \times 10^7 \text{ s year}^{-1}) \\ &= 2.58 \times 10^{22} \text{ J} \\ &= 2.45 \times 10^{19} \text{ Btu} \\ &= 2.45 \times 10^4 \text{ Quad} \end{aligned}$$

This compares with total global commercial energy use of about 430 Quad. Radiative forcing adds 57 times more energy to the Earth system over the course of a year than the total contributed by commercial energy.

---

interval, 100 years for example. The advantage of the concept is that it allows for a comparison of the climate-altering potential of different gases, the impact of emission of a given quantity of species X for example as compared with emission of a different quantity of species Y. The global warming potential (GWP) of an individual gas depends thus on a combination of the contribution the gas can make to instantaneous radiative forcing weighted by the persistence of the gas in the atmosphere. The GWP for a species with relatively modest radiative forcing that can survive in the atmosphere for an exceptionally long period of time may be greater as a consequence than the GWP of a different species whose lifetime in the atmosphere may be relatively brief even if the radiative forcing potential for this second species is much larger.

The Kyoto Protocol, formulated at the third meeting of the Conference of the Parties to the United Nations Framework Convention on Climate Change (the formal title of the meeting), was designed to curb future growth in the concentration of greenhouse gases. It targeted four species (CO_2, CH_4, N_2O, and SF_6) and two classes of compounds (hydrofluorocarbons and perfluoro-carbons), accepting that a number of additional greenhouse agents (members of the chlorocarbon and bromocarbon families of species) had been regulated previously under measures taken by the international community to deal with the threat to stratospheric ozone. The GWP concept was adopted as the standard with which to compare emissions of these different gases.

A summary of GWPs for a variety of greenhouse effective gases is presented in Table 13.2 (adopted from IPCC, 2007, Table 2.14). The table includes

Table 13.2 Lifetimes, radiative efficiencies, and direct (except for CH$_4$) GWPs relative to CO$_2$. From IPCC (2007)

Industrial Designation or Common Name (years)	Chemical Formula	Lifetime (years)	Radiative Efficiency (W m^{-2} ppb^{-1})	Global Warming Potential for Given Time Horizon			
				SAR ‡ (100-yr)	20-yr	100-yr	500-yr
Carbon dioxide	CO$_2$	See below[a]	[b]1.4 × 10^{-5}	1	1	1	1
Methane[c]	CH$_4$	12[c]	3.7 × 10^{-4}	21	72	25	7.6
Nitrous oxide	N$_2$O	114	3.03 × 10^{-3}	310	289	298	153
Substances controlled by the Montreal Protocol							
CFC-11	CCl$_3$F	45	0.25	3800	6730	4750	1620
CFC-12	CCl$_2$F$_2$	100	0.32	8100	11,000	10,900	5200
CFC-13	CClF$_3$	640	0.25		10,800	14,400	16,400
CFC-113	CCl$_2$FCClF$_2$	85	0.3	4800	6540	6130	2700
CFC-114	CClF$_2$CClF$_2$	300	0.31		8040	10,000	8730
CFC-115	CClF$_2$CF$_3$	1700	0.18		5310	7370	9990
Halon-1301	CBrF$_3$	65	0.32	5400	8480	7140	2760
Halon-1211	CBrClF$_2$	16	0.3		4750	1890	575
Halon-2402	CBrF$_2$CBrF$_2$	20	0.33		3680	1640	503
Carbon tetrachloride	CCl$_4$	26	0.13	1400	2700	1400	435
Methyl bromide	CH$_3$Br	0.7	0.01		17	5	1
Methyl chloroform	CH$_3$CCl$_3$	5	0.06		506	146	45
HCFC-22	CHClF$_2$	12	0.2	1500	5160	1810	549
HCFC-123	CHCl$_2$CF$_3$	1.3	0.14	90	273	77	24
HCFC-124	CHClFCF$_3$	5.8	0.22	470	2070	609	185
HCFC-141b	CH$_3$CCl$_2$F	9.3	0.14		2250	725	220
HCFC-142b	CH$_3$CClF$_2$	17.9	0.2	1800	5490	2310	705
HCFC-225ca	CHCl$_2$CF$_2$CF$_3$	1.9	0.2		429	122	37
HCFC-225cb	CHClFCF$_2$CClF$_2$	5.8	0.32		2030	595	181
Hydrofluorocarbons							
HFC-23	CHF$_3$	270	0.19	11,700	12,000	14,800	12,200
HFC-32	CH$_2$F$_2$	4.9	0.11	650	2330	675	205

HFC-125	CHF_2CF_3	29	0.23	2800	6350	3500	1100
HFC-134a	CH_2FCF_3	14	0.16	1300	3830	1430	435
HFC-143a	CH_3CF_3	52	0.13	3800	5890	4470	1590
HFC-152a	CH_3CHF_2	1.4	0.09	140	437	124	38
HFC-227ea	CF_3CHFCF_3	34.2	0.26	2900	5310	3220	1040
HFC-236fa	$CF_3CH_2CF_3$	240	0.28	6300	8100	9810	7660
HFC-245fa	$CHF_2CH_2CF_3$	7.6	0.28		3380	1030	314
HFC-365mfc	$CH_3CF_2CH_2CF_3$	8.6	0.21		2520	794	241
HFC-43-10mee	$CF_3CHFCHFCF_2CF_3$	15.9	0.4	1300	4140	1640	500
Perfluorinated compounds							
Sulphur hexafluoride	SF_6	3200	0.52	23900	16,300	22,800	32,600
Nitrogen trifluoride	NF_3	740	0.21		12,300	17,200	20,700
PFC-14	CF_4	50,000	0.10	6500	5210	7390	11,200
PFC-116	C_2F_6	10,000	0.26	9200	8630	12,200	18,200
PFC-218	C_3F_8	2600	0.26	7000	6310	8830	12,500
PFC-318	$c\text{-}C_4F_8$	3200	0.32	8700	7310	10,300	14,700
PFC-3-1-10	C_4F_{10}	2600	0.33	7000	6330	8860	12,500
PFC-4-1-12	C_5F_{12}	4100	0.41		6510	9160	13,300
PFC-5-1-14	C_6F_{14}	3200	0.49	7400	6600	9300	13,300
PFC-9-1-18	$C_{10}F_{18}$	>1000[d]	0.56		>5500	>7500	>9,500
trifluoromethyl sulphur pentafluoride	SF_5CF_3	800	0.57		13,200	17,700	21,200
Fluorinated ethers							
HFE-125	CHF_2OCF_3	136	0.44		13,800	14,900	8490
HFE-134	CHF_2OCHF_2	26	0.45		12,200	6320	1960
HFE-143a	CH_3OCF_3	4.3	0.27		2630	756	230
HCFE-235da2	$CHF_2OCHClCF_3$	2.6	0.38		1230	350	106
HFE-245cb2	$CH_3OCF_2CHF_2$	5.1	0.32		2440	708	215
HFE-245fa2	$CHF_2OCH_2CF_3$	4.9	0.31		2280	659	200
HFE-254cb2	$CH_3OCF_2CHF_2$	2.6	0.28		1260	359	109
HFE-347mcc3	$CH_3OCF_2CF_2CF_3$	5.2	0.34		1980	575	175

continued

Table 13.2 Continued

Industrial Designation or Common Name (years)	Chemical Formula	Lifetime (years)	Radiative Efficiency (W m^{-2} ppb^{-1})	SAR ‡ (100-yr)	Global Warming Potential for Given Time Horizon		
					20-yr	100-yr	500-yr
HFE-356pcc3	$CH_3OCF_2CF_2CHF_2$	0.33	0.93		386	110	33
HFE-449sl (HFE-7100)	$C_4F_9OCH_3$	3.8	0.31		1040	297	90
HFE-569sf2 (HFE-7200)	$C_4F_9OC_2H_5$	0.77	0.3		207	59	18
HFE-43-10pccc124 (H-Galden1040x)	$CHF_2OCF_2OC_2F_4OCHF_2$	6.3	1.37		6320	1870	569
HFE-236ca12 (HG-10)	$CHF_2OCF_2OCHF_2$	12.1	0.66		8000	2800	860
HFE-333pcc13 (HG-01)	$CHF_2OCF_2CF_2OCHF_2$	6.2	0.87		5100	1500	460
Perfluoropolyethers							
PFPMIE	$CF_3OCF(CF_3)$ $CF_2OCF_2OCF_3$	800	0.65		7620	10,300	12,400
Hydrocarbons and other compounds – Direct effects							
Dimethylether	CH_3OCH_3	0.015	0.02		1	1	<<1
Methylene chloride	CH_2Cl_2	0.38	0.03		31	8.7	2.7
Methyl chloride	CH_3Cl	1.0	0.01		45	13	4

estimates for the lifetimes of individual gases, their radiative forcing efficiencies, and their GWPs computed for time scales of 20,100, and 500 years. GWPs are reported here relative to CO_2. On a 100-year basis, the GWP of CH_4 is calculated to be 25 times greater than that of CO_2. N_2O is 298 times more effective than CO_2 and 11.9 times more effective than CH_4. Inclusion of sulfur hexafluoride (SF_6) in the Kyoto Protocol reflects the fact that even a modest emission of this gas to the atmosphere can have an important impact on climate. The high value for its GWP reflects both the unusual efficiency of SF_6 in radiative forcing (37,143 times greater than CO_2) and its exceptionally long lifetime (3200 years).

13.5 THE INSTRUMENTAL RECORD OF GLOBAL CHANGE

Gabriel Fahrenheit's invention of the thermometer in 1714 made it possible for scientists to carry out relatively straightforward measurements of temperature for the first time. Early applications of Fahrenheit's invention to measure air temperatures were relatively spotty. A useful record of changes in surface air temperatures can be reconstructed however dating back to about 1850 drawing on archival data from a wide range of stations around the world. The database includes not only measurements on land but also measurements for surface waters of the ocean, the latter taken by drawing water on board ship (originally by lifting a bucket of seawater onto the ship's deck, more recently by sampling water drawn on board as coolant for the ship's engines) and recording the resulting measurements in the ship's log.

A summary of trends for both global and hemispherically averaged surface temperatures is illustrated in Fig. 13.4 (IPCC Fig. 3.6). The upper panel refers to the globe as a whole, the middle panel to the northern hemisphere, the bottom one to the southern hemisphere. The data are presented in terms of departures from the average of temperatures recorded between 1961 and 1990. Temperatures for the northern hemisphere, the southern hemisphere, and for the globe were relatively constant between 1850 and 1910. They increased by about 0.4°C between 1910 and 1940 with indications for a small drop (by about 0.1°C) between 1940 and 1970. The decrease between 1940 and 1970 may have reflected the influence of an increase in emissions of sulfur from coal burning (the negative radiative forcing discussed earlier) offsetting the positive forcing due to the increase in the concentration of greenhouse gases. The long-term increase in temperatures, which resumed after 1970, may be attributed to the increasing relative importance of the positive radiative forcing associated with the continuing increase in the concentration of greenhouse gases combined with steps taken at least in developed countries to limit emissions of sulfur. The 12 years between 1995 and 2006 include eleven of the warmest years in the entire 156-year record (only 1996 fails to make this list). The upward trend in global surface temperatures since 1979 to 2005 corresponds to an increase of more than 0.16°C per decade.

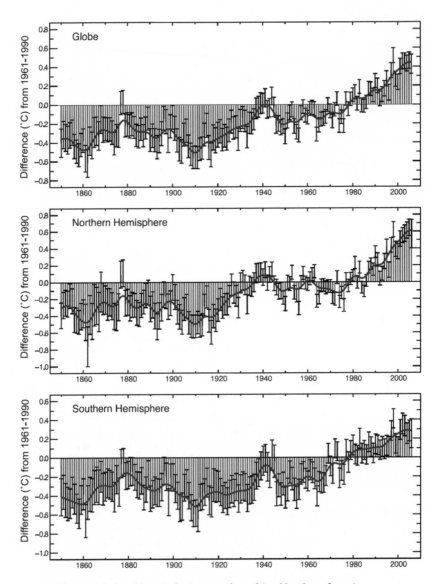

Figure 13.4 Global and hemispheric annual combined land–surface air
temperature and SST anomalies (°C) for 1850 to 2006 relative
to the 1961 to 1990 mean, along with 5 to 95% error bar ranges.
The smooth curves show decadal variations.
From IPCC (2007).

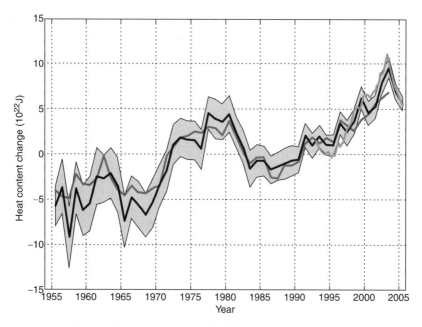

Figure 13.5 Time series of global annual ocean heat content (10^{22} J) for the 0 to 700 m layer. Individual and red curves denote the deviation from the 1961 to 1990 average and the shorter curve denotes the deviation from the average for the period 1993 to 2003. From IPCC (2007).

A record of the change in ocean heat content as observed since 1955 is presented in Fig. 13.5 (IPCC Fig. 5.1). The data displayed here refer to the upper 700 m of the global ocean. They are presented in terms of departures with respect to the 1961 to 1990 average. The increase in ocean heat content indicated here is consistent with the view expressed in the introduction to this chapter: that as much as 80% of the extra heat added to the earth system due to the past imbalance between energy absorbed and energy emitted to space is stored in the ocean, an indication of the important lag between current radiative forcing and its ultimate impact on climate.

The IPCC (2007) sought to place the recent record of global average surface temperatures in the context of the longer record available from a variety of studies of past climate. They concluded as follows:

Paleoclimatic information supports the interpretation that the warmth of the last half-century is unusual in at least the previous 1300 years. The last time the polar regions were warmer than present for an extended period (about 125,000 years ago), reductions in polar ice volume led to 4 to 6 m of sea level rise.

If the warming trend is permitted to continue, we may need to be concerned not just with the immediate challenge of having to live with a different, and shifting, climate but also with the potentially more catastrophic consequences of a significant rise in global sea level.

13.5 FUTURE PROSPECTS

The response of global average surface temperature to a presumed doubling of the concentration of CO_2 offers a convenient metric with which to assess the sensitivity of the climate system to a sustained increase in radiative forcing. Considering an instantaneous increase in the concentration of CO_2 from its pre-industrial value of 275 ppm to 550 ppm, summarizing results from a wide range of models, IPCC (2007) concluded that global average surface temperature would increase eventually, most likely, by about 3°C. Results from the models they considered varied from a low of 2°C to a high of 4.5°C. Radiative forcing associated with the presumed doubling of CO_2 amounts to 3.7 W m^{-2}, implying a sensitivity of equilibrium global average surface temperature to radiative forcing of 0.8°C per watt per square meter (0.8°C W^{-1} m^2).

Searching for an empirical estimate for the equilibrium response of global surface temperature to radiative forcing, James Hansen and colleagues (Hansen et al., 1984, 2005) considered the change in climate that occurred as the Earth made the transition from the last ice age to the present interglacial epoch. Radiative forcing associated with the increase in the concentrations of CO_2, CH_4, and N_2O was calculated as +3 W m^{-2}. The additional forcing contributed by the decrease in planetary albedo resulting from the demise of the continental ice sheets was estimated at +3.5 W m^{-2} implying a value for total radiative forcing of +6.5 W m^{-2}. They assumed that global average surface temperature increased by 5°C between glacial and interglacial time. The sensitivity to radiative forcing was calculated thus as 0.77°C W^{-1} m^2, in excellent agreement with the value recommended by IPCC (2007).

Trends in global mean surface temperature projected using a variety of climate models are summarized in Fig. 13.6 (IPCC, 2007), which includes also a comparison with the trend observed over the past century. The models explored four different scenarios for future emissions of CO_2, CH_4, N_2O, and SO_2. Scenario A2 envisages a continuing growth in emissions of CO_2, CH_4, and N_2O over the twenty-first century with a decrease in emissions of SO_2 after 2050. Net radiative forcing in Scenario A2 increases to about 8 W m^{-2} by 2100. Scenario A1B assumes that aggressive steps are taken to stabilize emissions of CO_2 in the latter half of the century with a net reduction in emissions of CH_4 and SO_2. Despite these steps, net radiative forcing in Scenario A1B increases to close to 6 W m^{-2} by 2100. Scenario B1 envisages more limited growth in emissions of CO_2 in the early part of the century with a net reduction after 2050 with similar initiatives to limit emissions of CH_4 and SO_2. Radiative forcing in Scenario B1 increases more slowly than in either of Scenarios A2 or A1B, stabilizing at a level of about 4 W m^{-2} in the latter half of the century. The fourth scenario assumes that the concentrations of all of the radiatively active gases are held constant and is included only by way of reference, to indicate the inertia of the climate system, the fact that temperatures will continue to increase (although slowly) even if radiative forcing is held fixed. Details of the different scenarios are summarized in Chapter 10 of the IPCC (2007) report (Meehl et al., 2007).

Concentrations of CO_2 rise to levels of close to 800 ppm, 700 ppm, and 550 ppm in Scenarios A2, A1B, and B1 respectively. The global average temperature in Scenario A2 increases over the next 100 years by about 4°C relative to the 1980–2000 mean, by about 3°C for Scenarios A1B and B1. Relative to 1900, the increase in global average temperature projected for Scenario A2 is comparable to the increase associated with the transition from glacial to interglacial conditions 20,000 years ago.

Spatial distributions of surface temperatures predicted for the three scenarios are displayed in Fig. 13.7 (IPCC 10.8). Results are presented for three time intervals: 2011–2030, 2046–2065, and 2080–2099. Increases in temperature are global in scope, greatest however at high latitudes, particularly in winter. As discussed, annual mean temperatures at high northern latitudes are projected with Scenario A2 to increase by as much as 4°C by the end of this century. Under this circumstance we might expect that the Arctic Ocean would be ice-free for much of the year—a very different world.

Models for future climate are in general agreement that there should be at least a modest increase in total global precipitation. The spatial and temporal pattern of precipitation is likely to change however. Precipitation is expected to increase at higher latitudes and most probably also in the tropics, offset by a potentially significant decrease in the sub-tropics. Transport of heat from the tropics to the sub-tropics is effected primarily by a circulation loop known as the Hadley circulation named after the English meteorologist (George Hadley) who first postulated its existence more than 265 years ago. Hadley's challenge was to provide an explanation for the persistent northeasterly direction of what are known as the trade winds in the northern sub-tropics. He correctly surmised that air would rise in response to intense solar heating in the tropics, move northward aloft (southward in the southern hemisphere), sinking eventually to the surface, completing a loop (turning west as the Earth turned underneath it, accounting thus for the prevailing northeasterly direction of the trade winds) as it returned to the tropics. He was wrong in that he thought that this circulation pattern would extend to high latitudes: in fact it reaches only to about ±30. As the air rises in the tropics, it loses most of its entrained moisture to precipitation. When it returns to the surface it is exceptionally dry and exceptionally hot (as dry air sinks, its temperature increases by about 10°C for every kilometer of descent due to compression). The landscape underlying the rising portion of the Hadley circulation is associated with lush tropical vegetation—rain forests. Beneath the descending loop, we find the great deserts of the world. In a warmer world, the Hadley circulation is expected to extend to higher latitudes, accounting for the anticipated decrease in sub-tropical precipitation, accompanied most likely by a latitudinal expansion of the existing sub-tropical deserts (think of the Sahara Desert extending into Southern Europe with a corresponding northward migration of the Southwestern deserts of the U.S.).

The atmosphere in a warmer world is expected to contain on average a higher concentration of water vapor. The implication is that when it rains (or snows) in a warmer world, precipitation is likely to be more intense. Higher rates of precipitation in some regions will be offset most likely by lower rates in others. The likely result: increased incidence of floods in some regions, more frequent droughts in others.

13.6 CONCLUDING REMARKS

The prospect of future global climate change is a matter of serious concern. Climate change that affected human civilizations in the past was for the most part triggered by natural forces. What is unique about the present situation is that we humans are responsible. As the late distinguished oceanographer Roger Revelle[3] famously remarked, we are in the midst of a great global experiment. For the first time in the history of our species we have developed the capacity to alter our environment globally. The long-term consequences are difficult to forecast but assuredly momentous.

We have attempted in this chapter to provide a primer on the science of climate change, skipping technical details but at the same time seeking accurately to present the essentials. There is no doubt that the Earth is currently out of radiative energy balance. The increase in the concentration of the principle greenhouse gases (unprecedented at least over the past 650,000 years, most probably longer) is assuredly responsible for the condition in which the Earth is now absorbing more energy from the sun than it is radiating back to space. A significant fraction of this excess energy is being stored in the ocean, ready to promote future climate change even if the growth in the concentration of climate altering gases (and particulate matter) were to be immediately halted (an unlikely possibility). Pending drastic and immediate action to reduce emissions of the primary greenhouse gases, notably CO_2, the end product of combustion of coal, oil, and natural gas, the global average surface temperature is likely to rise over the next century by as much as it did when the Earth made the transition from the last ice age to the present interglacial interlude. Unanticipated regional changes in climate are a likely result, seriously constraining our ability to cope.

Civilizations in the past, as recounted in Chapter 2, were also exposed to the challenges of climate change. On more than a number of occasions, they succumbed. What is unique about the present is that the human population is greater now than it ever was in the past. It has limited ability to adjust to a collapse in the function of essential local environmental services by moving elsewhere: international borders are increasingly impenetrable. At the same time, international economies are inextricably interconnected. Hurricane Katrina wreaked havoc on the lives of people in Louisiana. But the effects were not confined to Louisiana, or even to the southern states of the U.S. They were experienced worldwide in response to an immediate rise in the price of important globally traded commodities such as oil and natural gas but extending also to encompass more immediate life-essential resources such as corn, wheat, and soybeans.

To avoid the hazards of a potentially disruptive future climate change will require a C-change in how we currently use energy. If we are to continue to rely on fossil fuels, we need to find ways to capture and sequester CO_2 before it is released to the atmosphere (Chapter 14). We can seek to use biofuels as a substitute for fossil fuels—ethanol for example produced from a combination of corn, sugar cane, and/or cellulose, where the carbon emitted represents carbon recycled by photosynthesis from the atmosphere (Chapter 15). Or we can seek to develop an entirely new energy economy based on a combination of conservation, reduced reliance on fossil fuels, increased exploitation of nuclear power (Chapter 9), and perhaps, most promising, exploitation of renewable resources derived from the sun, wind, and the reservoir of heat emanating from the Earth's interior (Chapter 17).

NOTES

1 (http://cdiac.ornl.gov/trends/emis/em_cont.htm, read December 11, 2007).

2 (ftp://ftp.cmdl.noaa.gov/ccg/co2/trends/co2_annmean_mlo.txt, read December 11, 2007).

3 (http://en.wikipedia.org/wiki/Roger_Revelle, read October 27, 2008).

REFERENCES

BP Statistical Review of World Energy (2008). June 2008 http://www.bp.com/liveassets/ bp_internet/globalbp/globalbp_uk_english/reports_and_publications/statistical_ energy_review_2008/STAGING/local_assets/downloads/pdf/statistical_review_of_ world_energy_full_review_2008.pdf (read May 8, 2009).

Hansen, J. et al. (1984). Climate sensitivity: Analysis of feedback mechanisms. In: *Climate Processes and Climate Sensitivity* (eds. J.E. Hansen and T. Takahashi), pp. 130–163. AGU Geophysical Monograph 29, Maurice Ewing Vol. 5. American Geophysical Union.

Hansen, J. et al. (2005). Efficacy of climate forcings. *Journal of Geophysical Research*, **110**, D18104. doi:10.1029/2005JD005776.

IPCC (2007). *Intergovernmental Panel on Climate Change. Climate Change 2007: The Physical Science Basis. Working Group 1 Contribution to the Fourth Assessment.* Cambridge: Cambridge University Press.

Marland, G., Boden, T., and Andres, R.J. (2007). National CO2 Emissions from Fossil Fuel Burning, Cement Manufacture and Gas Flaring: 1751-2004, August 17, 2007.

Meehl, G.A. et al. (2007) Global climate projections. Chapter 10 from IPCC (2007).

McElroy, M.B. (2002). *The Atmospheric Environment: Effects of Human Activity.* Princeton: Princeton University Press.

McElroy, M.B. and Wang, Y.X. (2005). Human and animal wastes: implications for atmospheric N_2O and NO_x. *Global Biogeochemical Cycles*, **19**, GB2008. doi:10.1029/ 2004GB002429.

14

Prospects for Carbon Capture and Sequestration

14.1 INTRODUCTION

Coal, the most abundant of the fossil fuels, is also regrettably the most polluting. Conventional sources of pollution from coal burning include emissions of sulfur oxides, nitrogen oxides, mercury, and a variety of different forms of particulate matter, and—for purposes of this chapter—large quantities of the important greenhouse gas CO_2.

As discussed in Chapter 5, sulfur and nitrogen oxides contribute to the problem of acid rain. A portion of the sulfur emitted as a consequence of burning coal is converted in the atmosphere to small particles of sulfate that can serve as condensation nuclei (centers for condensation of water vapor) with implications for the formation of clouds. An increase in emissions of sulfur has the potential to increase cloud cover over regions subject to these emissions with consequences for local and regional climate. An increase in cloud cover would imply that a greater fraction of sunlight incident on a particular region would be reflected back to space resulting in regional cooling of the surface.

Particulate matter emitted either directly by combustion of coal or produced (as in the case of sulfur) by secondary processes in the atmosphere can have serious negative implications for public health (responsible for a variety of pulmonary problems when ingested by humans). Emissions of nitrogen oxides contribute to the production of ozone in the atmosphere with implications not only for humans (increased pulmonary problems) but also for the growth of plants as a result of damage to their stomata. And elemental mercury, with a relatively long lifetime in the atmosphere, can travel over large distances before depositing on the surface where it can be converted to chemical forms, methyl mercury for example (a serious neurotoxin), that can accumulate in the biota with consequences not only for humans consuming mercury-contaminated food stocks (fish in particular) but also for the health of the environment more generally.

Technological options are available (at a cost both in terms of energy consumption and economic resources required to supply the necessary capital equipment) to reduce, or even eliminate, most of these conventional emissions. The coal can be treated (washed for example) before it is consumed. The carbon of the coal can be converted prior to combustion to energy-rich gaseous forms such as CO and H_2 with the offending chemicals (the sulfur, nitrogen, mercury, and mineral particulate matter) removed in the process. Or a variety of means are available (again at a cost) to reduce emissions of particulates, sulfur, nitrogen, and mercury by treating the effluent from coal burning plants post combustion but prior to emission to the atmosphere. It is more difficult to eliminate (or reduce) emissions of CO_2.

The energy content of coal is present primarily in the form of chemically reduced carbon. Harnessing this energy involves necessarily the production of large quantities of CO_2. Combustion of oil and gas also contribute important sources of CO_2. A fraction of the energy released by burning oil and natural gas is associated, however, with oxidation of the hydrogen in these fuels to H_2O. The hydrogen content of natural gas is greater than that for oil (the hydrogen to carbon ratio is approximately 4 for natural gas which is composed mainly of methane, CH_4, as compared to about 2 for oil). It follows, judged solely by its contribution to emission of CO_2, that coal is the most polluting of the fossil fuels, followed by oil, with natural gas the most benign.

As discussed in Chapter 5, the energy content of the world's current reserves of recoverable coal is estimated conservatively at 230,000 Quad, sufficient to supply present global demand for energy (412 Quad per year) for at least 500 years, potentially longer given anticipated future advances in mining technology. Combustion of coal is responsible for approximately 40% of the total worldwide current production of CO_2 associated with combustion of fossil fuels. Prevailing patterns in energy use suggest that the contribution from combustion of coal to emissions of CO_2 is likely to increase in the future due in large measure to trends in energy use in large developing economies such as China (China is now the world's largest emitter of CO_2 having recently surpassed the U.S. for this dubious honor: coal provides the dominant source of China's energy and is likely to continue to do so for the foreseeable future).

A strategy designed to capture CO_2 produced by burning coal before it is emitted to the atmosphere and to bury (or sequester) it in some long-lived geological reservoir has the advantage that most of the coal consumed today is deployed in large stationary sources—electric power plants, chemical factories, and factories involved in energy-intensive industries such as the production of iron and steel. There are similar, though fewer, opportunities to capture CO_2 produced by consumption of oil and natural gas.

Approximately 70% of oil used in the U.S. is consumed by the transportation sector (cars, trucks, buses, trains, and aircraft), responsible for multiple, individually small, mobile sources of CO_2 that would be difficult to capture. Residential and commercial uses of natural gas account for approximately 36% of gas consumed in the U.S. Given the distributed nature of these sources and the fact that individually their contributions to emissions of CO_2 are relatively small, it is unlikely that these sources would be targeted in any early program designed to capture and sequester CO_2. Approximately 26% of natural gas usage in the U.S., however, is associated with electric power generation with about 4% used to produce nitrogen fertilizer (production of nitrogen fertilizer accounts for a much larger fraction of natural gas use in China, close to 70%). Coal accounted for 49% of electricity generated in the U.S. in 2006, natural gas 20% with oil responsible for only 1.6%. With respect to emissions of CO_2 from the electric power sector, coal was responsible for 82% of total emissions in 2006, gas for 16% with oil contributing a relatively minor 2% (note than non-fossil, carbon-free, sources, mainly nuclear, were responsible for approximately 29% of total U.S. power generation in 2006).

It is probable that an early program to capture CO_2 should focus on large industrial sources—electric power plants consuming various combinations of coal, oil, or natural gas and specific industries (iron and steel, cement factories, and chemical fertilizer plants for example) for which the magnitude of associated emissions might justify the expense of the equipment required to capture and dispose subsequently of the related sources of CO_2. Flue (or stack) gases emanating from such facilities are typically hot, characterized by temperatures of about 100°C with pressures close to the pressure of the atmosphere (the higher temperature accounts for the buoyancy that allows the stack gases to rise following their emission to the atmosphere). If the oxidant for the fuel is provided by air, as is the case for almost all current facilities, CO_2 accounts for less than about 14% of the gases produced by the combustion process (see Table 14.1), limited by the O_2 content of the air (21% by volume) that supplies the oxidant for the fuel.

We can conceive of potential future power plants (and fossil fuel fired factories) for which the oxidant could be supplied in the form of pure O_2 rather than air. This option is referred to as oxyfuel. Capture of CO_2 in this case could be designed in from the outset. The abundance of CO_2 in the stack gases of such a facility could be as great as 98%. The difficulty of capturing this CO_2 would be reduced significantly. Energy and capital would have to be expended, however, to provide the necessary source of pure O_2. Treatment of effluent gases and particulates would still be required. It would be essential also in the case of pure oxygen supplied facilities to dilute the supply of the O_2 oxidant by addition (through recirculation) of a portion of the CO_2 included in the exhaust gases. Otherwise, the temperature of the combustion chamber (boiler) would rise to unacceptably high values, sufficient to melt steel (addition of recirculated CO_2 allows the energy released by the combustion process to be shared by a greater number of molecules reducing the temperature of the resulting gaseous mix). This would require an additional demand not only for energy but also for capital.

Table 14.1 Properties of candidate gas streams that can be inputted to a capture process (from IPCC, 2005)

Source	CO_2 concentration % vol (dry)	Pressure of gas stream (atm)	CO_2 partial pressure (atm)
CO_2 from fuel combustion			
• Power station flue gas:			
Natural gas fired boilers	7–10	1.00	0.07–0.1
Gas turbines	3–4	1.00	0.03–0.04
Oil fired boilers	11–13	1.00	0.11–0.13
Coal fired boilers	12–14	1.00	0.12–0.14
IGCC: after combustion	12–14	1.00	0.12–0.14
Oil refinery and petrochemical			
plant fired heaters	8	1.00	0.12–0.14
CO_2 from chemical transformation + fuel combustion			
• Blast furnace gas			
Before combustion	20	2–3	0.4–0.6
After combustion	27	1.00	0.27
• Cement kiln off-gas	14–33	1.00	0.14–0.33
CO_2 from chemical transformation before combustion			
• IGCC: synthesis gas after gasification	8–20	20–70	1.6–14

Molecular nitrogen, the dominant constituent of the atmosphere, is not surprisingly the most abundant component of the stack gases of conventional, fossil fuel powered, factories and electric power plants. In addition to N_2, CO_2, and trace quantities of O_2, the effluent produced by coal-fired facilities includes a variety of trace organic and inorganic species and a significant quantity of particulate matter in addition to variable quantities of CO, H_2O, SO_2, NO, HCl, HF, mercury, and other metals. Effluent streams produced by oil- and gas-fired factories and power plants are less complex (particularly so for facilities fueled by natural gas), but the problem persists: CO_2 must be separated from a complex mixture of gases and particulate material and must be concentrated and pressurized before it can be made available for transfer to a suitable storage reservoir. All of this processing requires significant inputs, not only of capital to supply the necessary equipment but also of energy to fuel its operation.

We begin in Section 14.2 with an account of the technologies available for capture of CO_2. Options for burial or sequestration of CO_2 are discussed in Section 14.3. Futuristic possibilities for capturing CO_2 from the atmosphere are treated in Section 14.4. Summary comments are presented in Section 14.5.

14.2 OPTIONS FOR CAPTURE OF CO_2

The most promising approach to capturing CO_2 from a fossil fuel burning factory or power plant involves use of a solvent capable of selective absorption of CO_2. A solution composed of liquid water and mono-ethanolamine (MEA, C_2H_7NO), approximately 26% by weight mono-ethanolamine, offers one such possibility.

The absorption process for MEA may be summarized by the reaction

$$CO_2 + C_2H_7NO + H_2O \rightarrow HCO_3^- + C_2H_8NO^+ \qquad (14.1)$$

The CO_2 absorbed in reaction (14.1) may be released subsequently by driving reaction (14.1) from right to left rather than from left to right, i.e. by reversing the original absorption process. The absorption process is exothermic (energy is released when the reaction takes place). Energy must be supplied to reverse the process, to liberate the CO_2 absorbed in reaction (14.1) (the reverse reaction is endothermic).

Components of a conceptional CO_2 capture facility are illustrated schematically in Fig. 14.1. The key element consists of a column in which CO_2 is extracted from flue gases and incorporated in the solvent. This column is referred to as the absorber. Solvent is admitted to the absorber at its upper end and is allowed to flow under gravity to the bottom from whence it can be transferred (pumped) to the top of a second column, referred to as the stripper. Solvent, prior to incorporation of CO_2, is referred to as lean. Following loading with CO_2, the solvent is described as rich. The purpose of the stripper is to reverse the reaction that took place in the absorber, releasing a concentrated stream of close to pure CO_2. Lean solvent is reconstituted in the process. Exiting the bottom of the stripper, the reconstituted solvent, replenished with fresh supplies to replace solvent depleted in its passage through the absorber and stripper, is transferred back to the absorber where it is available once again to take up CO_2. The solvent is selected to ensure: (a) high efficiency for absorption of CO_2 from flue gases in the absorber; (b) minimal need for replacement; and (c) a minimum requirement for energy required to release CO_2 from the enriched solvent.

Flue gases enter the absorber at the bottom, propelled upward by a fan. The CO_2 content of the flue gases is reduced steadily as the gases rise in the absorber—as CO_2 is selectively incorporated in the downward moving solvent. Flue gases, depleted in CO_2, composed at this point primarily of N_2, are emitted to the atmosphere at the top of the absorber column (the exhaust gas). The flue gases are cooled prior to admission to the absorber in order to increase the rate at which CO_2 is incorporated in the solvent. To ensure that the major constituent incorporated in the solvent is CO_2, it is important that acidic impurities, notably sulfur and nitrogen oxides, be removed from the flue gas prior to its admission to the absorber. Care must be taken also to eliminate particulate matter that could otherwise clog the absorber.

The gaseous products exiting the stripper are composed primarily of CO_2 and H_2O. This gas stream is cooled to eliminate water vapor (by condensation followed by removal of the resulting liquid) avoiding thus the formation of acidic compounds (produced by reaction of CO_2 with H_2O) that would otherwise require use of special corrosion-resistant materials in transferring CO_2 to

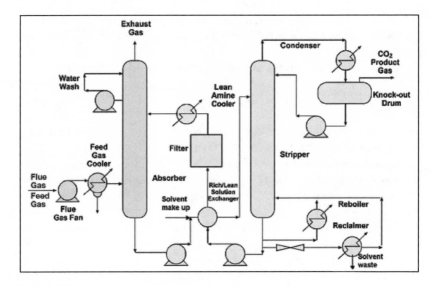

Figure 14.1 Process flow diagram for CO_2 recovery from flue gas by chemical absorption (from IPCC, 2005).

its ultimate storage reservoir. The product CO_2-rich gas must be pressurized prior to transfer (by pipe, truck, train, or ship) to its eventual storage reservoir.

All of these steps require significant inputs of energy. Energy is expended in removing sulfur and nitrogen oxides, particulates, and other impurities from the initial flue gas stream, in cooling the resulting gas mix prior to transfer to the absorber, in powering the blower involved in this transfer, in heating the enriched solvent in the stripper to liberate CO_2 and H_2O, in removing this water, in pressurizing the resulting CO_2, and in running the pumps employed to transfer solvent between the absorber and stripper. Estimates of penalties incurred for a variety of electric power generating systems are summarized in Fig. 14.2. Typically they average about 20%. That is to say, it is necessary to consume 20% more fuel (coal or natural gas for the cases included in Fig. 14.2) to produce a given quantity of electricity. The figure is adopted from the recent report by the Intergovernmental Panel on Climate Change on Carbon Dioxide Capture and Storage, Chapter 3 of that report (IPCC, 2005).

The abbreviation IGCC in the figure refers to a plant using an approach known as integrated gas combined cycle (IGCC) to generate electricity. A coal-fired IGCC plant converts the carbon of the coal initially to CO and H_2, with the CO converted subsequently to yield additional H_2 by reaction with H_2O (known as the water shift reaction). The advantage of the IGCC system is that it provides a concentrated stream of CO_2 from the outset but at a significantly higher cost for the necessary capital equipment (and for energy) as compared

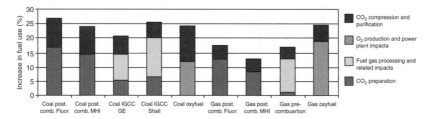

Figure 14.2 Percentage increase in fuel user per kWh of electricity due to CO_2 capture, compared to the same plant without capture (from IPCC, 2005).

to a conventional coal-fired power plant which typically burns pulverized coal—coal ground into small particles facilitating efficient combustion. Oxyfuel refers to plants in which the oxidant is supplied as pure oxygen rather than air. Fluor, MHI, and GE refer to the companies responsible for specific plant designs: the Fluor Corporation, Mitsubishi Heavy Industries, and General Electric.

IPCC (2005) offered estimates for the cost and energy increments required to install and operate carbon capture systems on eight new pulverized coal powered plants. Six of these plants were assumed to use super-critical steam (water at temperatures above 374°C at pressures of more than 220 atmospheres for which only one phase of water can exist) produced by burning bituminous coal. The other two employed sub-critical steam, making use of lower grade coal. The increment in energy required to operate the super-critical plants (with carbon capture) ranged from 24% to 40% relative to the energy required to operate the plants in the absence of carbon capture equipment. For the two sub-critical plants, the energy penalties were estimated at 40% and 42%. The increase in capital costs required to install the carbon capture equipment on the super-critical plants ranged from 44% to 74%: 75% and 87% for the sub-critical plants. Increases in cost for electricity delivered by the super-critical plants varied from 42% to 66% (1.8 and 3.4 cents per kWh): 77% and 81% for the sub-critical plants (3.8 cents per kWh). Expressed in terms of costs incurred for carbon emissions avoided, the costs for the super-critical plants ranged from $23–35 per ton of CO_2, $31 per ton of CO_2 for the sub-critical plants. Costs for electricity generated with the super-critical plants were estimated at between 6.2 and 8.6 cents per kWh, 7.7 and 8.7 cents per kWh for the sub-critical plants, greater in all cases than costs quoted in Table 9.3 for conventional nuclear plants, nuclear plants incorporating advanced design elements, and for electricity generated using wind. The data quoted here are summarized for convenience in Table 14.2.

There remains the problem of finding a suitable reservoir to store the captured CO_2, the costs and energy required to deliver it to the reservoir, and the costs to monitor its stability in the reservoir to guard against unexpected and potentially hazardous release, a topic addressed in the next section.

Table 14.2 CO$_2$ capture costs: new pulverized-coal power plants using current technology (from IPCC 2005)

Study assumptions and results	Parsons	Parsons	Simbeck	IEA GHG	IEA GHG	Rubin et al.	Range		NETL	Rao and Rubin	Stobbs and Clark
	2002b	2002b	2002	2004	2004	2005	min	max	2002	2002	2005
	Super-critical units / bituminous coals								Subcrit units / low bank coals		
Reference plant emission rate (tCO$_2$ MWh^{-1})	0.774	0.736	0.76	0.743	0.747	0.811	0.74	0.81	0.835	0.941	0.883
CO$_2$ emission rate after capture (t MWh^{-1})	0.108	0.101	0.145	0.117	0.092	0.107	0.09	0.15	0.059	0.133	0.060
CO$_2$ captured (Mt yr^{-1})	1.83	2.35	2.36	4.061	4.168	3.102	1.83	4.17	2.346	2.58	2.795
CO$_2$ reduction per kWh (%)	86	86	81	84	88	87	81	88	93	86	93
Reference plant COE (US$ MWh^{-1})	51.5	51.0	42.9	43.9	42.8	46.1	43	52	42.3	49.2	44.5
Incremental COE for capture (US$ MWh^{-1})	34.1	31.4	28	18.5	20.2	28	18	34	37.8	37.8	29.8
% Increase in COE (over ref. plant)	66	62	65	42	47	61	42	66	81	77	67
Cost of CO$_2$ avoided (US$/tCO$_2$)	51	49	43	29	31	40	29	51	43	47	36

14.3 OPTIONS FOR STORAGE OF CO$_2$

A variety of environments, as illustrated in Fig. 14.3, have been suggested as potential sites for storage (or sequestration) of CO$_2$. These include oil and gas fields (Stevens et al., 2001), coal beds (Gale and Freund, 2001), saline aquifers (Nordbotten et al., 2005), injection into the deep ocean (Drange et al., 2001; Herzog, 2001) and burial in ocean sediments (House et al., 2006). An additional possibility contemplates conversion and storage of CO$_2$ in the form of stable minerals (Lackner, 2002).

The potential for enhanced oil recovery (EOR) from largely depleted oil reservoirs by injection of CO$_2$ was discussed briefly in Chapter 6. This option is particularly effective for fields in which the residual oil is present at depths in excess of about 1200 m. The injected CO$_2$ mixes with the oil causing it to swell reducing its viscosity thus improving its mobility (Stevens et al., 2001). The CO$_2$ is injected through dedicated drill holes with the oil recovered at preexisting production wellheads. Stevens et al. (2001) point out that for present EOR projects in the Permian Basin of the Southwestern U.S. (think of Texas) the largest current expense is associated with acquisition of an adequate supply of high pressure CO$_2$: the price is currently close to $12 per ton of CO$_2$. Approximately 80% of CO$_2$ deployed in existing EOR projects is derived from naturally occurring geological deposits: that is to say, they employ a source of CO$_2$ that would not otherwise be released to the atmosphere. EOR accounts for about 4% of

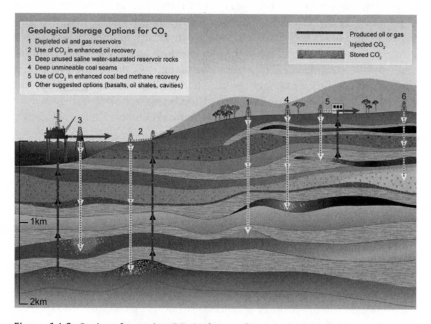

Figure 14.3 Options for storing CO$_2$ in deep underground geological formations (from IPCC, 2005).

present domestic production of oil in the U.S., a contribution that is predicted to double over the next decade. Clearly, the potential exists to make at least limited use of the CO_2 captured from large fossil fuel burning facilities for purposes of EOR, subsidizing at least to some extent the costs involved in capture and transport of the associated CO_2. Note, however, that the costs for capture of CO_2 from present and potential future pulverized coal fueled power plants (Table 14.2) are significantly higher than costs for CO_2 deployed in existing EOR projects ($23–35 as compared to $12 per ton of CO_2).

The ancillary economic benefits from injecting of CO_2 into depleted natural gas fields are significantly lower than those for EOR. Under normal operating procedures, as much as 90% of the natural gas from a gas field can be extracted without resort to extraordinary measures (Stevens et al., 2001). Injection of CO_2 may be effective in mobilizing a small fraction of the remaining natural gas in the reservoir. This may be accomplished, however, at the expense of a higher concentration of CO_2 in the gas brought to the surface, a contaminant that would have to be removed before the gas could be brought to market. If the drill holes were capped, following what is judged as the maximum economically feasible exploitation of a particular gas field, the reservoir could provide a convenient long-term storage medium for CO_2. IPCC (2005) estimates the combined worldwide potential of depleted oil and gas reservoirs for storage of CO_2 at between 675 and 900 billion tons of CO_2.

A further economically constructive possibility for sequestration of CO_2 involves injection into deep coal resources to stimulate release of methane (the dominant component of natural gas). This strategy is referred to as enhanced-coal-bed-methane recovery (ECBM). Methane is present both in the pore spaces of the coal and attached to its surface. Its concentration is generally greatest for the higher-grade coals that have been exposed to high temperatures and pressures during their formation. Injecting CO_2 into these coal seams would result in displacement of the methane. Drilling elsewhere into the coal could allow the methane to be released to the surface. The CO_2 would be retained in the coal seam and, so long as the coal resource is not subsequently mined, this process could provide a potential long-term storage reservoir for CO_2. ECBM accounted for over 7% of natural gas production in the U.S. in 1999 (Stevens et al., 2001), only a small fraction of which was accomplished however by injection of CO_2. IPCC (2005) estimates the global potential of ECBM for carbon sequestration in deep coal seams (resources that are unlikely to be mined in the future) at a minimum of 3–15 billion tons of CO_2, possibly as much as 200 billion tons of CO_2.

Burial of carbon in deep saline deposits offers an opportunity for sequestration of CO_2 potentially much greater than the combined resources offered by depleted oil and gas fields and deep coal deposits. IPCC (2005) estimates the minimum potential for carbon storage in worldwide saline deposits at 1000 billion tons of CO_2, potentially as much as ten times greater than this.

To put all of these numbers in context (the combined potential for storage in depleted oil and gas fields, deep coal seams, and saline deposits), the Energy

Table 14.3 Storage capacity for several geological storage options.
The storage capacity includes storage options that are not
economical (from IPCC, 2005)

Reservoir type	Lower estimate of storage capacity (GtCO$_2$)	Upper estimate of storage capacity (GtCO$_2$)
Oil and gas fields	675	900
Unminable coal seams		
(ECBM)	3–15	200
Deep saline formations	1000	Uncertain, but possible 10^4

Information Administration (EIA, 2002) projected growth in worldwide energy consumption of 2.3% per year between 1999 and 2020. If the relative contribution of fossil fuels to this growth should remain comparable to what it is today (close to 30 billion tons of CO_2 per year), cumulative emissions of CO_2 over the next century would amount to more than 8500 billion tons of CO_2 (most likely significantly greater than this given current trends favoring increasing use of coal in the face of higher prices for oil and gas). The IPCC (2005) data quoted here for the potential to sequester CO_2 in depleted oil and gas fields, deep coal seams, and saline deposits are summarized in Table 14.3. The authors point out, however, that not all of these opportunities are economically viable.

Approximately one-half of the net amount of CO_2 released to the atmosphere from the combination of burning fossil fuels and from global deforestation up to present has been transferred to the ocean. Uptake of CO_2 by the ocean is limited by the abundance of carbonate ion (CO_3^{--}) present in the seawater that comes into contact with the atmosphere. The carbonate ion serves to partially neutralize the acidic properties of CO_2. The net reaction is represented by

$$CO_2 + CO_3^{2-} + H_2O \rightarrow 2HCO_3^- \tag{14.2}$$

Over the past several hundred years, since the industrial revolution and the beginning of the modern rise in the abundance of atmospheric CO_2, approximately 10% of the total volume of ocean water has been exposed to the atmosphere. It takes close to a thousand years for the entire ocean to physically turn over, a consequence of the extended lifetime of the cold waters that dominate conditions in the ocean at depth. The sluggish rate at which the ocean turns over and the relatively slow rate at which CO_3^{2-} is supplied to the surface accounts for the comparative inefficiency of the ocean as a sink for CO_2. One way around this limitation would be to inject the fossil source of CO_2 directly into the deep ocean, below the thermocline (the zone in the ocean marking the transition between the relatively warm, low density, waters of the upper ocean and the cold, stable, waters of the abyss) where the CO_2 could be sequestered, isolated from the atmosphere, for hundreds of years or even longer (taking advantage of the sluggish nature of the circulation of seawater below the thermocline).

For typical temperatures of the ocean below the thermocline (3–10°C), CO_2 can exist either as a gas or as a liquid depending on pressure (as indicated by the phase diagram for CO_2 presented in Fig. 14.4). The liquid phase is stable for pressures greater than about 50 atm corresponding to a depth in the ocean of about 500 m (pressure increases by approximately 1 atm for every 10 m increase in ocean depth). If CO_2 were injected at depths greater than 2500 m (pressures greater than about 250 atm), the density of the resulting CO_2 liquid would be greater than the density of the seawater it displaced and the liquid would tend to sink. If the droplets containing the liquid CO_2 were large enough, they could survive to the bottom to form a coating of liquid CO_2 on the surface of the ocean sediment (a lake if the release were large enough). For injection of CO_2 at shallower depth, above 2500 m but below 500 m (where CO_2 would still be present primarily in liquid form), the density of the resulting liquid would be less than the density of seawater. Droplets would rise in this case, dissolving eventually in the water column with conversion of their constituent liquid CO_2 to dissolved CO_2 and subsequently to HCO_3^-, with an associated reduction in the abundance of CO_3^{2-} accompanied by an increase in ocean acidity.

The decrease in the concentration of CO_3^{2-} in seawater resulting from the dissolution of CO_2 present initially in the liquid phase may be expected to increase the alkalinity of ocean water by triggering dissolution of $CaCO_3$ present in deep ocean sediments. $CaCO_3$ is supplied to sediments in the form of the

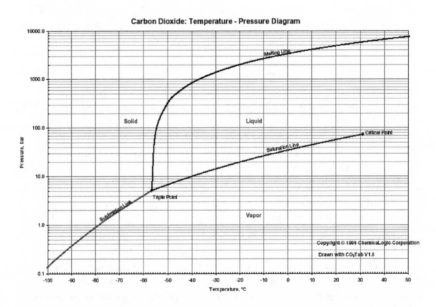

Figure 14.4 Phase diagram for CO_2 (from http://www.chemicalogic.com/download/co2_phase_diagram.pdf, read Oct 17, 2007).

shells or hard parts of organisms that once lived in the water column. The concept of alkalinity is defined in Table 14.4. The net reaction corresponding to the dissolution of the $CaCO_3$ in sediments may be represented by

$$CaCO_3 \rightarrow Ca^{2+} + CO_3^{2-} \tag{14.3}$$

in which the product calcium and carbonate ions are released into the overlying water column. The resulting increase in alkalinity, with the associated increase in the concentration of CO_3^{2-} (offsetting the reduction in CO_3^{2-} associated with the initial dissolution of the injected CO_2), would further enhance the capacity of the ocean to take up CO_2 from the atmosphere, as indicated by reaction (14.2).

Concentrations quoted in Table 14.4 are expressed in moles per liter $(mol\, l^{-1})$. A mole of substance X defines the number of molecules of the substance required to supply a mass of 1 gram. A liter corresponds to a volume of $10^3\, cm^3$. Entries in the table under the heading charge eq l^{-1} (charge equivalent per liter) identify the magnitude of the charge expressed in $mol\, l^{-1}$ contributed by the individual species. It is equal to the number density $(mol\, l^{-1})$ for species carrying a single charge (Na^+ and Cl^- for example): twice as large for species carrying a double charge (Ca^{2+} and SO_4^{2-} for example). Alkalinity is defined by the net positive charge contributed by the cations indicated in the table, offset by the negative charge represented by the anions, for a net balance of 0.002 charge eq l^{-1}. This (positive) charge is balanced by the net negative charge contributed by

Table 14.4 Concentrations of the major positively and negatively charged species that determine the alkalinity of seawater (from McElroy, 2002)

Positive charge

Cation	mol l^{-1}	charge eq l^{-1}
Na^+	0.470	0.470
K^+	0.010	0.010
Mg^{2+}	0.053	0.106
Ca^{2+}	0.010	0.020
Sum		0.606

Negative charge

Anion	mol l^{-1}	charge eq l^{-1}
Cl^-	0.547	0.547
SO_4^{2-}	0.028	0.056
Br^-	0.001	0.001
Partial sum		0.604
$HCO_3^- + 2CO_3^{2-}$		0.002
Sum		0.606

HCO_3^- and CO_3^{--}. Concentrations of H^+ and OH^- in seawater are too small to significantly affect the overall charge balance for seawater. A pH of 8.0 for example would correspond to a concentration of H^+ of only 10^{-8} mol l^{-1}.

Eventually, a portion of the CO_2 injected into the deep ocean will be released to the atmosphere (assuming that the strategy to reduce the build-up of CO_2 in the atmosphere was successful in the interim). The time scale for this release would depend on a number of factors: the time for CO_2 injected in the liquid phase to dissolve in ocean waters; the extent to which the alkalinity of ocean waters was increased by dissolution of sedimentary $CaCO_3$; and the time for CO_2-enriched waters to return to the surface. The delay involved would extend most likely to centuries, possibly to as long as several thousand years, given the slow turn over of waters in the deep ocean. The CO_2 could be isolated for much longer times if it were injected into the sediment of the ocean rather than the water column.

Possibilities for essentially permanent sequestration of CO_2 in ocean sediments were discussed by House et al. (2006). They proposed that CO_2 should be injected in liquid form into sediments of the ocean at depths greater than about 3000 m. The density of liquid CO_2 in the upper regions of the sediment would be greater in this case than the density of the sediment pore water. The density of liquid CO_2 would decrease however with depth in the sediment in response to the increase in temperature resulting from the geothermal gradient (temperature increases by about 0.03°C per meter as a function of depth). A few hundred meters below the sea floor, the density of liquid CO_2 would be equal to the density of the pore water. At greater depths, the density of liquid CO_2 would fall below the density of the pore water. CO_2 injected into this zone would be buoyant as a result (its density would be less than the density of the fluid it displaced) and would tend to rise following injection.

The buoyant liquid CO_2 would arrive eventually at a level at which ambient conditions would favor formation of CO_2 hydrates. Hydrates consist of a crystalline (ice) phase in which the CO_2 molecules are trapped in hydrogen-bonded cages of H_2O molecules (5 to 6 molecules of H_2O for every molecule of CO_2). The pore spaces would be clogged by the resulting icy hydrates inhibiting further upward motion of liquid CO_2. Through time, CO_2 would dissolve from the liquid phase into the aqueous phase represented by the pore water. The resulting CO_2-rich pore water would be denser than the original pore water and would tend to sink. Eventually the hydrates would also dissolve. CO_2-rich pore water would sink to the point where dilution with ambient pore water would eliminate the density difference with respect to waters in the pore environment. CO_2 would be essentially permanently trapped in this case. Upward migration of CO_2 could proceed subsequently only as a result of negligibly slow molecular diffusion. This possible sequence is illustrated in Fig. 14.5.

Uptake of CO_2 by the ocean could be enhanced by an artificially induced increase in ocean alkalinity. This would result in a decrease in the build-up of the concentration of CO_2 in the atmosphere for any given emission of CO_2 resulting from combustion of fossil fuels or deforestation (without the need to

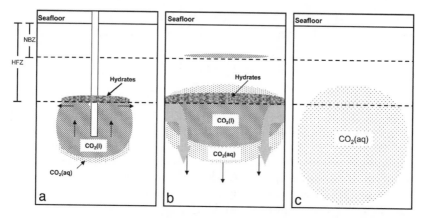

Figure 14.5 The long-term evolution of the injected CO_2: (A) On the injection time scale (~1 year), small amounts of hydrate form as the top of the plume enters the hydrate formation zone (HFZ). The hydrate that forms is expected to impede the upward migration of $CO_2(l)$ and force the $CO_2(l)$ to flow laterally; (B) After about 10^2 years, most of the CO_2 will have reached the bottom of the HFZ, and we expect the self-forming hydrate cap will have expanded laterally and trapped substantial quantities of $CO_2(l)$ below it. Simultaneously, the CO_2-saturated pore fluid will sink away from the HFZ by buoyancy-driven advection; (C) Eventually the $CO_2(l)$ and CO_2 hydrates will have dissolved and formed a $CO_2(aq)$ solution. The solution will percolate through the porous matrix until it has mixed with a large enough quantity of water to become neutrally buoyant. Once the solution is neutrally buoyant, further solute migration will only occur by diffusion (from House et al., 2006).

capture these gases prior to their release to the atmosphere). An intriguing mechanism that could accomplish this objective was suggested by House et al. (2009).

Their proposal involves as a first step electrochemical conversion of salt in seawater (or salt mined from salt deposits) to form sodium hydroxide, gaseous hydrogen, and gaseous chlorine (a standard industrial process enabled by application of energy supplied in the form of electricity):

$$NaCl(aq) + H_2O(l) \rightarrow NaOH\ (aq) + \tfrac{1}{2}\ Cl_2\ (g) + \tfrac{1}{2}\ H_2\ (g) \qquad (14.4)$$

Here the parentheses denote the phase of the indicated substance, either aqueous (aq), liquid (l), or gas (g). The hydrogen and chlorine gases would be combined subsequently to form either aqueous or gaseous HCl:

$$\tfrac{1}{2}\ Cl_2\ (g) + \tfrac{1}{2}\ H_2\ (g) \rightarrow HCl \qquad (14.5)$$

Introduction of NaOH to ocean waters would result in an increase in ocean alkalinity. Enhanced uptake of CO_2 could proceed then through the net reaction

$$CO_2(g) + NaOH\ (aq) \rightarrow HCO_3^- + Na^+ \qquad (14.6)$$

House et al. (2009) point out that release of NaOH (aq) directly into ocean water would result in a local increase in the pH of the seawater receiving this input (as a consequence of reaction of the added OH^- with H^+). To offset the possible negative impact of this change on the ocean biota, they suggested that it might be preferable to neutralize the NaOH (aq) prior to release to the ocean by reacting it with CO_2 externally to form either $NaHCO_3$ or Na_2CO_3. A number of approaches were suggested for disposal of the HCl produced in reaction (14.5) including the possibility of reacting it with silicate rocks, a process that would be equivalent to an acceleration of the natural rate for weathering of these rocks.

House et al. (2009) did not propose that their scheme for enhanced sequestration of CO_2 should be applied immediately. They pointed out that "technologies currently exist that can offset or eliminate emissions from large point sources more cost effectively than the process described here." They noted, however, that "if anthropogenic climate change is expected to remain a serious threat despite the fullest practical deployment of these schemes, then the process [they] discuss could provide the additional mitigation necessary to avoid further damage from climate change." Similar caveats should apply, it would appear, to proposals to dispose of CO_2 by reacting it either with $CaCO_3$ (Caldeira and Rau, 2000) or with silicate rocks (Lackner, 2002). Large quantities of rocks would have to be mined in either case to provide an adequate feedstock to accommodate these options, comparable or even greater than the quantity of CO_2 designated for such treatment. And there would be further challenges associated with environmentally, and economically, acceptable disposal of the products and residues.

In the near term, the best opportunities for carbon capture and sequestration will involve most likely injection of CO_2 into depleted oil reservoirs to enhance secondary recovery of oil, EOR. The next option in terms of preference, should it prove feasible, would be to insert CO_2 into deep, economically inaccessible, coal seams to recover methane that would otherwise be unavailable, ECBM. Option three might involve depositing CO_2 in stable, deep, saline deposits or in depleted gas reservoirs. The economic return in this case would be minimal (although there might be some benefit in the latter case in terms of increased recovery of methane). Option four could require conversion of CO_2 to a stable mineral form such as $CaCO_3$. Economic returns in this case would also be minimal and this procedure is likely to be more expensive than any combination of options 1–3. Use of the deep ocean, preferably ocean sediments, could represent a fifth option, one that is unlikely to be exercised, however, until the potential for options 1–3 has been exhausted. As discussed above, instituting procedures to increase the alkalinity of the ocean, enhancing thus the capacity of the ocean to take up CO_2 from the atmosphere, could provide a sixth option. This is likely, however, to draw opposition from constituencies concerned with the potential for unanticipated consequences associated with such a deliberate alteration of an important component of the global ecosystem. Indeed, if carbon capture and sequestration should become a major activity in the future, it is likely that there will be opposition to all of the choices noted here with the possible exception of options 1 and 2.

A summary of the costs estimated by IPCC (2005) for capture and sequestration of CO_2 produced by three representative new fossil fuel powered plants is presented in Table 14.5. Note in all cases the increase in fuel required to generate a given amount of electricity, ranging from a low of 11% for the natural gas combined cycle plant to a high of 40% for the pulverized coal plant. Assuming that the CO_2 captured is transported and deposited in a geological reservoir (absent potential offsets associated with its use for EOR), the cost for electricity would be increased by a minimum of 21% (under optimal conditions for the integrated coal gasification combined cycle plant), potentially by as much as 91% (for the case of the pulverized coal plant). Estimates for the cost of CO_2 emissions avoided range from $14 to $91 per ton of CO_2. The analysis conducted by IPCC (2005) assumed costs for transporting CO_2 from source to depository of $0–5 per ton of CO_2, with additional expense of $0.6–8.3 per ton of CO_2 for storage and monitoring (to ensure safety, guarding against inadvertent release). These estimates are most likely conservative (on the low side) given lack of experience with actual operating systems. If the captured CO_2 is deployed for purposes of EOR, costs both for electricity generation and carbon sequestration would be reduced as indicated in Table 14.5, benefiting from a subsidy estimated at between $10 and $16 per ton of CO_2, a subsidy assumed to be paid by the operator of the EOR facility for access to the associated source of CO_2.

It is clear from the foregoing that the economically optimal opportunity for sequestration of CO_2 will involve most likely its use for enhanced recovery of oil from depleted oil fields (EOR). As discussed earlier, however, CO_2 employed for this purpose in the American Southwest today is obtained largely from geological deposits (with some contribution from the separation of CO_2 from natural gas) at a cost of approximately $12 per ton. The costs for capture of CO_2 quoted in Table 14.5 reflect costs for capturing CO_2 from new plants for which the capture equipment is incorporated in the design and construction of these plants from the outset. Costs for capturing CO_2 from existing plants, for which the capture equipment must be installed retroactively, will be necessarily much higher. These costs will depend on the time horizon over which the expense for this equipment can be amortized. It is important in this context to note that the average age of existing coal-fired power plants in the U.S. is close to 40 years (17 years for plants fueled by natural gas).

Absent a sizable tax on emissions of CO_2 (or a comparable policy initiative to discourage such emissions), operators of existing fossil fuel powered plants would appear to have little incentive to install the expensive equipment required

Table 14.5 Range of costs for CO_2 capture, transport, and geological storage based on current technology for new power plants (from IPCC, 2005)

	Pulverized Coal Power Plant	Natural Gas Combined Cycle Power Plant	Integrated Coal Gasification Combined Cycle Power Plant
Cost of electricity without CCS (US$ MWh⁻¹)	43–52	31–50	41–61
Power plant with capture			
Increased Fuel Requirement (%)	24–40	11–22	14–25
CO_2 captured (kg MWh⁻¹)	820–970	360–410	670–940
CO_2 avoided (kg MWh⁻¹)	620–700	300–320	590–730
% CO_2 avoided	81–88	83–88	81–91
Power plant with capture and geological storage[6]			
Cost of electricity (US$ MWh⁻¹)	63–99	43–77	55–91
Electricity cost increase (US$ MWh⁻¹)	19–47	12–29	10–32
% increase	43–91	37–85	21–78
Mitigation cost (US$/t$CO_2$ avoided)	30–71	38–91	14–53
Mitigalion cost(US$/tC avoided)	110–260	140–330	51–200
Power plant with capture and enhanced oil recovery[7]			
Cost of electricity (US$ MWh⁻¹)	49–81	37–70	40–75
Electricity cost increase (US$ MWh⁻¹)	5–29	6–22	(–5)–19
% increase	12–57	19–63	(–10)–46
Mitigation cost (US$/t$CO_2$ avoided)	9–44	19–68	(–7)–31
Mitigalion cost(US$/tC avoided)	31–160	71–250	(–25)–120

to capture their emissions. For it to make sense economically for them to do so, even given the opportunities to deploy the captured CO_2 for EOR, it would appear that the tax would have to amount to at least \$100 per ton of CO_2, arguably more. Whether it will make sense for utilities to build new fossil fuel powered plants incorporating CO_2 capture equipment from the outset will also depend on the magnitude of this tax. The choice in this case will involve an assessment of the relative merits of investing in new fossil fuel powered plants incorporating CO_2 capture or pursuing other opportunities (wind and nuclear for example) to supply the projected future market for power.

14.4 PROSPECTS FOR CAPTURING CO_2 FROM THE ATMOSPHERE

Prospects for capturing CO_2 from the atmosphere were discussed by Zeman and Lackner (2004) and by Keith et al. (2005). Thermodynamic considerations indicate that the minimum energy required to extract CO_2 from a gas stream in which it is present at a pressure of p_1 and to concentrate it to a pressure of p_2 should be proportional to the natural logarithm of the ratio of pressures, p_2/p_1. The partial pressure of CO_2 in the atmosphere today is about 4×10^{-4} atm as compared to about 10^{-1} atm for CO_2 in the flue gases produced by a fossil fuel power plant (Table 2.1). It follows that to concentrate CO_2 from the atmosphere to a pressure of 1 atm in preparation for transfer to a suitable geological depository would require an energy investment greater by at least a factor of 3.4 (7.82/2.30) than the energy required to accomplish the same task using the more concentrated CO_2 present in the flue gases of a power plant (Keith et al., 2005). The investment would be reduced if the decision was made to capture less than 100% of the CO_2 present in the air.

Both Zeman and Lackner (2004) and Keith et al. (2005) propose to capture CO_2 from the air by subjecting it to a spray composed of an aqueous solution of sodium hydroxide (NaOH). The CO_2 is converted in the process to CO_3^{2-} which is reacted subsequently with lime (CaO) to form solid calcium carbonate ($CaCO_3$) with reconstitution of the original NaOH solution. CO_2 would be released by heating the product $CaCO_3$ with concurrent recovery of the CaO(s). The key steps in this sequence are summarized in Table 14.6.

Table 14.6 Key steps in the procedure proposed by Keith et al. (2005) for capturing CO_2 from the atmosphere

1. CO_2 dissolved in a solution of NaOH:
   $$CO_2\,(g) + 2Na^+ + 2OH^- \rightarrow CO_3^{2-} + 2Na^+ + H_2O\,(l)$$

2. Addition of lime to the solution:
   $$CaO\,(s) + H_2O\,(l) \rightarrow Ca^{2+} + 2OH^-$$

3. Formation of solid calcium carbonate:
   $$CO_3^{2-} + Ca^{2+} \rightarrow CaCO_3\,(s)$$

4. Heating of $CaCO_3$ to release CO_2 (g) and reconstitute CaO (s):
   $$CaCO_3\,(s) \rightarrow CaO\,(s) + CO_2\,(g)$$

The approach is basically similar to the procedure described earlier for capture of CO_2 from flue gases using MEA. The objective in both cases would be to optimize the ease with which CO_2 was incorporated in the solvent, to minimize requirements for replacement of solvent, and to minimize the energy required to extract CO_2 from the solvent. In the case where the solvent used is NaOH, steps (1) and (2) in Table 14.6 are exothermic (they proceed without requirements for extraneous energy). Step (3) is effectively energy neutral while energy must be supplied to enable step (4).

Keith et al. (2005) envisaged a tower 120 m in height, 110 m in diameter, in which an upward directed air stream would interact with a downward moving spray of sodium hydroxide droplets. They estimated the capital cost for construction of such an integrated capture system at $18 million with annual operating costs of $400,000. The facility could capture, they suggested, as much as 280,000 tons of CO_2 per year at an average cost of about $140 a ton. They recognized that considerable uncertainty should be attached to this cost estimate, however, pointing out that further study could lead to a more efficient and consequently less expensive design. While not cheap, the cost for capturing CO_2 directly from the atmosphere may not be much greater than the cost for capturing CO_2 from an aging coal-fired power plant, especially given the limited time available to amortize the necessary capture equipment, and the fact that in many cases it may be difficult if not impossible to install the capture equipment on existing facilities given constraints with respect to the space available for its placement.

The scheme described by Zeman and Lackner (2004) for capturing CO_2 from the atmosphere is similar to that outlined by Keith et al. (2005). Zeman and Lackner discussed a potential cost for capturing CO_2 from the atmosphere of between $25 and $75 per ton of CO_2, somewhat lower than the figure quoted by Keith et al. (2005). Given the large uncertainties involved in both analyses, however, and the fact that relevant technology has not yet been deployed on an industrial scale, the differences in cost estimates are probably not significant. Further research is clearly indicated to refine our understanding of the ultimate potential (and cost) of a strategy that might emphasize capture of CO_2, not just from smoke stacks but prospectively also directly from the atmosphere.

14.5 CONCLUDING REMARKS

As discussed, capturing CO_2 from large plants burning fossil fuel prior to its emission to the atmosphere is technically feasible. A number of options exist for its subsequent disposal ensuring that it could be isolated from the atmosphere for times ranging from centuries to millennia or longer. The critical question is one of cost. Absent a tax on CO_2 emissions (or a comparable policy initiative), it is unlikely that the capture-sequestration option will be implemented, at least on any large scale, in the near future. A possible exception concerns the potential use of the CO_2 captured from an industrial facility to enhance recovery of oil from depleted oil reservoirs (EOR). CO_2 isolated from methane during processing of natural gas is already being used for this purpose in Texas.

Capturing CO_2 from the atmosphere is also possible. There are advantages in this case in terms of the wide range of choices available for location of the capture facility (or facilities). CO_2 is distributed relatively uniformly throughout the atmosphere (with the exception of a small excess in the northern hemisphere where most of the anthropogenic CO_2 is emitted and of course significant excesses in the immediate vicinity of large sources). A further advantage of targeting the atmosphere is that it would provide an opportunity to capture CO_2 not just from major point sources such as power plants and industrial facilities producing commodities such as iron, steel, cement, and chemical fertilizer but also from distributed sources associated for example with the consumption of fossil fuels in domestic and commercial settings and their use in powering cars, trucks, trains, and airplanes. Again, cost will be a determinative factor. It is not possible at this time to estimate the cost (or tax) that could provide the tipping point to encourage this option.

Photosynthesis offers a natural means for capture of CO_2 from the atmosphere (the energy is supplied in this case by the sun). The trees and plants that absorb this CO_2 provide, however, only a temporary storage reservoir for this carbon. As the plants die, their carbon is returned to the atmosphere as CO_2. In a steady state, uptake and release of carbon may be expected to strike a balance (see discussion in Chapter 3). If the carbon incorporated in biomass by photosynthesis is employed, however, as a fuel (adding wood chips to coal-burning power plants for example) and if the CO_2 produced as a consequence is captured and sequestered, this could provide a potentially significant, additional, path for net removal of CO_2 from the atmosphere. It would be important, though, to ensure that the biomass harvested for this purpose should be replaced by regrowth. And it would be critical to account fully for the CO_2 produced and emitted in conjunction with harvesting the vegetation, in transporting it to the industrial facility where it would be consumed, and in replanting to ensure a continuing source of this biofuel (accounting for example for CO_2 emitted in conjunction with the manufacture of chemical fertilizer that might be required to facilitate regeneration of the related plant material).

The future of carbon capture and sequestration will be determined ultimately by the magnitude of the penalty imposed on emissions of CO_2 to the atmosphere. The higher the tax on carbon emissions, the greater the incentive to introduce carbon-free alternative sources of energy, notably nuclear, wind, and potentially solar.

REFERENCES

Caldeira, K. and Rau, G.H. (2000). Accelerating carbonate dissolution to sequester carbon dioxide in the ocean: geochemical implications. *Geophys. Res. Lett.* 27 no. 2 pp 225–228.

Drange, H., Alendal, G., and Johannessen, O.M. (2001). Ocean release of fossil fuel CO_2. *Geophysical Research Letters*, **28** (12), 2637–2640.

EIA (2002). International Energy Outlook 2002, DOE/EIA-0484 (2002), Energy Information Administration, Office of Integrated Analysis Forecast, U.S. Department of Energy.

Gale, J. and Freund, P. (2001). Coal-bed methane enhancement with sequestration worldwide potential. *Environmental Geosciences*, 8 (3), 210–217.

Herzog, H. (2001). What future for carbon sequestration? *Environmental Science and Technology*, 35 (7), 148A–153A.

House, H.Z., Schrag, D., Harvey, C., and Lackner, K. (2006). Permanent carbon dioxide storage in deep sea sediments. *Proceedings of the National Academy of Sciences*, 103 (33), 12291–12295.

House, H.Z., House, C.H., Schrag, D.P., and Aziz, M.J. (2009). Electrochemical acceleration of chemical weathering as an energetically feasible approach to mitigating anthropogenic climate change. *Energy Environ. Sci.*, DOI. 1039: 2009.

IPCC (2005). *Intergovernmental Panel on Climate Change, Carbon Dioxide Capture and Storage*. Cambridge: Cambridge University Press.

Keith,D.W., Ha-Duong, M., and Stolaroff, J.K. (2005). Climate strategy with CO_2 capture from the air. *Climate Change*, 74. pp 17–45.

Lackner, K.(2002). Carbonate chemistry for sequestering fossil carbon. *Annual Review of Energy and the Environment*, 27 (1), 193–232.

McElroy, M.B. (2002). *The Atmospheric Environment: Effects of Human Activity*. Princeton University Press.

Nordbotten, J., Celia, M., and Bachu, S. (2005). Injection and storage of CO_2 in deep saline aquifers: analytical solution for plume evolution during injection. *Transport in Porous Media*, 58(3), 339–360.

Stevens, S.H., Kuuskraa, V.A., Gale, J. and Beecy, D. (2001). CO_2 injection and sequestration in depleted oil and gas fields and deep coal seams: worldwide potential and costs. *Environmental Geosciences,* 8, no.3, pp 200–209.

Zeman, F. and Lackner, K. (2004). Capturing carbon dioxide directly from the atmosphere. *World Resource Review*, 16 (2), 157–171.

15

Ethanol from Biomass: Can It Substitute for Gasoline?

15.1 INTRODUCTION

Present day annual consumption of gasoline in the United States amounts to close to 150 billion gallons, approximately 500 gallons for every man, woman, and child in the country. With gasoline prices up by almost a third in 2006, the annualized bill for gasoline for a typical U.S. family of four was close to $6000, a burden that fell disproportionally on those least equipped to bear it. Not surprisingly, there has been a political reaction. Leaders of the major oil companies were called to testify in Congress and there were calls for a windfall profits tax. The price of gasoline is linked inevitably, however, to the price of oil and there is little Congress or the oil companies can do about that, at least in the short term. Geopolitical considerations, notably the instability in the Middle East, and international market conditions (increased demand from China and India, political uncertainties in Russia and Venezuela, the general state of the world economy), determine the price of oil, at a then all-time high of more than $75 for a 42-gallon barrel of crude in 2006, hitting a peak of close to $150 a barrel in mid 2006 before falling back to about $50 a barrel in mid 2009. But there is a solution, some would claim.

Why not replace gasoline with ethanol, the stuff that adds zip to your beer and your gin and tonic, a fuel produced from homegrown corn? After all, more than 40% of the world's corn is grown in the U.S. and the U.S. can legitimately lay claim to its status as the world's most efficient agricultural economy. Corn grows by drawing carbon dioxide from the atmosphere through photosynthesis. That should offset concerns about increasing levels of greenhouse gases and consequences for global warming, should it not?

Brazil has emerged in recent years as an ethanol success story. The feedstock in this case is sugar cane rather than corn. The seeds of Brazil's success date back to the oil crises of the 1970s. The military government in power at that time made a decision to subsidize production of ethanol from sugar cane to reduce their dependence on expensive imported oil. They provided generous subsidies and tax breaks to owners of sugar mills encouraging them to switch from refining sugar to producing ethanol, developing at the same time a distribution system to ensure that the product was readily available to consumers. In 1975, they ordered that all gasoline sold in Brazil should be mixed with 10% ethanol, a percentage increased subsequently to between 20 and 25%. Cars capable of running on ethanol only were introduced in the late 1970s, fruits of a research program funded by the government at a Brazilian Air Force research laboratory. The program fell on hard times, however, in 1990 when the combination of a poor sugar cane harvest and high sugar prices led to a serious shortage of ethanol prompting drivers to switch back to gasoline. The program is now back on track thanks to an innovation that allows computers installed in modern Brazilian cars to be programmed (or reprogrammed) at minimal cost to calculate the ethanol to gas mixture present in the tank of a car at any given time and to adjust the operation of the engine accordingly.

Today, more than 80% of all non-diesel new cars sold in Brazil are flex-fuel. With access to either ethanol or a gasoline–ethanol blend (gasohol, containing up to 25% ethanol) at filling stations, motorists in Brazil have a choice. They can opt for either gasoline or ethanol, or a combination, basing purchasing decisions simply on considerations of price and personal preference. The supply of ethanol is insufficient, however, to totally supplant current demand for gasoline.

Ethanol belongs to the class of chemical compounds known as the alcohols. Molecules in this family are characterized by an OH group bonded to a hydrocarbon framework. Ethanol (CH_3CH_2OH), for example, is equivalent to ethane (CH_3CH_3) with one of the hydrogen atoms replaced by OH. Referred to also as ethyl alcohol, ethanol is a liquid at ambient temperatures with a boiling point of 78°C. The energy content of a gallon of ethanol is equal to 76,000 Btu as compared to about 115,000 Btu for a gallon of conventional gasoline. It follows that to replace the energy equivalent of a gallon of gasoline we would need approximately 1.5 gallons of ethanol or, to put it another way, a gallon of ethanol can contribute the energy equivalence of 0.66 gallons of gasoline.

Some 3.4 billion gallons of ethanol were produced from corn in the United States in 2004, sold as a blend with gasoline accounting for about 2% of total gasoline sales in that year by volume, or 1.3% of sales by energy content (Davis and Siegel, 2004). By 2005, production had risen to 3.9 billion gallons accounting for 2.8% of sales by volume or 1.9% by energy content., reaching a level of 4.8 billion gallons per year by mid-2006, with plant under construction expected to add an additional 2.2 billion gallons of capacity over the next few years. Current federal policy in the United States sets a goal for up to 7.5 billion gallons of so-called renewable fuel to be used as an additive to gasoline by 2012 (Farrell et al., 2006). President Bush, in his State of the Union address on January 31,

2006 expressed an objective "to replace more than 75% of our oil imports from the Middle East by 2025."

Brazil is currently the world's largest producer of ethanol (the U.S. is number 2). Some 4 billion gallons were produced in 2004 in Brazil as compared to 3.4 billion gallons in the U.S. The bulk of this production was consumed domestically, more than 80% in 2004. India is currently Brazil's largest customer for ethanol exports, just ahead of the U.S. with significant exports also to Venezuela and South Korea and prospectively (assuming significant growth in future Brazilian domestic production) to Japan (Lynch, 2006). Domestic demand is growing rapidly, however, in Brazil driven in part by the high price of oil, in part by the growing number of flex-fuel vehicles. Not surprisingly, prices have risen accordingly, by a record 14% in the month of March 2006 alone, raising questions as to the quantity of ethanol that might be available in the immediate future for export. Brazil, like the U.S., has plans for important future expansion of its ethanol production capacity.

Production of ethanol from either corn or sugar cane poses a dilemma: whether the feedstock should be devoted to food or fuel. With increasing use of corn and sugar cane for fuel, a rise in related food prices would seem inevitable (and indeed such an increase took place in 2007 and 2008). A potential future option that could avoid this dilemma would involve production of ethanol from cellulose, the ubiquitous component of indigestible grass and wood. Optimists foresee a future where currently idle land could be devoted to cultivation of fast growing grasses (prairie grasses for example) and trees (poplars and willows are mentioned) that could be harvested to produce cellulose to feed a new generation of ethanol factories capable of supplanting as much as 50% of current gasoline use with important savings in terms of emission of greenhouse gases—a great, new, domestically based, energy industry. This would require of course a major commitment of land, perhaps as much as 75% of the land currently devoted to crops in the U.S.

Production of ethanol from corn is discussed in Section 15.2. Production from sugar cane is discussed in Section 15.3, with prospects for cellulose treated in Section 15.3. Concluding summary remarks are presented in Section 15.4.

15.2 ETHANOL FROM CORN

Some 73.4 million acres of land were harvested for corn in the United States in 2004, accounting for approximately 23% of the nation's total cultivated land area (the total planted area amounted to 80.9 million acres). The United States is responsible for production of more than 40% of the world's corn. Most of this corn is fed to animals. Corn is traded in units known as bushels: a bushel consists of 25.4 kg or 56 lb of kernels with a moisture content of about 15% composed of approximately 72,800 kernels. U.S. production of corn in 2004 amounted to 11.8 billion bushels corresponding to an average yield of 160.8 bushels per acre (an 18% increase relative to 2003: variations in weather conditions play an important role in determining year-to-year variations in productivity).

The fraction of the corn crop used to manufacture ethanol amounted to about 12% of total production in 2004 (1.4 billion bushels).[1] Combining the value reported for the total quantity of corn employed in manufacturing ethanol in the United States in 2004 (1.4 billion bushels) with the total quantity of ethanol produced (3.4 billion gallons), we may conclude that 2.4 gallons of ethanol were produced for every bushel of corn consumed by ethanol factories in 2004. Domestic production of ethanol in the U.S. in 2004 fell short of demand by about 160 million gallons (4.5%). Imports—86 million gallons from Brazil, 39 million from Jamaica, 25 million from Costa Rica, and 6 million from El Salvador—made up the deficit.

Use of corn as a feedstock to produce ethanol has been a subject of considerable controversy. Much of the debate has focused on what is referred to as the net energy value (NEV) of corn-based ethanol. This is defined conventionally as a measure of the difference between the energy contained in the final ethanol product as compared to the fossil energy consumed in its production. However, even the definition of NEV has been controversial. The question is how one should treat the energy content of products obtained coincidentally in conjunction with the manufacture of ethanol, what is referred to in the literature as the co-product credit (see for example the discussion in Farrell et al., 2006). Important co-products of ethanol production include dried distiller grains, corn gluten, and corn oil. Should one simply credit the ethanol stream with the caloric energy content of these products? Or would it be more appropriate to estimate first the energy that would be consumed if these products were manufactured in the most efficient manner and to credit the ethanol stream then with the energy costs avoided by substitution of the ethanol co-products for the more conventionally produced alternatives? In estimating the energy input to ethanol production, should one account for the energy embedded in the facility in which the ethanol is produced—the energy consumed to make the steel, concrete, and other materials involved in constructing the ethanol processing plant? Should we account also for the energy embedded in the farm machinery employed to plant, tend, and harvest the corn in the field or the trucks used to transport the corn to the factory after harvest? Not surprisingly, there has been a tendency for proponents of ethanol to stack the numbers one way—to put their case in its most optimistic light—and for opponents to take the opposite tack. Normally, one might expect this not to matter too much. But in the case of the corn–ethanol issue the differences are large and have come to play an important role in the ensuing public debate. One side claims you have to put in more energy than you get back in the final ethanol product: the other disagrees. We shall attempt in what follows to clarify the essential elements of the controversy.

We shall distinguish from the outset between the energy required to produce the corn and the energy employed subsequently to convert the corn to ethanol. Approximately 30% of the total fossil energy employed in the production of ethanol is associated with planting, growing, and harvesting the corn (we exclude from this consideration the solar energy used to grow the corn, the truly renewable input). To achieve the impressive yields of corn realized in the U.S. in 2004 (more than 160 bushels per acre) farmers had to apply copious quantities of fertilizer and significant quantities of pesticides. On an energy basis, nitrogen is the most expensive of the fertilizer inputs. Nitrogen is readily available in the atmosphere. But to be incorporated in living tissue (growing corn for example) the triple bond linking the atoms of N in N_2 (the dominant form of nitrogen in the atmosphere) must be broken. Nitrogen incorporated in living tissue is referred to as fixed nitrogen. Examples of fertilizers delivering fixed nitrogen include ammonia (NH_3), urea ($CO(NH_2)_2$), and ammonium nitrate ($NH_4 NO_3$).

To produce fixed nitrogen from atmospheric N_2 we need to supply energy. Natural gas (methane) is the most common source of the chemical energy used to fix nitrogen. The energy investment required to produce a pound of fixed nitrogen in the U.S. averages 24,500 Btu per pound of N according to Shapouri and McAloon (2004), a little more if we allow for the energy expended in granulating, packaging, and transporting the fertilizer. Typical recommended rates for application of nitrogen fertilizer range between about 120 and 140 lb N per acre, or, assuming the yield for corn achieved in the U.S. in 2004, between 0.8 and 0.9 lb N per bushel. If we assume that 0.85 lb of N were applied for every bushel of corn produced in the U.S. in 2004, the energy investment in nitrogen required to produce a bushel of corn would have amounted to 20,825 Btu. Assuming a yield of 2.4 gallons of ethanol per bushel, the energy investment in nitrogen required to produce a gallon of ethanol (corresponding to a nitrogen input of 0.35 lb N) would be equal to about 8677 Btu corresponding to a little more than 11% of the energy incorporated eventually in the ethanol. Shapouri and McAloon (2004) quote a figure of 23,477 Btu for the average expenditure in nitrogen required to produce a bushel of corn in the U.S. (reflecting presumably somewhat higher applications of N per bushel than assumed above). With the ethanol yield assumed here, this would imply that the energy investment in nitrogen required to produce a gallon of ethanol would be equal to 9782 Btu or about 13% of the ultimate energy yield in the ethanol. We shall adopt in what follows the numbers recommended by Shapouri and McAloon (2004).

Growing corn requires an application of not only nitrogen but also phosphate and potassium. To achieve maximum yields we need in addition to apply pesticides and, in some cases, lime to increase the fertility of the soil. Accounting for these added energy investments, the energy associated with the application of N should be increased by about 6534 Btu per bushel or 2723 Btu per gallon of ethanol product according to Shapouri and McAloon (2004). We should allow also for the energy consumed by farm machinery, for the energy content of the seed, for the energy expended in drying and hauling the harvest, and for the energy deployed in irrigation (the energy used to pump water from below ground), the latter especially important when the corn is grown in dry environments such as Nebraska. Reflecting these additional expenditures, adopting nationally averaged figures recommended by Shapouri and McAloon (2004), the energy invested in corn production should be increased by a further

19,744 Btu per bushel, or 7975 Btu per gallon of product. Combining these various components would imply a total (U.S. average) energy cost for production of a bushel of corn in the U.S. in 2004 of 23,477 + 6534 + 19,744 = 49,755 Btu, an advance payment of 20,731 Btu for the feedstock converted subsequently to a gallon of ethanol (assuming a yield of 2.4 gallons per bushel) corresponding to about 27% of the energy incorporated eventually in the fuel. The estimate of the fossil energy invested in producing a bushel of corn given by Shapouri and McAloon (2004) is almost a factor of 2 lower than the value reported by Pimentel and Patzek (2005), 94,693 Btu. A breakdown of assumptions implicit in the two analyses is presented in Table 15.1.

The estimates of energy expenditures by Pimentel and Patzek (2005) are systematically higher than those reported by Shapouri and McAloon (2004). Their value for nitrogen, for example, is larger than Shapouri and McAloon's recommendation by 21%. Farrell et al. (2006) suggested that the Pimentel and Patzek (2005) value was based on a numerical error introduced in translating the result presented earlier for nitrogen by Patzek (2004). It should be reduced, they suggest, by a factor of 1.23. Applying this correction, Pimentel and Patzek's value for nitrogen would be in good agreement with Shapouri and McAloon's result. Farrell et al. (2006) were critical also of the application rate for herbicides (6.2 kg per hectare) assumed by Pimentel and Patzek (2005), suggesting that this value may be too high by as much as a factor of 2. They disputed also Pimentel and Patzek's approach to estimating the energy embedded in farm machinery and dismissed out of hand their decision to charge for the energy included in the food consumed by farm laborers. On this latter point, we would agree without question: after all, farm workers have to eat whether or not they are involved in cultivating and harvesting corn! We concur also with their criticism of Pimentel and Patzek's estimate for the energy embedded in farm machinery. Their value is only slightly less than the total energy consumed directly on an annual basis by the machinery. Somewhat arbitrarily, we assign to the energy embedded in farm machinery a value of 2000 Btu, equal to about 10% of Pimentel and Patzek's result for the annual energy consumed in the form of diesel and gasoline and in pumping water for irrigation. An amended version of Pimentel and Patzek's energy budget is included in Table 15.1.

The investment in energy required to produce a bushel of corn in our revised version of the Pimentel and Patzek budget is still larger than the value suggested by Shapouri and McAloon (2004)—by 38%—although the discrepancy is reduced significantly in this case relative to the 90% excess associated with the unamended Pimentel and Patzek budget. In an earlier study, Shapouri et al. (1995) reported a value of 59,765 Btu for the energy required to produce a bushel of corn, a value intermediate between the more recent Shapouri and McAloon conclusion and our revised version of the Pimentel and Patzek analysis included in Table 15.1. For present purposes, we shall tentatively adopt a value of 55,000 Btu per bushel, slanted to the lower range of the values listed in the Table 15.1.

Table 15.1 Energy (Btu) required to produce a bushel of corn. Comparison of estimates by Shapouri and McAloun (2004) and Pimentel (2005)

Shapouri and McAloon (2004)		Pimentel and Patzek (2005)		Pimentel and Patzek (2005) (revised)	
Inputs	Energy (BTU)	Inputs	Energy (BTU)	Inputs	Energy (BTU)
Seed	603	Seed	6005	Seed	6055
Nitrogen	23,477	Nitrogen	28,507	Nitrogen	23,500
Potash	1899	Potassium	2923	Potassium	2923
Phosphate	1631	Phosphate	3144	Phosphate	3144
Lime	63	Lime	3668	Lime	3668
Diesel	7491	Diesel	11,680	Diesel	11,680
Gasoline	3519	Gasoline	4716	Gasoline	4716
LGG	2108	Electricity	396	Electricity	396
Electricity	2258	Irrigation	3726	Irrigation	3726
Natural gas	1846	Chemicals**	10,481	Chemicals**	5000
Chemicals	2941	Transport	1968	Transport	1968
Other*	1919	Machinery	11,855	Machinery	2000
		Labor	5380	Labor	0
Total	49,753	Total	94,500	Total	68,776

*Other includes contributions from custom work (1581 Btu), purchased water (136 Btu), and input hauling (202 Btu).

**Includes contributions from herbicides (7220 Btu) and insecticides (3261 Btu).

The first step in converting corn to ethanol involves separation of starch from other components of the corn. Starch accounts for a little more than 60% of the total mass of corn kernels (Graboski, 2002; Shapouri and McAloon, 2004). The chemical composition of starch may be represented by the formula $(C_6H_{10}O_5)_n$. Glucose is produced from starch by hydrolysis (addition of a water molecule to each unit of the starch):

$$C_6H_{10}O_5 + H_2O \rightarrow C_6H_{12}O_6 \tag{15.1}$$

Ethanol is formed subsequently by fermentation, summarized by the net reaction:

$$C_6H_{12}O_6 \, (1 \text{ molecule of glucose}) \rightarrow$$
$$2C_2H_5OH \, (2 \text{ molecules of ethanol}) + 2CO_2 \tag{15.2}$$

The initial fermentation process yields an ethanol/water mixture with an ethanol content of about 8%. A series of as many as three distillation steps must be carried out subsequently to produce ethanol at a concentration of 95%. Further processing is required to obtain essentially pure ethanol (99.5%). Before shipping, ethanol is denatured (made unsuitable for drinking) by addition of 5% gasoline.

All of this processing requires important inputs of energy. Shapouri and McAloon (2004) estimate that 49,733 Btu are required on average to produce a gallon of ethanol from corn, approximately 80% in the form of thermal energy obtained from a combination of natural gas and coal (mainly the former) with the balance supplied by electricity (a little more than 1 kwh per gallon). Their reconstruction of the total energy budget is summarized in Table 15.2. According to their analysis, before accounting for the energy value of co-products, production of ethanol is associated with a positive energy return: an investment of 72,000 Btu yields a gallon of ethanol with an energy content of 76,000 Btu corresponding to a net gain of 5.2%. Shapouri and McAloon (2004) assumed a yield of 2.66 gallons of ethanol from a bushel of corn. As discussed earlier, the statistical data for 2004 suggests a somewhat lower yield of 2.4 gallons per bushel. If we adopt this lower yield, Shapouri and McAloon's energy balance for ethanol would be slightly negative: production of a gallon of ethanol would require an investment of 3799 Btu over and above the energy incorporated in the ethanol corresponding to a deficit of 5.0%. With the numbers recommended by Pimentel and Patzek (2005), the energy balance would be distinctly negative as indicated in Table 15.2: an additional 24,602 Btu would be required to produce a gallon of ethanol, a shortfall of 32%. Even taking into account the revisions to the Pimentel and Patzek budget proposed above, the energy balance is still negative, by 14,242 Btu per gallon or 19%. Note the relative agreement between Shapouri and McAloon (2004) and Pimentel and Patzek (2005) with respect to their estimates for the energy expended in converting the corn feedstock to ethanol. Both studies suggest an investment of about 55,000 Btu per gallon. The major discrepancies relate to the differences between their estimates for the energy expended in growing and harvesting the corn. There are important differences also in how they choose to compensate for the energy value of co-products.

Shapouri and McAloon (2004) begin by noting that only 66% of the corn mass (the starch component) is involved in producing ethanol. Most of the balance is converted to animal feed products. Accordingly, they reduce the energy invested in producing a gallon of ethanol by approximately 36%, from 72,052 Btu to 45,802 Btu (they charge only for the 66% represented by the starch component of the corn). The energy balance in this case is markedly positive, by 30,198 Btu per gallon, corresponding to a positive return on the energy investment of 66% (30,198/45,802). Accounting for the lower ethanol yield inferred here for 2004 (the second column in Table 15.2) the energy involved in producing a gallon of ethanol would be increased to 50,726 Btu (allowing for co-products as in Shapouri and McAloon). The return on the energy investment in this case, while still positive, is reduced to 50%. It is difficult, however, to justify Shapouri and McAloon's approach to estimating the energy savings associated with co-products. Implicit in their analysis is an assumption that obtaining animal feedstock by processing corn through an ethanol factory is energy efficient. This seems unlikely. It would seem preferable to estimate first the energy required to produce, by the most efficient (conventional) path, feedstock nutritionally equivalent to that represented by the ethanol co-products. The energy saved in the ethanol production stream should not exceed this limit. Estimating the energy value of co-products in this

Table 15.2 Energy required to produce a gallon of ethanol

Process	Shapouri and McAloon (2001)		Pimentel and Patzek (2005)	
	Energy[a] (Btu/gallon)	Energy[b] (Btu/gallon)	Energy[c] (Btu/gallon)	Energy[c,d] (Btu/gallon)
Corn production	18,713	20,725	37,870	27,51
Corn transport	2120	2348	4830	4830
Ethanol production	49,733	55,079	56,415	56,415
Ethanol distribution	1487	1487	1487	1487
Total	72,052	79,639	100,602	90,242
Net energy value	+3948	−3639	−24,602	−14,242
Percent gain or loss	+5.2%	−5.0%	−32%	−19%

[a]Corresponding to a yield of 2.66 gallons per bushel.

[b]Assuming a yield of 2.4 gallons per bushel rather than the 2.66 gallons per bushel assumed by Shapouri and McAloon (2004).

[c]Estimate of energy consumed in ethanol distribution taken from Shapouri and McAloon (2004).

[d]Revisions to the Pimentel and Patzek (2005) budget as discussed in the text.

manner is referred to in the literature as the displacement method. Accepting the estimate of the displacement value reported by Farrell et al. (2006), 14,738 Btu per gallon, the net energy investment required to produce a gallon of ethanol in the Shapouri and McAloon model should be raised to 57,314 Btu, representing a surplus of 18,686 Btu relative to the energy incorporated ultimately in the gallon of ethanol (a positive energy yield of 33%).

Results for net energy yields (or deficits) corresponding to the variety of models discussed here are summarized in Table 15.3, with co-product energy values applied as indicated. Considering the uncertainties inherent in all of these analyses, we conclude that the energy balance for corn-produced ethanol is marginally positive: the energy captured in the ethanol is greater than the fossil energy employed in its production by about 20 to 30%. The bulk of this fossil energy is supplied in the form of coal and natural gas, 51% and 38% respectively, with petroleum accounting for as little as 6% according to the Ethanol Today model of Farrell et al. (2006). As indicated earlier, the energy content of a gallon of ethanol is equivalent to the energy content of 0.66 gallons of gasoline. We can think of production of a gallon of ethanol from corn therefore as an opportunity to forgo use of about 0.63 gallons of gasoline (allowing for the petroleum consumed in producing the ethanol). This obviously could be considered a benefit in the United States in terms of reducing demand for imported oil. A separate question is whether substitution of ethanol for gasoline makes sense economically. And we need also to assess the significance of the tradeoffs in terms of the environment, in particular the implications of a switch from gasoline to ethanol for emission of greenhouse gases.

The wholesale price for neat (pure) ethanol in November 2005 was between $2.06 and $2.16 a gallon (Yacobucchi, 2006). Ethanol in the United States is

Table 15.3 Estimate of net energy gains or losses associated with production
of a gallon of ethanol from corn allowing for the energy value
of co-products

	Case A	Case B	Case C	Case D	Case E	Case F
Energy before allowing for co-products (Btu/gallon)	72,052	79,799	72,052	79,799	90,242	74,417
Co-products allowance (Btu/gallon)	26,250	26,250	14,738	14,738	14,738	14,738
Net energy cost (Btu/gallon)	45,802	53,549	57,314	65,061	75,504	56,679
Net gain or loss (Btu/gallon)	+30,198	+22,451	+18,686	+10,939	+496	+16,321
Percent gain or loss	+66%	+42%	+33%	+17%	+0.7%	+27%

Case A   Refers to the original Shapouri and McAloon (2004) budget.

Case B   Refers to the Shapouri and McAloon (2004) budget assuming ethanol yield of 2.4 gallons
per bushel.

Case C   Same as Case A but with co-product allowance reduced as recommended by Farrell et al. (2006).

Case D   Same as Case B with Farrell et al. (2006) value for co-product allowance.

Case E   Revised Pimentel and Patzek (2005) budget as indicated in Table 15.2 but allowing for energy
benefit of co-products following Farrell et al. (2006).

Case F   The Ethanol Today model of Farrell et al. (2006).

used as a blend with gasoline, referred to as E10 (10% ethanol, 90% gasoline).
Use of ethanol as a blend with gasoline is subsidized in the United States at a
level of 51 cents a gallon of ethanol (the subsidy was 52 cents prior to 2004).
The subsidy accrues in the form of an income tax rebate to the company that
blends the ethanol with gasoline. The effective wholesale price for a gallon of
ethanol was reduced therefore to between $1.44 and $1.45 in November 2005,
roughly equal on a per gallon basis to the wholesale price of gasoline that prevailed at the same time (between $1.44 and $1.45 a gallon). By May 2006, however, it had risen to $2.65 a gallon. We must allow, however, for the fact that the
energy content of a gallon of ethanol is much less than the energy content of a
gallon of gasoline: it takes 1.5 gallons of ethanol to provide the energy punch of
a gallon of gasoline. The subsidized price for a gallon of ethanol translates
therefore on an energy basis to an equivalent wholesale price for a gallon of
gasoline of $2.17 to $2.19 in November 2005. The customer was obliged consequently to pay an extra $0.90 to $1.06 per gallon of gasoline equivalent on an
energy basis as a hidden subsidy for ethanol when filling up his or her car in
November 2005. The inequity was even more pronounced a few months later.
The wholesale price of gasoline had risen in April 2006 to close to $2.00 a gallon.
With ethanol selling at $2.65 a gallon, the energy equivalent price of ethanol
was then close to $4.00 a gallon, almost twice the price of gasoline reflecting
primarily an increase in demand for ethanol, driven mainly by moves to substitute ethanol for MTBE (methyl tertiary-butyl ether) as an antiknock additive

to gasoline (use of MTBE is currently being phased out in the U.S. due to concerns about the chemical's persistent carcinogenic contamination of groundwater). By July 2006 the wholesale price of ethanol had risen to $3.10 a gallon in corn-producing states such as Illinois, touching close to $4.00 a gallon in California.

As noted earlier, approximately 4.4% of ethanol consumed in the United States in 2004 was supplied by imports. Of this, 54% came from Brazil. The United States imposes a tax of 54 cents per gallon on imported ethanol in addition to a duty equal to 2.5% of the value of the imported product. The price of ethanol loaded on a ship in Brazil averaged 87 cents a gallon in 2004. Allowing for shipping costs the price per gallon for ethanol arriving at a port in the United States was $1.01. Adding the tax and import duty, the price climbed to $1.58 per gallon, competitive with prevailing, subsidized, prices in the U.S. at the same time. Ethanol imported into the U.S. in 2004 that did not originate in Brazil came mainly from countries in the Caribbean. There is a simple explanation for this. The U.S. Congress in 1984 passed the Caribbean Basin Initiative (CBI) designed to promote economic development in 24 Caribbean and Central American countries. The initiative included a provision that allowed ethanol made or processed in these countries to be imported duty free to the U.S., capped at a limit equal to 7% of U.S. ethanol consumption. In addition to this, a further 35 million gallons can be imported duty free if it is produced using 30% or more of local (CBI country) sugar cane. An additional 35 million gallons can be imported duty free if produced using 50% or more of locally grown sugar cane. U.S. companies are moving aggressively to take advantage of the loopholes available under the CBI. Cargill, for example, was negotiating in 2004 to build a dehydration facility in El Salvador that would convert Brazilian ethanol to fuel grade product (by removing the residual water) to be imported subsequently duty free to the U.S. Would it not be better to eliminate all of these barriers to trade at a time when ethanol is in short supply in the U.S. and when prices are at an all time high despite the 51cents a gallon subsidy provided to domestic producers by the U.S. tax payer? To do so would provide a welcome break for the U.S. consumer, and at the same time a boon to the economies of developing countries such as Brazil. Everybody would win (except possibly Cargill and other similarly positioned companies). Why have we not done it? Put it down to U.S. domestic politics and the power of special interests.

Table 15.4 provides a list of the top-ten ethanol producers by capacity in the United States in 2006 (Renewable Fuels Association, January 2006). You should not be too surprised therefore when you see advertisements on television from Archers Daniels Midland trumpeting the merits of ethanol as a clean fuel that reduces our dependence on foreign oil, offering at the same time a potential solution to the problem of climate change. Is ethanol a good deal economically for the consumer? The answer is unequivocally no. Senator John McCain (2003) summed it up thus: "Plain and simple, the ethanol program is highway robbery perpetrated on the American public by Congress." Archers Daniel Midland and the other companies listed in Table 15.4 are the winners. Succinctly

Table 15.4 Major producers of ethanol in the U.S., 2006

Company	Capacity (millions of gallons per year)
Archer Daniel Midland	1070
VerSun Energy Corporation	230
Aventine Renewable Energy	207
Hawkeye Renewables	200
ASAlliances Bioenergy Corporation	200
Abenoga Bioenergy Corporation	198
Midwest grain Processors	152
U.S. BioEnergy Corporation	145
Cargill	120
New Energy Corporation	102
All Others	3658
Total	6258

Source: Renewable Fuels Association, U.S. fuel Ethanol Industry Plants and Production Capacity, January 2006. Includes capacities both for plants in production and under construction. Quoted from Yacobucci (2006).

put, ethanol from corn offers an opportunity to convert coal and natural gas to a liquid fuel that can substitute partially for petroleum. But at the moment the price is unacceptably high. The consumer and the taxpayer are obliged to foot the bill. And the contribution of corn-based ethanol as a solution to the climate problem is marginal at best. The benefits from ethanol produced from sugar cane are more significant as discussed below.

Combustion of the fossil fuels involved in production of ethanol from corn is associated inevitably with emission of CO_2, the most abundant anthropogenic greenhouse gas. Application of nitrogen fertilizer in growing the corn leads to emission of a second greenhouse gas, nitrous oxide (N_2O), which is even more effective as a greenhouse agent than CO_2, by a factor of 296 on a molecule per molecule basis according to the Intergovernmental Panel on Climate Change (2001). Allowing for emissions of N_2O in addition to CO_2, Farrell et al. (2005), in their Ethanol Today model, concluded that the climate impact of a corn-based ethanol–gasoline substitution was comparable to that of a gasoline-only system (a 13% reduction). The impact was somewhat more negative with their CO_2 Intensive model (this model accounts for plans to ship corn from Nebraska to a coal-fired ethanol plant in North Dakota): greenhouse emissions were increased in this case by 2 % relative to the gasoline-only standard.

15.3 ETHANOL FROM SUGAR CANE

On an energy basis, sucrose accounts for approximately 30% of the photosynthetic product of sugar cane, with the balance distributed more or less equally between leafy material normally left in the field after harvest (if not previously burnt) and bagasse, the fibrous matter that contains the sucrose (Wikipedea, 2006). The traditional practice for harvesting the cane involves first setting fire to the field in which it is growing. This has a two-fold purpose: first it acts to remove much of the extraneous vegetation; second, it acts to soften the cane, making it easier to harvest the crop. Harvesting cane in the traditional manner is a backbreaking, labor-intensive job. The cane in this case is cut by hand by laborers wielding machetes, not much different from the technique used by the slaves who supplied the labor for the first sugar plantations in the Caribbean. The sucrose content of the cane is concentrated near the base increasing the backbreaking nature of the work involved in its harvest. The worker engaged in this traditional harvesting practice is expected to harvest up to 10 tons of cane per day (Rohter, 2006) for a fraction of the wages earned by agricultural workers in the U.S. Increasingly though, the harvesting process is being mechanized in Brazil eliminating the need to burn the fields prior to harvest, providing higher paying jobs (but much fewer).

After it is harvested, the cane is transported to one of several hundred ethanol factories (distilleries) scattered across the cane-producing region of the country (concentrated in the south near Sao Paulo and in the northeast where climatic conditions are favorable for cultivation of the sugar cane plant). Once there, the cane is crushed through a series of rollers much like the rollers used of old to wring water from clothes after they had been washed. The juice extracted in this manner consists of a mixture of water and sucrose. The sucrose is concentrated by boiling off the water and is converted subsequently to ethanol by fermentation following the same procedures used to process corn to ethanol after the starch of the corn has been converted to sugar as discussed above. Burning bagasse left behind when the sucrose is separated from the cane provides the energy needed to boil off the water in the cane juice and to promote subsequent fermentation and concentration of the ethanol. Ethanol factories in Brazil are generally energy self-sufficient. Electricity requirements for the factories are met typically on site with turbines driven by steam produced by burning the bagasse, with a surplus available for sale to the national electricity grid. More than 1300 MW of electricity were generated in this manner in Brazil in 2001 with the excess of a little more than 10% (150 MW) sold to utilities (Wikipedia, 2006).

Ethanol as produced in Brazil today is clearly positive in terms of its implications for emission of greenhouse gases. The bulk of the CO_2 released either in the process of producing the ethanol or in its ultimate consumption represents CO_2 recycled by photosynthesis from the atmosphere. Fossil energy is consumed in producing the nitrogen fertilizer required to grow the sugar cane, in harvesting it (to the extent that harvesting is mechanized), in transporting the cane to the processing facility, and in delivering the ethanol product to market. Emission of greenhouse gases (N_2O as well as CO_2) from the combination of these activities is more than offset, however, by the CO_2 saved by substituting ethanol for gasoline and the additional savings associated with the substitution of electricity produced by combustion of (photosynthetic) bagasse for electricity that would be produced otherwise by burning fossil fuels with associated emission of excess CO_2 to the atmosphere.

Today, approximately 50% of Brazil's sugar cane is used to make sugar with the balance devoted to production of ethanol. The decision on whether to make sugar or ethanol depends on prevailing prices for sugar and gasoline (ethanol's competition in the market). It is anticipated that in the future the balance will switch significantly in favor of ethanol (although that could change again as it did in the early 1990s depending on future trends in prices for sugar and oil). Current plans call for an investment in ethanol facilities of as much as $9 billion over the next 6 years with the aim of doubling the amount available for export by 2010 (Lynch, 2006). Annex 1 Parties to the Kyoto Climate Protocol (mainly countries of the European Union in addition to Canada and Japan—the United States and Australia declined to ratify the Protocol) are obligated to reduce emissions of greenhouse gases and clearly have an incentive to supply at least part of this capital. The Clean Development Mechanism (CDM) incorporated in the Protocol (Article 12) encourages investments by Annex 1 Parties in developing countries that can lead to net reductions in global emissions of greenhouse gases. Investing Parties are permitted under terms of the Protocol to assume credit for reductions certified under CDM. Credits so earned can be applied to offset obligations to which they are committed nationally under conditions of the Protocol.

An acre of land planted to sugar cane in Brazil today yields approximately 640 gallons of ethanol, an increase by more than a factor of 3 over yields achieved 25 years ago according to Luhnow and Samor (2006). This may be compared with the yield of ethanol obtained from corn in the U.S. in 2004, about 380 gallons per acre as discussed above. Production of 4 billion gallons of ethanol in Brazil in 2004 with a yield of 640 gallons per acre would have required processing sugar cane harvested from approximately 6.25 million acres, or 25,291 km^2 equal to about 0.3% of the total land area of Brazil (9.37 billion km^2). In practice, the national average yield of ethanol was probably less than 640 gallons per acre (yields are appreciably lower in the northeast as compared to the south) and the estimate for the amount of land required to cultivate the necessary sugar cane should be increased accordingly. Wikipedia (2006) quotes a figure of 45,000 km^2 for land devoted to cultivation of sugar cane for purposes of ethanol production in 2000. This would correspond to a little more than 0.5% of the total land area of the country and represents probably a more reasonable estimate of the actual commitment. By way of comparison, the fraction of the land area of the lower 48 states of the U.S. used to cultivate corn for ethanol is equal to about 0.4 %, not very different from the fractional commitment of land to sugar cane ethanol in Brazil. The fraction of available solar energy converted to energy that can be employed as ethanol in the transportation sector is evaluated in Box 15.1.

A field planted with sugar cane can provide multiple crops. After the initial harvest, so long as the roots are not disturbed, the cane will regenerate. Up to ten harvests are possible as a result of a single planting. The yield of sugar cane from subsequent crops decreases with time, however, providing an incentive for early replanting. The high yield quoted by Luhnow and Samor (2006) may refer to experience relating to the early harvests following initial planting.

Box 15.1

Calculate the fraction of incident solar energy converted to ethanol by processing sugar cane under optimal conditions in Brazil. The average rate at which solar energy is absorbed by the Earth, as evaluated in Box 3.2, is equal to 241 W m^{-2}. Assume that solar energy available on average in the sugar cane growing regions of Brazil is approximately 25% higher than this, 300 W m^{-2}, reflecting the relatively low latitude of the cane growing region and the fact that the region is cloud free to a greater extent than the global average. Assume further that sugar cane grows for 10 months a year (the harvest period lasts approximately two months). Assume a yield of ethanol of 640 gallons per acre as discussed above.

Solution:

As indicated earlier in this chapter, 1 W is equivalent to an energy source of 9.48×10^{-4} Btu s^{-1}. The total energy, E, available from the sun over a 10-month period, assuming power input of 300 W m^{-2}, is given by

$$E = (3 \times 10^2)(9.48 \times 10^{-4})(2.62 \times 10^7) \text{ Btu m}^{-4}$$
$$= 7.45 \times 10^6 \text{ Btu m}^{-2}$$

where 2.62×10^7 is the number of seconds corresponding to a time interval of 10 months.

One acre corresponds to an area of 4.05×10^3 m^2. Hence the total solar energy available over a 10-month period per acre, F, is given by

$$F = (7.24 \times 10^6)(4.05 \times 10^3) \text{ Btu}$$
$$= 3.02 \times 10^{10} \text{ Btu}$$

The energy content of a gallon of ethanol is equal to 7.6×10^4 Btu. The energy content of 650 gallons of ethanol, G, is given then by

$$G = (7.6 \times 10^4)(6.5 \times 10^2) \text{ Btu}$$
$$= 4.94 \times 10^7 \text{ Btu}$$

It follows that the fraction, f, of the available solar energy converted to ethanol is given by

$$f = \frac{G}{F} = \frac{4.94 \times 10^7}{3.02 \times 10^{10}} = 1.6 \times 10^{-3}$$

Conclusion: A little less than 0.2% of the available solar energy is converted to chemical energy in the form of ethanol, a reasonably impressive return given that sucrose accounts for only about 30% of the energy content of above-ground sugar cane biomass.

15.4 ETHANOL FROM CELLULOSE

Cellulose, composed of long chains of glucose molecules, is the most abundant molecular component of the biosphere. Together with lignin, it is responsible for the structural integrity of plants—trees and grasses (trees are able to stand tall largely because of their cellulose and lignin). The primary cell walls of plants are composed mainly of cellulose with variable quantities of lignin included in the cell secondary walls. While cellulose is similar in elemental composition to starch, $(C_6H_{10}O_5)_n$, it is structurally quite distinct. Cellulose is an extended straight chain polymer: starch exhibits a characteristic coiled structure. Starch can be broken down more easily than cellulose to isolate its glucose components.

Cellulose cannot be digested directly by either humans or animals: ruminants (cattle sheep and goats) are able to live on cellulose but they do so through a symbiotic relationship with bacteria present in their stomachs (rumens). The grass the ruminant eats is processed and decomposed by the bacteria. The products of this decomposition provide the nutrient for the animal host providing at the same time nourishment for the bacteria (that is what we mean by a symbiotic relationship: profitable for both parties).

Several steps are required to produce the sugars from cellulose that can be fermented to make ethanol. First, the cellulose must be separated from the lignin. The preferred method in this case involves first breaking down the plant material into smaller units, elements no more than a few millimeters in size. The resulting pellets are treated subsequently by exposing them either to a dilute acid at high temperature (in excess of 230°C) or a more concentrated acid at lower temperature (100°C). The relatively large surface area relative to volume of the pellets as compared to the parent material promotes the efficiency with which the acid is able to access the plant material and isolate the cellulose. The advantage of the concentrated acid/low temperature treatment is that most of the sugar content of the cellulose is conserved, in contrast to the high temperature case where as much as 50% of the cellulose may be lost. The disadvantage is that the treatment takes a much longer time (hours) (Badger, 2002).

Further processing is required to release the fermentable sugars of the cellulose. The primary objective of current research is to find an economically efficient means to accomplish this task (Detchon, 2006). Genetic engineering of naturally occurring organisms offers a potential solution. Costs for production of the necessary enzymes have been reduced significantly (by a factor of 30) over the past 5 years as discussed by Detchon (2006). Ethanol is produced by subsequent fermentation following the methods discussed earlier for treating corn and sugar cane. A second possibility is to combine the two processes into a single procedure: to find (or more likely engineer) an organism that can both break down the cellulose and promote the fermentation required subsequently to produce the ethanol. This latter option is referred to as consolidated bioprocessing or CBP (NRDC, 2004).

There can be little doubt, in principle at least, that plant material can provide a viable source of ethanol. The challenge is to find the most efficient means to accomplish this objective, to assess its cost and future potential and to identify preferred sources of the necessary plant material, whether wood or grass or both. Current debate has focused on the potential of purposefully grown crops such as switchgrass or fast growing trees such as poplars and willows. It is clear that there is an opportunity also to use a variety of otherwise surplus or waste products including discarded paper, sawdust, and the stocks of crops such as corn and wheat (Detchon, 2006; Lave et al., 2001; NRDC, 2004).

Switchgrass is a perennial grass native to the American prairies (perennial means that after mowing, the grass will regrow: it does not need to be replanted). Since it is native, it is relatively pest resistant. Also, its demand for fertilizer, notably nitrogen, is significantly less than that for corn, by as much as a factor of 2 or 3 (NRDC, 2004; Pimentel and Patzek, 2005). It follows that the requirements for fossil energy inputs in growing switchgrass are significantly lower than those for corn (see the discussion in Section 15.2), particularly so if the energy stored in lignin is used to fuel the ethanol-producing plant. Emission of greenhouse gases (allowing for the different climate-altering potential of the different gases) is reduced as a consequence, by close to 86% according to Farrell et al. (2006) (comparing their "Ethanol Today" and "Cellulosic" models).

The stated objective of the U.S. Department of Energy is for ethanol production in the U.S. to rise to the point where it can displace 60 billion gallons of gasoline annually by 2030 (U.S. DOE, 2006). Given the lower energy content of ethanol relative to gasoline, this would require an annual production capacity for ethanol of 90 billion gallons. Is this realistic? How much land would it take?

The total land area of the U.S. amounts to a little less than 2.3 billion acres, 1.9 billion acres in the lower 48 states. The United States Department of Agriculture (USDA) classifies agricultural land under four categories: cropland, grassland pasture and range, forest use land, and land devoted to other special purposes (including highway, road and railroad rights-of way, national and state parks, wilderness areas, urban areas, and land restricted for defense or industrial purposes). Cropland is further divided into land actively planted with crops, land that could be planted with crops but is currently idle (including land enrolled in the federal Conservation Program),[2] and cropland used only for pasture. The total land area devoted to agriculture in the U.S. in 2002 amounted to 1.17 billion acres, almost all in the lower 48 states. Of this, cropland accounted for 442 million acres, grassland 584 acres with forested land used in part for grazing (generally forested areas occupied in part by grass) contributing an additional 134 million acres. Some 1.09 billion acres were classified as non-agricultural land, 723 million acres of this in the lower 48 states. A summary of U.S. land use is presented for 2002 in Table 15.5. Trends over time are illustrated in Fig. 15.1.

As discussed earlier, 2.4 gallons of ethanol were produced from a bushel of corn kernels in 2004 (see Section 15.2). The mass of a bushel of corn is equal to 25.4 kg. It follows that approximately100 gallons of ethanol were produced from a ton of corn kernels in 2004. We shall assume as a point of departure that the yield of ethanol from biomass is similar to that for corn (1 ton biomass: 100 gallons of ethanol). Production of 90 billion gallons of ethanol would require therefore a source of 900 million tons of biomass. The yield of corn kernels from an acre of land in 2004 was equal to approximately 160 bushels, implying that the yield of biomass in the form of corn kernels from an acre of land in that year was equal to approximately 4 tons. If the biomass yield of cellulose per acre was comparable to that realized for corn kernels in 2004, it follows that we

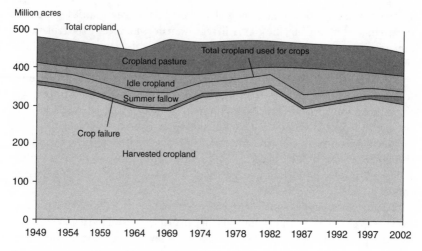

Million acres

Figure 15.1 Major uses of U.S. cropland.

would require 225 million acres of land to replace 60 billion gallons of gasoline—more than six times the cropland acreage idled under the Federal Conservation Reserve Program (CRP), 60% of the land currently under active crop cultivation or 39% of all grassland and range. To displace 50% of current gasoline consumption we would need more than seven times the amount of cropland idled under CRP, 75% of all of the cropland currently in use or 49% of all grassland and range, a formidable commitment especially if the goal is at the same time to maintain current output from the agricultural sector.

A little more than half of the price of ethanol produced from corn in the U.S. is represented by the cost of the corn feedstock (Pimentel and Patzek, 2005). The cost of cellulose is unlikely to exceed that of corn, but it could be comparable, especially so if marginal lands had to be developed (requiring extensive irrigation for example) to supply the necessary cellulose. The expense in producing ethanol from cellulose will be greater of necessity than the expense involved in producing it from corn. It is impossible, however, to assess the magnitude of these costs until the details of the manufacturing strategy are defined. It is clear though that for cellulosic ethanol to be viable it must be able to compete economically with ethanol from corn and both must be competitive with gasoline. The challenge here is that the price of gasoline is linked inextricably to the international price of oil and, as recent history has demonstrated, this can change rapidly, with price swings in both upward and downward directions. A prospective ethanol industry could be stillborn if oil prices were to drop markedly below $50 a barrel (as they did in the early months of 2009). Given the significant sources of capital required to develop a large-scale, national,

Table 15.5 Agricultural and nonagricultural uses of land, United States, 2002

Land use	Acreage		Proportion of total	
	48 States	**U.S.**	**48 States**	**U.S.**
	Million acres		**Percent**	
Agricultural				
Cropland:				
Cropland used for crops[1]	340	340	18.0	15.0
Idle cropland	40	40	2.1	1.8
Cropland used only for pasture	62	62	3.3	2.7
Grassland pasture and range	584	587	30.8	25.9
Forest-use land:				
Forest land grazed	134	134	7.1	5.9
Special uses:				
Farmsteads, farm roads	11	11	0.6	0.5
Total agricultural land[2]	1171	1174	61.8	51.8
Nonagricultural				
Forest-use land:				
Forest-use land not grazed[3]	425	517	22.4	22.8
Special uses:				
Transportation uses[4]	27	27	1.4	1.2
Recreation and wildlife areas[5]	100	242	5.3	10.7
National defense areas[6]	15	17	0.8	0.8
Urban land	59	60	3.1	2.6
Miscellaneous other land[7]	97	228	5.1	10.1
Total nonagricultural land[2]	723	1,091	38.2	48.2
Total land area[2]	**1894**	**2264**	**100.0**	**100.0**

[1]Cropland harvested, crop failure, and cultivated summer fallow.

[2]Breakdown of land uses may not add to totals due to rounding.

[3]Excludes 98 million acres of forest land in parks and other special uses.

[4]Rural highways, roads and railroad rights-of-way, and rural airports.

[5]National and State parks and related recreational areas, national and State wildlife refuges, and national wilderness and primitive areas.

[6]Federal land administered by the Department of Defense for military purposes and land administered by the Department of Energy.

[7]Includes miscellaneous uses not inventoried, and areas of little surface use such as marshes, open swamps, bare nock areas, desert, and tundra.

Sources: Estimates are based on reports and records of the Bureau of the Census (HUD/BOC. 1992, 2002, 2003) and Federal, State, and local land management and conservation agencies, including DOI/BLM, 2003; DOT/BTS, 2004; DOT/FAA, 2002; DOT/FHWA, 2002; DOT/FRA, 2004; USDA/FS 1989, 1998; DOI/FWS, 2001; GSA, 2001; GDT, 2000; USDA/NASS, 2004a, 2004b, 2005; DOI/NPS, 2002; USDA/NRCS, 2000, 2004a; and Wl, 2002.

ethanol-manufacturing capacity, investors will need assurance for a reasonable return on their investments before committing their money.

This assurance could be provided by introducing a tax on gasoline linked to the production price for ethanol on an energy equivalent basis. The tax should increase in this case in response to a drop in the price of gasoline and would be unnecessary if the price of gasoline were to rise above some predetermined threshold. Given the political opposition to "new taxes," it would be important that such a tax should not be perceived as an additional source of government revenue. Revenue raised should be applied to reduce other government expenditures, ideally to compensate those most impacted by a gasoline surcharge—specifically the least advantaged members of our society. It could be applied for example to reduce the tax rate for lower income wage earners, or to decrease the taxes they pay for social security, or for a myriad of other, socially-desirable, revenue-neutral, purposes. It would fall on politicians to work out the details and convince citizens of the merits of any such gasoline/ethanol tax proposal.

Ethanol from cellulose has the potential to markedly reduce our dependence on imported oil, even though the goal to replace 50% of current domestic use of gasoline may be unrealistic (given the requirements for land and the competition with land required for agriculture). Replacing some portion of gasoline use with cellulose-produced ethanol can provide for an important reduction in emission of greenhouse gases. To realize the full environmental and fossil energy-saving potential of cellulose-produced ethanol, it will be important, however, to make use of the lignin associated with the cellulose in manufacturing the ethanol. Whether this happens in practice will depend inevitably on considerations of cost. In principle, corn stocks could be harvested in combination with corn kernels and the energy content of the stocks could be employed to provide the heat needed to fuel the boilers of the ethanol plants and to generate or at least supplement requirements for electricity (as with bagasse in Brazil). In practice, though, operators of corn-based ethanol plants find it more economical in the U.S. to use commercially available coal and electricity to supply their needs for energy.

Farrell et al. (2005) estimate that the savings of oil in energy terms for cellulose-produced ethanol used as a substitute for gasoline could amount to as much as 93% with reductions in greenhouse gas emissions (on a climate-altering equivalent basis) as large as 88 %, clearly a positive on both scores. Until the source of the cellulose, the means for its production, and the details of its subsequent conversion to ethanol are clear, it will be difficult to assess definitively the accuracy of their conclusions.

15.5 SUMMARY AND CONCLUDING REMARKS

We provided a detailed discussion in Section 15.2 of the complex arguments relating to the use of corn-based ethanol as a substitute for gasoline in the U.S. The reader may have been understandably confused by the plethora of numbers quoted in this context. Hopefully, though, the tables included in the section provide a convenient summary of the facts.

We addressed a number of issues. The first concerned the energy balance of corn-based ethanol—inputs and outputs—associated with production of ethanol from corn. How does the fossil energy expended in producing ethanol compare with the energy recovered in the ethanol? Proponents, notably Shapouri and McAloon (2004), argue that the energy balance is distinctly positive (you invest much less than you gain by way of return). Opponents, Pimentel and Patzek (2005) for example, reach the opposite conclusion. We reviewed the arguments in Section 15.2 and reached the conclusion that the truth lies somewhere in the middle. As best we can tell with current data, the energy balance is mildly positive, by 20% or so, in agreement with conclusions reached earlier by Farrell et al. (2005).

There is a net saving of gasoline and therefore imported oil associated with the use of ethanol produced from corn. The fossil fuel inputs employed in production of the ethanol under current practice in the U.S. involve mainly coal and natural gas. We concluded that a gallon of ethanol could displace approximately 0.63 gallons of gasoline.

The savings in greenhouse gas emissions associated with corn-derived ethanol are minimal when we consider the entire life cycle of the product, from the field used to grow the corn to the factory used to covert it to ethanol. The carbon transformed to CO_2 and released to the atmosphere when the ethanol is consumed may be considered simply as a return of carbon to the atmosphere, replacement of carbon taken up earlier by the corn plant when it grew using photosynthesis. But along the way, fossil energy was consumed—in planting, growing, harvesting, and processing the corn—and extra CO_2 was added to the atmosphere—the carbon bound up for hundreds of millions of years in coal, oil, and natural gas. In addition, microbial processing of nitrogen applied as fertilizer in the cornfield is responsible for emission of a greenhouse gas, N_2O, that is almost 300 times more efficient on a molecule per molecule basis than CO_2 as a climate-altering agent. The net result is a wash: the climate impact of using ethanol produced from corn is about the same as for gasoline when we take into account the detailed life cycles of the two products.

A recent General Motors advertisement, lauding the potential of corn-based ethanol, begins with the slogan: "Live green, go yellow." The advertisement was taken to task by no less than the influential Consumer Reports magazine in its October 2006 issue. In an article titled "The ethanol myth," the magazine reports on a test conducted using a 2007 Chevrolet Tahoe Flexible Fueled Vehicle (FFV) running on 85% ethanol, 15% gasoline (E85). The retail price of E85 at the pump in August 2006 averaged $2.91 a gallon. They found that the mileage of the car they tested running on E85 decreased from an already low 14 miles per gallon (mpg) on gasoline to 10 mpg on E85. To realize the same mileage as they would with gasoline (accounting for the lower energy content of ethanol) they concluded that the price of $2.91 for a gallon of E85 was actually equivalent to a price of $3.99 for a gallon of gasoline: clearly not a good deal. The price of ethanol may be artificially high at present reflecting a temporary mismatch between supply and demand. And it is possible that it could decrease significantly in the future with anticipated expansion of production capacity. In the long term, though, competition with the demand for corn as a feedstock for animals in the U.S. and as a valuable commodity for export poses a dilemma. The advantages for the environment, specifically the contribution corn-based ethanol could make towards mitigating the problem of global climate change, are limited and would not appear to justify the cost of a major expansion of corn-based ethanol in the foreseeable future.

Prospects for ethanol as produced from sugar cane in Brazil are more promising. The yield of ethanol per acre planted with sugar cane is greater than for corn and there is potential for significant expansion of future production. Demand for ethanol in the U.S., in Europe, in Japan, and elsewhere is likely to grow significantly over the next few years: Brazil may be best equipped to meet this demand. A potential problem could arise, however, if expansion of sugar cane production in Brazil were to increase pressure for conversion of land currently occupied by rain forest to purposes of agriculture, compensating for example for pasture lost in the subtropics, displaced as the result of an increased commitment to sugar cane. A large fraction of the world's above ground carbon is stored in tropical rain forests. Current studies suggest that deforestation in the tropics is responsible for addition at present of as much as 2 billion tons of carbon every year to the atmosphere, 30% of the contribution from global consumption of fossil fuels (McElroy, 2002). From an environmental point of view, it would be unfortunate if savings in greenhouse gas emissions associated with substitution of biomass produced ethanol for gasoline were to be offset by destruction of a critical global ecological resource, the tropical rain forests, site of so much of the world's irreplaceable biological diversity (Wilson, 1992).

In the long run, ethanol produced from cellulose may hold the greatest promise, as an environmentally constructive alternative to fossil fuel derived gasoline. Vast reservoirs of cellulose are stored in the boreal forests of Canada and Russia. And there is potential for significant production from switchgrass and other sources in the U.S. and elsewhere (Argentina for example). Whether cellulose lives up to its promise as a source of fuel competitive with gasoline will depend, however, on results from the ongoing research program and on considerations of economics that are difficult to predict given uncertainties in what the future cost of ethanol produced from cellulose will be, in addition to uncertainties in the future price of oil. Creative public policy initiatives may be required to ensure a level playing field for ethanol and may be justified ultimately by considerations of the implications for global security and for the health of the global environment. But, as we shall argue later, there may be more attractive alternative options to achieve the same end.

NOTES

1 Data quoted here were taken from http://www.ethanolmarket.com/corngrains. html (March 28, 2006). See also Vesterby, M. and Krupa, K.S. (2004). Amber waves, U.S. Department of Agriculture: Economic Research Service, November 2004.

2 The Conservation Reserve Program (CRP) was established in 1985 to compensate farmers who committed voluntarily to retire land considered environmentally sensitive. Commitment periods under the program range from 10 to 15 years. Funds devoted to the program in 2002 amounted to $1.6 billion. Cropland retired under the program amounted to 34 million acres in 2002 (Lubowski et al., 2002).

REFERENCES

Badger, P.C. (2002). Ethanol from cellulose: a general review. In: *Trends in New Crops and New Uses* (eds. J. Janick and A. Whipkey), pp 17–21, ASHS Press, Alexandria, Va.

Consumer Reports, October, 2006.

Davis, S. and Siegel, S. (2004). *Transportation Energy Data Book*. Oak Ridge, Tenn: Oak Ridge National Laboratory.

Detchon, R. (2006). *The Biofuels Source Book*. http://www.energyfuturecoalition.org/biofuels/acknowledgements.htm (read July 1, 2006).

Farrell, A.E. Plevin, R.J., Turner, B.T., Jones, A.D., O'Hare, M., and Kammen, D.M. (2006). Ethanol can contribute to energy and environmental goals, *Science*, **311**, 27 January 2006, 506–508.

Graboski, M.S. (2002) Fossil energy use in the manufacture of corn ethanol, National Corn Growers Association. Available at www.ncga.com/ethanol/main.

Intergovernmental Panel on Climate Change (2001). *Climate Change 2001: The Scientific Basis, Contribution of Working Group 1 to the Third Assessment*. Cambridge: Cambridge University Press.

Lave, L.B., Griffin, W.M., and Maclean, H.(2001). The ethanol answer to carbon emissions, issues in Science and Technology on line, Winter, 2001. (http://bob.nap.edu?issues/18.2/lave.html.

Lubowski, R.N., Vesterby, M., Bucholtz, S., Baez, A., and Roberts, M.J. (2002). Major uses of land in the United States, 2002, USDA Economic Research Service, EIB-14.

Luhnow, D. and Samor, G. (2006). As Brazil fills up on ethanol, it weans off energy imports. The *Wall Street Journal*, January 16, 2006.

Lynch, D.J. (2006). Brazil hopes to build on its ethanol success. *USA Today*, March 28, 2006.

McCain, J. (2003). Statement on the Energy Bill. Press Release November 21, 2003.

McElroy, M.B. (2002). *The Atmospheric Environment: Effects of Human Activity*. Princeton: Princeton University Press.

NRDC, National Resources Defense Council. (2004). Growing energy: how biofuels can help end America's oil dependence. December, 2004. (http://www.nrdc.org/air/energy/biofuels.pdf).

Patzek, T.W. (2004). Thermodynamics of the corn-ethanol biofuel cycle. *Critical Reviews in Plant Sciences*, **23**(6), 519–567.

Pimentel, D. and Patzek, T.W. (2005). Ethanol production using corn, switchgrass, and wood; biodiesel production using soybean and sunflower. *Natural Resources Research*, **14**(1) 65–76 March 2005.

Renewable Fuels Association. (2006). U.S. fuel ethanol industry plants and production capacity, January 2006. Data quoted by Yacobucci (2006).

Rohter, L. (2006). With big boost from sugar cane, Brazil is satisfying its fuel needs. The *New York Times*, April 10, 2006.

Shapouri, H., Duffield, J.A., and Graboski, M.S. (1995). Estimating the net energy balance of corn ethanol. Agriculture Economic Report No. 721, U.S.

Department of Agriculture, Economic Research Service, Office of Energy, July 1995.

Shapouri, H. and McAloon, A. (2004). The 2001 net energy balance of corn ethanol, U.S. Department of Agriculture, 2004. Also available at www.usda.gov/oce/oepnu.

U.S. Department of Energy, Office of Science, fact sheet (2006). A scientific roadmap for making cellulosic ethanol a practical alternative to gasoline. http//www.er.doe.gov/News Information/News/2006/Biofuels/factsheet.htm, 2006.

Vesterby, M. and Krupa, K.S.(2004). Amber waves. U.S. Department of Agriculture: Economic Research Service, November 2004.

Yacobucci, B.D. (2006). Fuel ethanol: background and public policy issues, Congressional Research Service report for Congress, Order Code RL33290, March 3, 2006.

Wikipedia, The Free Encyclopedia. Ethanol fuel in Brazil, http://en.wikipedia.org/wiki/Ethanol_fuel_in_Brazil, Read May 3, 2006.

Wilson, E.O. (1992). *The Diversity of Life*. Cambridge: The Belknap Press of Harvard University Press.

16

Current Patterns of Energy Use

16.1 INTRODUCTION

This chapter presents a summary of current patterns of energy use for four representative national economies, the U.S. (Section 16.2), the UK (Section 16.3), Canada (Section 16.4), and China (Section 16.5). We conclude in Section 16.6 with some perspective on the differences in the challenges faced by these countries as they strive to develop their economies while seeking at the same time to minimize the impact of this development on energy security and on the acceptable function of the global climate system.

The discussion underscores the extent to which not only national but also global economies currently depend on access to and consumption of the primary fossil fuels, coal, oil, and natural gas. As discussed earlier in this volume, coal replaced wood as the energy source of choice for much of the developed world during the early part of the last century, supplemented more recently by oil and natural gas. Coal, in the developed world today, is deployed largely as the energy source for production of electricity. Its use for cooking and for space heating has been supplanted for the most part in such societies either by electricity or by a combination of either oil or natural gas, largely in response to concerns over the implications for air quality (both indoor and outdoor). Coal continues to play an important role as a source of fuel for domestic applications in large developing countries such as China and India. Where locally available, coal offers generally the least cost option for energy. Where coal is not locally accessible, as is the case for many poor countries, biomass (wood, plant material, and animal waste) is often the only source of available fuel.

Demand for oil has accelerated rapidly in recent years. Oil provides the essential energy source for long-range transport not only of goods but also of people. The mobility of cars, trucks, ships, planes, and the railroad depends on reliable and affordable access to oil and its various secondary products (diesel, gasoline, and jet fuel). To a growing extent, demand is outstripping globally available sources of supply. Prices are rising accordingly, subject to rapid and unpredictable swings. Disruptions in supply, even when temporary, can threaten the health of global economies posing serious risks to national and even international security. The current situation is clearly unsustainable.

Supplies of natural gas are generally greater than those of oil. The U.S. for example is largely self-sufficient and likely to remain so at least for the immediate future, particularly so given the significant resources available not only in the lower 48 states but also in the American Arctic. The situation for Europe is less propitious. Continuity of supply in this case is increasingly dependent, as discussed in Chapter 7, on sources in Eastern Europe (in Russia and countries of the former Soviet Union), in the Middle East, and North Africa. Demand for natural gas in Japan and China is accommodated largely by imports of LNG (liquefied natural gas) from Southeast Asia and West Africa.

Consumption of fossil fuels, as elaborated in Chapter 13, is primarily responsible for the changes in the composition of the atmosphere implicated in the current threat to the global climate system. This chapter is intended to provide a perspective on the specific activities (applications of fossil energy) that contribute to the relevant emissions, notably to emissions of CO_2.

16.2 ENERGY USE IN THE U.S.

Figure 16.1 presents an overview of the sources of energy that fuel the U.S. economy. The data displayed here refer to calendar year 2002. The relative contributions of nuclear, hydro, biomass/other (other includes wind and solar), natural gas, coal, and oil to the primary energy supply in the U.S. in 2002 were equal to 8.4%, 2.7%, 3.3%, 23.9%, 23.2%, and 39.3% respectively. Total consumption of primary energy in 2002 amounted to approximately 97 Quad with fossil fuels accounting for 86.4% of this total. Of the total (97 Quad), 35.2 Quad was deployed in the performance of useful work, with 56.2 Quad (57.9 % of the total input) rejected as waste heat (a portion of the oil consumed was not combusted but was converted to products—plastics for example).

The efficiency with which primary energy was converted to electricity amounted on average to 31.2%. The efficiency with which it was employed to perform useful work in the transportation sector (to turn the wheels of cars, trucks, ships, and trains, or to provide the motive force for planes) was even lower, 19.9%. We return later (in Chapter 17) to a discussion of the relative inefficiency of the internal combustion engine and the opportunities for significant savings of energy (and CO_2 emissions) that could be realized by replacing cars driven by gasoline and diesel with cars propelled at least in part by

U.S. Energy Flow Trends – 2002
Net Primary Resource Consumption ~97 Quads

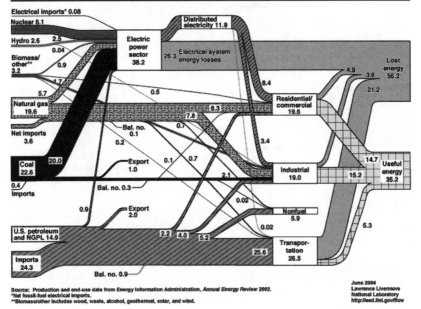

Figure 16.1 The energy flow chart showing relative size of net primary energy consumption in the United States in 2002. (https://eed.llnl.gov/flow/02flow.
php, read October 2, 2008).

electricity drawn from the commercial grid (plug-in hybrids for example). It is
of interest to note that the quantities of energy deployed ultimately to provide
useful services in the residential/commercial and industrial sectors are approximately equal, 41.8% and 43.2% respectively, as compared to 15.1% in the
transportation sector.

A summary of CO_2 emissions associated with the U.S. energy sector is presented in Fig. 16.2. The residential/commercial sector accounts for 38.5% of
the total, with 29.2% and 32.2% contributed by the industrial and transportation sectors respectively. Note the dominance of emissions from coal-generated
electricity, 1875 Mt CO_2 per year, comparable to the magnitude of the source
(1850 Mt CO_2 per year) from consumption of oil in the transportation sector.

Trends in energy use and CO_2 emissions for the U.S. over the past 25 years
are presented in Fig. 16.3. Note the long-term rise both in energy use and CO_2
emissions. Interruptions to the long-term trend are observed between 1980
and 1982, between 1990 and 1991, and to a lesser extent between 2000 and
2001. The reversals in all cases reflected retrenchments of the U.S. economy.
The gross domestic product (GDP) of the U.S. declined by 2.9% between 1981
and 1982 coincident with the spike in oil prices that ensued following the

U.S. 2002 Carbon Dioxide Emissions from Energy Consumption — 5,682* Million Metric Tons of CO$_2$**

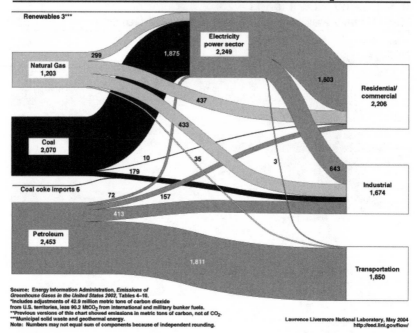

Source: Energy Information Administration, *Emissions of Greenhouse Gases in the United States 2002*, Tables 4–10.
*Includes adjustments of 42.9 million metric tons of carbon dioxide from U.S. territories, less 90.2 MtCO$_2$ from international and military bunker fuels.
**Previous versions of this chart showed emissions in metric tons of carbon, not of CO$_2$.
***Municipal solid waste and geothermal energy.
Note: Numbers may not equal sum of components because of independent rounding.

Lawrence Livermore National Laboratory, May 2004
http://eed.llnl.gov/flow/

Figure 16.2 Carbon dioxide emissions from U.S. energy consumption in 2002. (https://eed.llnl.gov/flow/carbon02.php, read October 2, 2008).

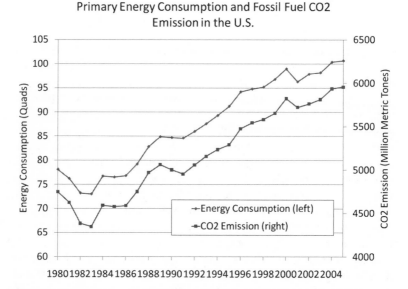

Figure 16.3 Energy consumption and CO$_2$ emission data reported by the U.S. Energy Information Administration (EIA): http://tonto.eia.doe.gov/country/index.cfm, read October 7, 2008.

374

Iranian hostage crisis of 1978 (cf. Fig. 6.4). The reversal in 1990–1991 was associated with the invasion of Kuwait by Iraq in 1990, which resulted in the first
Gulf War. The war ended in this case (in 1991) when the Iraqis were forced to
withdraw. Before doing so, however, they set fire to the oil wells in Kuwait. The
GDP of the U.S. declined by 1.3% between 1990 and 1991. The decrease in the
growth rate both of energy consumption and CO_2 emissions between 2000
and 2001 was associated with the mild relatively short-lived reversal in the
economy that ensued following the drop in the stock market prompted by what
is commonly referred to as the dot com bubble (the NASDAQ market on which
many of the technology stocks traded peaked on March 10, 2000). Note that
CO_2 emissions increased by more than 20% between 1990 and 2005. Had the
U.S. ratified the Kyoto Climate Treaty, it would have been obligated to reduce
its emissions of greenhouse gases by 7% between 1990 and 2008–2012. Clearly,
it has fallen far short of this target.

16.3 ENERGY USE IN THE UK

The pattern of energy use in the United Kingdom (UK) is summarized in
Fig. 16.4. The data presented here refer to calendar year 2007. The data in the
original report were quoted in terms of millions of tons of oil equivalent
(mtoe). For ease of comparison with the U.S. data in Fig. 16.1, the UK data are
restated here in units of Btu.

The width of the bands in Fig. 16.4 provides an indication of the magnitude
of the associated energy flows. An important difference between the presenta-
tions in Figs. 16.1 and 16.4 is that the data for energy use in the transportation
sector in the latter case refer to the quantity of energy deployed in the sector
rather than to the amount that was actually applied as a source of propulsion
(as useful work) as is the case for the U.S. data in Fig. 16.1. As indicated earlier,
approximately 80% of the energy consumed by the transportation sector is
released as waste heat. To compare data for energy end use in the UK transportation as presented in Fig. 16.4 with the corresponding entry for the U.S.
(Fig. 16.1), the value indicated in Fig. 16.4 should be reduced by about 80%.

Petroleum accounts for approximately 43% of the total supply of primary
energy in the UK, natural gas, 41%, coal, 6%, nuclear plus hydro, 9%, with the
balance represented by various renewable sources including wind and biomass.
The breakdown on energy inputs to electricity production in 2006 is as follows:
coal, 37.5%; natural gas, 36%; nuclear, 18%; oil, 1%; with the balance from a
variety of non-fossil sources. Data on trends in the UK energy sector as quoted
here and in what follows were taken mainly from official UK government
sources.[1]

The UK engages in significant trade in both energy and energy products.
The balance of energy trade expressed in monetary units was significantly positive until recently. It is now negative with the value of imports increasing rapidly relative to exports (the crossover occurred in 2004). Domestic production
of both oil and natural gas peaked in the UK in 2000 (sources in both cases

derived primarily from fields in the North Sea). Production of oil rose by 38% between 1990 and 2000. It decreased by 39% between 2000 and 2006. Production of natural gas increased by 138% between 1990 and 2000. It fell, however, by 26% between 2000 and 2006. Production of coal has declined steadily since 1990, by 80% between 1990 and 2006. The UK switched from being a net exporter to a net importer of coal in 2001. Since then, imports have been rising at an average annual rate of 15%.

Coal accounted for 67% of UK electricity production in 1990; the contribution from natural gas was then less than 1%. By 2007, the relative fraction of electricity produced from coal had dropped to 34% with natural gas accounting then for 43% of total power generation. The use of coal in the UK power sector has enjoyed, however, a recent resurgence. It is been deployed to an increasing extent both as a means to compensate for the decrease in production of electricity from nuclear sources (a trend that began in 2000) and as a substitute for increasingly expensive natural gas.

Trends in UK energy consumption over the past 25 years with related emissions of CO_2 are presented in Fig. 16.5. The decrease in emissions since 1990 is a result for the most part of the substitution of natural gas for coal over this period, primarily in the electric power sector as noted above. Emissions declined by about 6.4% between 1990 and 2006, in contrast to the significant increase in emissions reported over the same time interval for the U.S. (Fig. 16.2). The rise in the recent years is associated primarily with the increase in coal use. If current trends continue, it is likely that UK emissions will resume their long-term growth. Overall emissions of greenhouse gases from the UK over the interval 1990 to 2006 (expressed on a standard based on their potential impact on

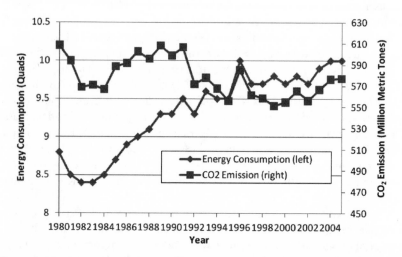

Figure 16.5 Energy consumption and CO_2 emission trends for the UK. Data from EIA http://tonto.eia.doe.gov/country/index.cfm, read October 7, 2008.

climate, a standard that attaches greater significance to CH_4 as compared to CO_2 and even greater weight to N_2O) decreased by 15.4% reflecting significant reported declines in emissions of CH_4 and N_2O (52% and 40% respectively in terms of the assessed climate impact of these gases). It is interesting to note in this context that consumption of oil in the transportation sector, and related emissions of CO_2, continued to increase over the entire period 1990 to 2006 despite significant substitution of more efficient diesel fuel for gasoline (referred to as petrol in the UK).

16.4 ENERGY USE IN CANADA

A flow diagram for energy use in Canada is presented in Fig. 16.6. The data displayed here refer to calendar year 2006. The Canadian energy economy differs from that of either the U.S. or the UK in that Canada is not only an important producer and consumer of primary energy, it is also an important net exporter. Exports of crude oil, mainly to the U.S., accounted for 67% of Canada's total production in 2006. More than half (59%) of the nation's production of natural gas was also exported, and again the U.S. was the primary customer. Canada's exports of crude oil and oil products, in addition to natural gas, coal, and electricity, were valued at $32 billion in 2006.[2]

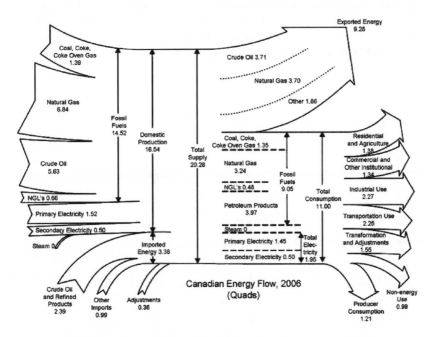

Figure 16.6 Energy flow chart for Canada 2006. Adapted from Report on Energy Supply–Demand in Canada, 2006, http://www.statcan.ca/english/freepub/57-003-XIE/2006000/technote4.htm, read October 8, 2008.

Hydropower accounted for 59% of Canada's total production of electricity in 2006. Nuclear ranked second contributing 16% of the total with most of the remainder derived from fossil sources. The contribution from wind, solar, and tidal sources accounted for only 0.5% of total power generation in 2006. Production from wind, however, is increasing rapidly at the present time. It more than doubled between 2005 and 2006, rising from 680 MW average power output in 2005 to 1460 MW in 2006.[3]

The entry in Fig. 16.6 designated as primary electricity refers to electricity generated from a combination of hydro, nuclear, wind, tidal, and solar. Secondary electricity refers to electricity produced by thermal plants fueled by a combination of oil, coal, and natural gas.

The categories used by the Canadian statistical authorities to report final energy use differ from those favored by the U.S. and UK authorities. The industrial sector is the largest single customer for energy in Canada accounting for approximately one-third of the total. The transportation sector is responsible for about 31% of the total with about 20% deployed in the combination of the residential/agricultural and commercial sectors. Important quantities of energy are consumed by the mining and oil/gas extraction industries specifically in the production of oil from the tar sand resources in Alberta (cf. Chapter 6).

Trends in energy use and CO_2 emissions since 1980 are summarized in Fig. 16.7. Overall consumption of energy in Canada increased by about 50% between 1980 and 2005 accompanied by about a 32% rise in emissions of CO_2. The increase in CO_2 emissions since 1990 (the base year for the Kyoto Climate Protocol) amounts to about 30%. It is unlikely under the circumstances that Canada can meet its commitment under the Kyoto Protocol for a 6%

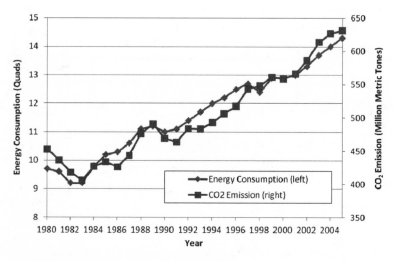

Figure 16.7 Energy consumption and CO_2 emission trend for Canada. Data from EIA http://tonto.eia.doe.gov/country/index.cfm, read October 7, 2008.

reduction in greenhouse gas emissions by 2008–2012 relative to 1990. As an interesting footnote, energy consumption (and CO_2 emissions) decreased in Canada by approximately 2% in 2006 relative to 2005. The reduction in energy use extended across the national economy and was attributed largely to warmer weather conditions that prevailed across the country in 2006.

16.5 ENERGY USE IN CHINA

An overview of energy use in China is presented in Fig. 16.8. The data displayed here refer to calendar year 2000 and were developed by economists at the Chinese Institute of Policy and Management at the behest of Ho and Nielsen (2007). The total supply of energy in China in 2000 amounted to 37.75 Quad with coal, petroleum, biomass, hydro, and natural gas accounting for 52%, 24%, 16%, 6%, and 2% respectively of the total. Nuclear power was responsible for less than 0.5% of the available primary energy. The industrial sector accounted for 50% of total consumption followed by the residential sector (30%), the combination of the agricultural and commercial sectors (11%) and transportation (9%).

The structure of the Chinese economy differs significantly from those of more developed countries such as the U.S., the UK, and Canada. China's economy is in a state of transition. Typical of such transitional economies, large quantities of energy are being deployed in China at the present time to establish

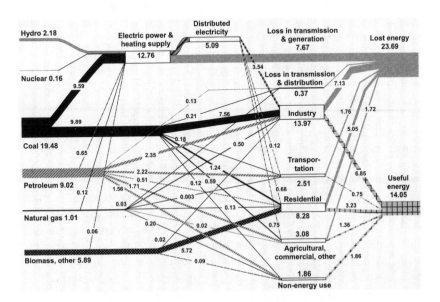

Figure 16.8 Energy flow chart for China 2000 (Ho and Nielsen, 2007).

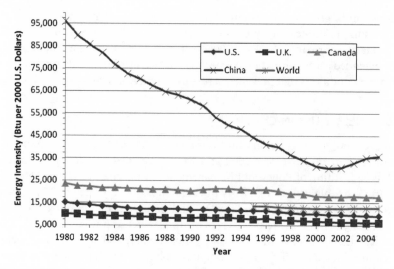

Figure 16.9 Total primary energy consumption per dollar of GDP based on purchasing power parity. Data from EIA, http://www.eia.doe.gov/iea/wecbtu. html, read October 7, 2008.

essential infrastructure, for the construction of buildings, factories, roads, and airports for example. The efficiency of the Chinese economy (defined as the quantity of energy applied to produce a given unit of GDP) differs significantly as a consequence from that of more developed economies such as the U.S., UK, and Canada as illustrated in Fig. 16.9. Note, however, that the efficiency of the Chinese economy is increasing rapidly, on a trend that is likely to bring it close to that of more developed economies within a few decades despite the interruption of the long-term trend observed over the past few years. The decrease in the energy efficiency over the past few years reflects most likely a temporary increase in the pace of investments in infrastructure.

Trends in energy use and CO_2 emissions for China over the past 25 years are presented in Fig. 16.10. Note the rapid increase in both energy use and CO_2 emissions since 2000. China is now the world's largest emitter of CO_2, having surpassed in U.S. in the early months of 2006. It is difficult to account for the change in slope both for Chinese energy use and CO_2 emissions in the late 1990s. It is generally thought that this may reflect an underestimate in official statistics for coal use over this interval. There is no doubt though that Chinese energy use is increasing rapidly and with it the country's emissions of CO_2. Coal continues to represent the dominant contribution to the Chinese energy economy. To reverse the upward trend in CO_2 emissions will require important structural changes in the Chinese economy, specifically a transition to less carbon intensive sources of energy. We return later in Chapter 17 to the options that might be available for China to successfully confront this challenge.

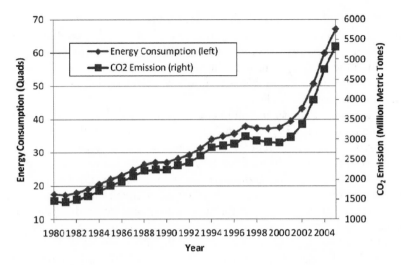

Figure 16.10   Energy consumption and CO_2 emission trend for China. Data from
EIA http://tonto.eia.doe.gov/country/index.cfm, read October 7, 2008.

16.6 CONCLUDING REMARKS

End use demands for energy for the three developed economies (the U.S., UK, and Canada) highlighted here are basically similar. Transportation accounts for between 31% and 41% of final demand, with Canada and the U.S. on the low and high ends of this range respectively. Requirements for industry range from 20% in the UK to 33% in Canada with a similarly narrow range for the fractions of energy deployed in the residential/commercial sectors (from about 20% in Canada to about 30% in both the U.S. and UK). The pattern is distinctly different for the developing economy of China. There, industry accounted for 50% of total energy consumption in 2000 with transportation responsible for only 9%. The structure of the Chinese economy is changing rapidly, however. Demand for oil to fuel the transportation sector is rising at a rate approaching 10% per year. China, like the U.S. and the UK, is now increasingly dependent on imports to meet this growing demand and the trend is unlikely to reverse, at least in the immediate future.

The U.S. relies on a broad range of sources to satisfy its requirements for primary energy (notably the fossil fuels—oil, coal, and natural gas—with a significant contribution from nuclear). Natural gas is now more important than coal in meeting demands for primary energy in the UK although as noted above this situation may be changing (coal is becoming more important). Coal represents the dominant source of primary energy for China. Its use continues to increase and is largely responsible for the rapid recent rise in the country's emissions of CO_2. Canada, in contrast to the U.S., UK, and China, is energy self-sufficient. Canada is a net exporter of coal, oil, natural gas, and electricity (primarily to the U.S.). Much of its recent supply of oil is derived, however, from the tar sands of Alberta. Extracting this resource, as discussed in Chapter 6, is an energy-intensive process. Significant quantities of natural gas are consumed in the process (to provide the heat needed to liberate the oil from the sand). Emissions of CO_2 are increasing accordingly, contributing to the trend in emissions displayed in Fig. 16.7.

Consumption of coal and oil in combination is largely responsible for the continuing rapid rise in global emissions of CO_2. In most of the world, coal is deployed mainly as the energy source used to generate electricity. Oil is employed primarily in the transportation sector, as the motive source to propel cars, trucks, trains, ships, and planes. To reduce emissions of CO_2, we need both to cut back on our reliance on coal for electricity production and to trim back our dependence on oil for transportation. There are good reasons, independent of the climate concern, to address the oil issue. When oil prices hit a record of close to $150 a barrel in mid-2008, the bill for imported oil in the U.S. was projected to rise to an annual level of more than $700 billion (comparable in 1 year to the entire cost of the military intervention in Iraq or to the funds made available controversially by the U.S. Congress in October, 2008, to address the crisis in the U.S. banking industry). Oil has been used on several occasions in the past by producing countries to impose their will on dependent consumers. Reducing dependence on uncertain supplies from potentially unreliable sources is an imperative not just for the U.S. and the UK and other developed economies but also prospectively for China. Not to do so would be to expose the economies of these countries, and indeed their national securities, to serious and unnecessary risk.

NOTES

1 (http://www.berr.gov.uk/whatwedo/energy/statistics/publications/in-brief/page17222.html, read October 9, 2008).

2 (http://www.statcan.ca/english/freepub/57-003-XIE/2006000/part1.htm, read October 8, 2008)

3 (http://www.statcan.ca/english/freepub/57-003-XIE/2006000/part1.htm, read October 8, 2008).

REFERENCE

Ho, M.S. and Nielsen, C.P. (2007). *Clearing the Air: The Health and Economic Damages of Air Pollution in China*. Cambridge, Mass: The MIT Press.

17

Vision for a Low-Carbon Energy Future

17.1 INTRODUCTION

Chapter 16 presented an overview of how energy is used in four representative economies, the U.S., UK, Canada, and China. It concluded that a responsible future energy policy should address two interrelated objectives: first to reduce the carbon intensity of global economies; second to scale back their dependence on unreliable sources of oil.

This chapter offers a vision for a low-carbon future, starting with relatively straightforward energy options. It begins in Section 17.2 with some general views on how practices emphasizing more efficient use of energy could contribute to a reduction in future demand, while providing at the same time an opportunity to reduce emissions of CO_2. Section 17.3 discusses a number of potential alternatives to imported oil as energy sources for transportation, notably methanol and CNG. Prospects for ethanol were discussed earlier in Chapter 15.

We move then to a more expansive view of a different energy future, one in which electricity is envisaged to assume an even greater importance than it does today, where it could be produced mainly from non-carbon sources, and where it could play a critical role not only in domestic and commercial applications but also in transportation.

We begin in Section 17.4 with an analysis of the prospects for wind as a potentially important source of electricity, extending the more general discussion in Chapter 8. Opportunities for energy from the sun to complement energy from wind are explored in Section 17.5. Prospects for geothermal sources are discussed in Section 17.6. The opportunity for electricity to substitute for fossil energy (oil) in the transportation sector is explored in

Section 17.7. Investments in the electric power transmission system that would be required to support an expanded role for renewable energy are discussed in Section 17.8 with concluding comments in Section 17.9.

17.2 ENERGY CONSERVATION

The obvious first step should be to increase the efficiency with which we use energy. Public education—enhanced energy literacy—could play an important role in advancing this objective. The business sector, industrial and commercial, has been generally ahead of the curve on this issue. It has been more difficult for the homeowner and for individuals to know what they should do that could make a difference. The central motivating factor for the individual could and should be to cut expense while at the same time preserving valued energy services. A first step would be to develop the capacity to carry out a personal energy audit. How do we use energy in our homes, in our workplaces, and in our everyday lives? What do we pay for these services? And what could we do that would both cut energy use and save money but would not require a significant change in current lifestyles?

For a start it would be good if we could compare directly the cost of the different energy sources we contract for in our everyday lives. As discussed in Chapter 3, electricity purchased in the U.S. is billed in units of kWh (kilowatt hours). Natural gas is priced in therms while home heating oil and gasoline are sold by the gallon (a measure of volume rather than energy as discussed in Chapter 3). In early October 2008, in Cambridge, Massachusetts, the retail charges (prices paid by the consumer including prices for delivery in the case of electricity, natural gas, and home heating oil) for these various fuels were as follows: electricity, 19.9 cents per kWh; natural gas, $1.46 per therm; home heating oil, $3.50 per gallon; and gasoline, $3.20 per gallon. Expressed in terms of cost per 100,000 Btu (per therm equivalent), the comparable figures for electricity, natural gas, home heating oil, and gasoline corresponded to $5.83 (electricity), $1.46 (natural gas), $2.52 (home heating oil), and $2.78 (gasoline).

There is a good reason why electricity on an energy basis costs more than energy provided by either natural gas or oil-based fuels. As noted in Chapter 16, the efficiency with which primary energy is converted to electricity in the U.S. averages about 31%. It takes 3.1 units of primary energy to produce a given unit of energy as electricity. The cost of electricity is correspondingly higher than the cost for competitive primary sources of energy such as oil or natural gas. Heating one's house with electricity is obviously more expensive (and less energy efficient) than doing so with either oil or natural gas. Given the mix of energy sources used to generate electricity in the U.S., it is also clearly wasteful, at least on a national average basis, in terms of emissions of CO_2.

Lighting accounts for about 9% of total household use of electricity in the U.S. Incandescent bulbs, the bulbs still in most common use, are extremely inefficient in converting electrical energy to light. As discussed in Chapter 11, to radiate light in the visible portion of the electromagnetic spectrum with sufficient intensity to be useful, the tungsten filaments of incandescent light bulbs must be raised to temperatures as high as 2500°C. The filament behaves as a blackbody (Chapter 3) consuming energy and emitting radiation in proportion to the fourth power of its temperature. Most of the radiation produced by the 2500°C filament, however, is emitted in the longer wavelength (invisible to the human eye) region of the spectrum. It is radiated as heat rather than light (adopting the common day use of this terminology).[1] Less than 2% of the energy consumed by an incandescent light bulb is emitted in the visible; more than 98% is released as waste heat. Clearly, there are significant opportunities to save energy by switching to more efficient light sources. Compact fluorescent bulbs are 3–5 times more efficient in producing visible light than incandescent bulbs. Were we to make the transition on a large-scale basis from incandescent lighting to compact fluorescence, the savings in electricity use could be as much as 7%. And with technological advances, there are prospects for even more efficient light sources in the future.

Table 17.1 presents a summary of electricity consumed by a variety of appliances in a typical American home. There are a number of important take-home messages. It is not a good idea to use electrical energy as a source of heat (note the high levels of consumption for clothes dryers, electric cookers, electric water heaters, and portable electric space heaters). Recall that up to 70% of primary energy is wasted as heat (it is released to the environment) in producing the electricity in the first place. In contrast, the use of electricity to run motors or devices that do not produce a great deal of heat is efficient (more than 90% of the electrical energy may be usefully deployed in this case and a comparable service could not be provided by tapping a primary source of

Table 17.1 Monthly consumption of electricity for a typical American home

Application	Energy (kWh)
Lighting	100
Refrigerator (old)	120
Refrigerator (Energy Star)	36
Freezer	195
Dishwasher	36
Clothes Washer	15
Clothes Dryer	150
Color TV (27")	27
Computer (monitor and printer)	26
Electric range and oven	127
Electric water heater	500
Dehumidifier	30
Window air conditioner	150
Portable electric heater	274
Miscellaneous appliances	25

Source: http://www.blachlylane.coop/customer_service/tips.php, read October 15, 2008.

energy such as coal, oil, or natural gas). For applications that require heat (cooking, space heating, or drying clothes for example), it would be better (more energy efficient) to use natural gas so long as it is available (or bottled propane—or a comparable alternative—where it is not). Of course if the electricity were produced from a non-carbon source (nuclear, wind, solar, and geothermal), it could be argued that it might be socially responsible to use this electricity in order to realize the consequent savings in emission of CO_2. With current prices, however, exercise of this social conscience would come at a significant price to the pocketbook.

Conservation can make an important contribution to energy savings. Use of a programmable thermostat could ensure that space heating and cooling is supplied only when needed. Turn the thermostat down when occupants are out of the house in winter or when demand for heat is reduced when they are in bed at night. Reverse the process in response to requirements for space cooling in summer: turn the thermostat up during the day when the house is unoccupied, and up at night. Consider adjusting targeted temperatures down by a few degrees when outside temperatures are low, up when they are high. Explore the advantages of an investment in improved insulation. There are services using infrared imaging that can identify areas where heat is leaking differentially from homes, or where selective insulation could significantly reduce demands for cooling in summer. Payoff for investments in insulation in many cases may be as quick as a year or even less and there may be subsidies from local utilities or local authorities to encourage such investments (and if not, there should be!). Consider drying clothes passively rather than using an energy intensive clothes dryer. Wash the dishes in the dishwasher but let them dry passively. Use the dishwasher and the clothes washer only when they are full. Turn off all electrical power in rooms when they are not in use. A central turn-on/turn-off switch could encourage this practice similar to the system that exists in many hotel rooms, specifically in Asia but also in parts of Europe. There, you need to insert a room key to turn on the power when you enter the room. When you leave, and inevitably recover the room key, the power is automatically switched off. And there are many more possibilities for savings without seriously impacting desired services.

In response to the first energy crisis of the early 1970s, Israel passed a law which required solar water heaters to be installed in all new homes. The technology is relatively straightforward. Sunlight absorbed by a black painted absorbing material mounted either on the roof of a building or on a sun facing wall can be used to heat circulating water directly or alternatively to heat an intermediate fluid that can deliver heat using a heat exchanger to water held in a storage tank. It is estimated that solar water heating in Israel has been responsible for a net savings in primary energy by as much as 3%. Solar water heating is widely used also in China. To date it has not enjoyed wide application in the U.S., largely because costs for retrofitting existing buildings are sufficiently high as to make such investments uneconomic. It should be considered however for new buildings, particularly in warmer sunnier regions.

The State of California has led the way in the U.S. in demonstrating how one can achieve significant savings in domestic consumption of energy without compromise to delivery of essential energy services. The initiatives taken by California date back to the early 1970s, to the aftermath of the first energy crisis. Rigorous statewide standards for energy efficiency of buildings were enacted in 1974. Laws were passed that provided economic incentives for electric utilities to reduce consumers' demand for energy. Since 1974, California has succeeded in holding its per capita energy consumption effectively constant while energy use per person in the U.S. as a whole has risen by more than 50%. Mufson (2007), in an informative article on the California energy experience, quotes Greg Kats, managing principal of the energy and clean technology advisory company Capital E, to the effect that the average California family now spends about $800 a year less on energy than would be the case had the State not implemented the energy savings initiatives it did over the past 20 years. The experience in California demonstrates that with education and public support, it should be possible to reduce domestic consumption of energy in the U.S. as a whole by at least 30–50% and there should be opportunities for similar savings elsewhere.

Conservation offers the best immediate opportunity to cut back on demand for energy while at the same time saving money for consumers and reducing emissions of CO_2. We stressed in this context the importance of an energy literate public. California succeeded in maintaining electricity use per capita constant for more than 30 years in part by instituting demanding energy efficiency standards for buildings. The problem with these standards is that they do not, in most instances, address the challenge of improving the energy efficiency of existing buildings. When a party is negotiating to purchase an existing house in the U.S. and wishes to negotiate a mortgage with a financial institution, they are required to furnish financial information with respect to assets and projected income. The bank makes a decision then as to what the purchasing party can afford to pay in terms of monthly principal, interest, taxes, and insurance on the property (or at least they used to before the current sub-prime mortgage crisis!). It also requires, customarily, a building inspection at the expense of the purchasing party to ensure that the building is structurally sound, that there are no problems that could militate against its safe, effective function in the future (and the security of the bank's mortgage). How about instituting a policy that would require as part of the bank's deliberative process an energy audit to determine the monthly expenses on energy that are likely to be incurred by the new owners? These expenses could be added to the costs projected otherwise to service the mortgage, taxes, and insurance. The purchaser and seller could then, with informed professional advice, determine which if any capital investments in energy savings would make sense and adjust the price of the property accordingly. Since mortgages are offered often for periods as long as 30 years, this would provide incentives for the parties to invest in energy savings even if payoffs for such investments should extend over a decade or longer. As a side benefit, such an arrangement would create opportunities for a new class of energy efficiency professional consultants, promoting an expertise that is currently lacking but much needed, providing a source of high paying jobs at a time when for much of the world this is judged to be a politically desirable objective. Implementing the recommendations of the energy consultants, when judged to be economically constructive, would provide important additional opportunities for employment.

17.3 ALTERNATIVE TRANSPORTATION FUELS

The Texas oil baron T. Boone Pickens is currently campaigning for a U.S. energy policy that would encourage a switch from oil-based fuels in the transportation sector to compressed natural gas (CNG).[2] He argues, correctly, that this could markedly reduce U.S. dependence on imported oil. The numbers quoted above suggest that at present prices such a switch could result in significant savings for consumers, by as much as 47% if the vehicle were driven by natural gas at ambient pressure. Driving one's car or truck with natural gas rather than gasoline would cost in this case, at current prices, the equivalent of $1.52 for a gallon of gasoline (the savings this suggests is in fact an important component of Pickens' campaign). However, it would be impractical to use gas at the relatively low pressures at which the product is supplied through the conventional distribution system (a little more than 1 atm). For it to serve as a useful transportation fuel, the pressure of the gas would need to be raised by at least a factor of 100 (to 100 atm or more) to convert the conventional supply to CNG. Energy would have to be expended to accomplish this upgrade and costs might be expected to rise accordingly. It is likely further if the Pickens plan were to be implemented, that vehicles would be fueled not from home supplies of gas but rather from commercial stations catering to this new demand supplying gas as CNG. Additional expense would be incurred to develop the infrastructure required to service this demand. And if CNG were to gain wide acceptance as an energy source for transportation in addition to its current use, it would be necessary to extend and expand the existing natural gas distribution system (Chapter 7). Again costs would rise. Economic benefits to the motorist might be expected to decline accordingly.

Vehicles using CNG would require larger, sturdier, and consequently heavier, storage tanks. This would mitigate most likely (at least so it seems to me) against wide spread use of CNG in compact personal automobiles where the overhead in terms of weight and volume to accommodate the bulky heavy tanks and other modifications to the combustion system might be prohibitive. There is no doubt though that there should be a useful market for expansion of current use of CNG in heavier vehicles such as trucks and buses and that this should certainly contribute to a reduced demand for imported oil and to a related decrease in emissions of CO_2 (as discussed in Chapter 13, combustion of natural gas results in production of less CO_2 per unit of energy recovered than combustion of either gasoline or diesel fuel).

A switch to methanol (CH_3OH) in the transportation sector rather than CNG offers an alternative to the Pickens' plan to reduce dependence on imported oil. Methanol is a liquid similar to ethanol. It has an octane rating higher than gasoline, suffering though from the fact that its energy density (the energy contained in unit volume) is only half that of gasoline (35% less than ethanol). A larger tank would be required as a result to store a given amount of energy. The chemical properties of methanol are similar, however, to those of ethanol. The distribution system required to accommodate a methanol industry should be comparable therefore to that now in place for ethanol suggesting that a widespread shift from gasoline to methanol (or use of methanol as an important additive) in the transportation sector should be less taxing than a shift to CNG.

The path for production of methanol in most common use today makes use of natural gas (CH_4) as a feedstock. Natural gas is reacted first with H_2O to produce a mixture of CO, H_2, and CO_2 referred to as syngas. Production of methanol is controlled subsequently by an equilibrium represented by

$$CO_2 + 3H_2 \leftrightarrow CH_3OH + H_2O \qquad (17.1)$$

As noted by Olah et al. (2006), modern methanol plants are capable of converting natural gas to methanol with an energy efficiency as high as 70% although additional energy must be consumed to convert the product methanol to its essentially pure form (it must be concentrated by passing it through one or more distillation columns to eliminate impurities, notably H_2O).

An interesting futuristic plan for production of methanol would be to react CO_2 captured from an oxyfuel power plant with H_2 formed by the electrolysis of H_2O:

$$H_2O \rightarrow H_2 + 1/2O_2 \qquad (17.2)$$

As indicated in Chapter 14, an oxyfuel plant is one that uses pure O_2 rather than air to convert a fuel such as coal to syngas. It is consequently easier to capture CO_2 from an oxyfuel plant than from a conventional power plant given the much higher concentration of CO_2 in the stack gas of the former (stack gases of a conventional power plant are dominated by N_2 since fuel is combusted in this case in the presence of air rather than pure O_2, Chapter 14). The oxygen for the oxyfuel plant could be supplied by reaction (17.2) which could be implemented with an efficiency as high as 80%. The carbon atom in the initial fuel would thus get a second chance before it had to be emitted to the atmosphere. It could be used first to produce electricity, enjoying then a second life as fuel for a car or truck. If the hydrogen in reaction (17.2) were produced from a non-carbon source (nuclear, wind, solar, or geothermal for example), aggregate consumption of carbon-based fuel would be reduced accordingly.

17.4 ELECTRICITY FROM WIND

As this book was delivered to the publisher (November 1, 2008), our research group at Harvard was engaged in a comprehensive analysis of the opportunities for generation of electricity on a worldwide basis using a combination of wind and solar energy resources. The data presented in this section are based on a preliminary report of results from the wind component of this work.

The study takes advantage of a database on global wind fields derived from retrospective analysis of meteorological observations using a state-of-the–art, global, weather/climate, model constrained by the full range of globally available measurements (the approach is described by Rienecker et al., 2007). Think of it as the best possible reconstruction of the state of the atmosphere with a temporal resolution of 6 h and with a spatial resolution of approximately 67×50 km (i.e. the database provides a record of the strength of winds not only in the near surface region but as a function of elevation every 6 h on a spatial grid of 67×50 km).

We envisaged deployment of a land-based network of 2.5 MW wind turbines in regions where these turbines would be capable of capturing wind energy on an annually averaged basis at more than 20% of the rated capacity of the turbines. Since wind strengths are necessarily variable at any given location, an individual turbine cannot be expected to live up to its full potential (its rated capacity) all the time in converting wind energy to electricity. A turbine operating full time at a power level of 2.5 MW would produce 2.2×10^7 kWh of electricity over the course of a year (the maximum power output multiplied by the number of hours in a year, 8760). If wind conditions allowed it to produce only 20% of its rated power level (this is referred to as the capacity factor, abbreviated as CF), the annual yield of electricity would be equal to 4.4×10^6 kWh. Assuming a market price for electricity of 10 cents per kWh, a turbine would generate in this case an annual revenue of $440,000. At present prices, the capital cost for a 2.5 MW turbine is approximately $4.5 million ($1800 per kW). It is expected that with proper maintenance this cost could be amortized over a period as long as 20 to 30 years (the anticipated operational life of the turbine). Turbines installed recently in the U.S. have been able to operate with capacity factors approaching 40% (a tribute to high quality manufacturing and careful attention to siting).

The electricity that could be generated on a worldwide, country-by-country basis, by deploying 2.5 MW turbines in regions where wind conditions could accommodate capacity factors of 20% or greater, is summarized in Fig. 17.1 (Lu et al., 2009). Specifically excluded in this analysis were regions classified as forested, areas occupied by permanent snow or ice cover (Greenland and Antarctica for example), environments covered by water (the Great Lakes for example) and regions identified as either developed or urban. Results are quoted here in units of annual petawatt hours (10^{15} Wh or 10^{12} kWh). Table 17.2 provides a breakdown of results for the ten countries identified as the largest global emitters of CO_2 together with a comparison of wind power potentials for

Table 17.2 Annual wind energy potential (in units of 1012 Watt-hour, TWh) for the top 10 CO2 emitting countries

No	Country	CO_2 (million tons)	Electricity consumption (TWh)	Potential wind energy (TWh)		
				Onshore	Offshore	Total
1	United States	5956.98	3815.9	74,400.8	14,474.6	88,875.4
2	China	5607.09	2398.5	39,072.4	4605.9	43,678.3
3	Russia	1696.00	779.6	116,193.0	22,811.2	139,004.2
4	Japan	1230.36	974.1	572.9	2672.1	3245.0
5	India	1165.72	488.8	2936.2	1057.8	3994.0
6	Germany	844.17	545.7	3189.3	940.3	4129.6
7	Canada	631.26	540.5	78,083.4	21,226.6	99,310.0
8	United Kingdom	577.17	348.6	4397.6	6234.4	10,632.0
9	South Korea	499.63	352.2	125.9	987.9	1113.8
10	Italy	466.64	307.5	247.7	163.6	411.3

Note: CO_2 emission and electricity consumption for 2005, data source from EIA (http://tonto.eia.doe.gov/country/index.cfm).

these countries with current rates of consumption. Table 17.2 includes also estimates for electricity that could be generated offshore within about 90 km (50 nautical miles) of the nearest shoreline in waters with depths less than 200 m, assuming in this case deployment of turbines with rated capacities of 3.6 MW.

The electricity consumption and CO_2 emission data quoted here refer to calendar year 2005. As noted earlier, China has now surpassed the U.S. as the world's largest emitter of CO_2. The conclusion from the results summarized in Table 17.2 is that land-based wind farms could account for a major fraction of current and reasonably projected future growth in electricity demand for all of the major CO_2 emitting countries with the possible exception of Japan where it might be necessary to exploit also the potential of resources available offshore. The entry for the U.S. in Table 17.2 includes also contributions from Alaska and Hawaii.

The potential for land-based generation of electricity in the U.S. on a state-by-state basis is illustrated for the lower 48 states in Fig. 17.2. Notable here is the high concentration of wind resources in the central plain region extending northward from Texas to the Dakotas, westward to Montana and Wyoming, and eastward to Minnesota and Iowa. Figure 17.3 offers a comparison of potential expressed as a ratio with respect to current state consumption. Potential exceeds demand by more than a factor of 400 in Montana and the Dakotas, and by as much as a factor of 33 in Texas. Wind power in the lower 48 states, according to this analysis, would be sufficient to accommodate up to 16 times current consumption of electricity, as much as 18 times current demand if resources available offshore were developed also, subject to the constraints noted above.

17.5 ELECTRICITY FROM THE SUN

The problem with wind power is that it is intermittent. It peaks typically in winter with a minimum in summer. It can vary in particular locations on time scales ranging from minutes to hours to days. Variability on short time scales can be accommodated to some extent by incorporating contributions from spatially separated regions into an integrated, interconnected, electric grid, as discussed in Section 17.8. Seasonal variations are more predictable. Demand for electricity in the U.S. is greatest in summer, expressed for the most part by requirements for air conditioning in the southern states.

The seasonal variation of wind power potential for the lower 48 states of the U.S. (monthly averages) is presented in Fig. 17.4, which includes also an illustration of the (monthly average) seasonal pattern of demand. Tapping energy from the sun, which peaks typically in summer, could provide a useful complement to energy derived from wind, particularly so if solar and wind resources could be combined and shared over an extensive integrated electric transmission grid.

A map of concentrated solar power (CSP) available for the contiguous states of the U.S. is presented in Fig. 17.5.[3] CSP refers to the opportunity to concentrate sunlight using mirrors and to focus this light on a target where it can be used to heat a fluid with the concentrated energy employed subsequently to generate electricity by means of a conventional turbine. In one such design, the light is focused on a receiver sited on the top of a tower in the middle of the mirror array. In another, curved mirrors deployed in a linear array are programmed to rotate tracking the sun. Again, the light concentrated in this system is used to heat a fluid. Oil is the medium of choice most often for

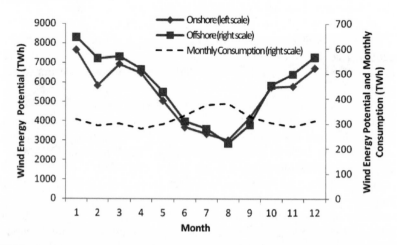

Figure 17.4 Monthly wind energy potential for the contiguous U.S. in 2006 (in units of 10^{12} Watt-hour, TWh) and the monthly electricity consumption (TWh) for the whole U.S. (from Lu et al., 2009).

this application. The focused sunlight is used to heat the oil, which flows to a central station where the energy it contains is extracted through a heat exchanger to be converted subsequently to steam. The steam is then used to drive a conventional turbine. Having delivered its heat to the heat exchanger, the oil flows back through the mirror array to be reheated and the sequence repeats.

Two power tower demonstration projects have been implemented in the U.S. The first, Solar One, with a rated capacity of 10 MW, was deployed near Barstow, California, in 1982. It operated from 1982 to 1988 and produced more than 38 million kWh of electricity. In this design, the energy concentrated by the mirror array was used directly to produce steam. The problem in this case was that the plant could generate electricity only during the day and only when skies were clear (not cloud covered). Solar Two involved a retrofit of Solar One. It used molten salt to store the energy harvested when the sun was shining. Use of a high heat capacity intermediate energy storage medium allowed Solar Two in one demonstration exercise to deliver power to the grid for 24 h a day for seven consecutive days before persistent cloud cover interrupted operation. There are a number of solar power systems either operating or in the course of construction also in Spain. Planta Solar 10 and Planta Solar 20 use water/steam systems with rated capacities of 11 and 30 MW respectively. Solar Tres is projected to function at a capacity of 15 MW with a facility for 24-h operation provided by use of molten salt as an intermediate energy storage reservoir.[4]

The National Renewable Energy Laboratory (NREL) of the U.S. Department of Energy is committed, in combination with the solar industry, to an objective to install 1000 MW of CSP capacity in the Southwestern U.S. by 2010. The goal is that with continuing research and with economies of scale in manufacturing, the cost for electricity produced using CSP could be lowered to about 7 cents per kWh by 2010.[5] By way of comparison, the current cost for power generated by wind is about 5 cents per kWh (Chapter 8). In terms of capital investment, Greenblatt (2008) estimates that the expense to install a watt of CSP capacity at present would be about $3.50, projecting that with growth in the industry this could drop to about $2.00 by 2030. The corresponding numbers he quotes for present day and for 2030 onshore wind capacity are $2.00 and $1.50 respectively. CSP is not currently competitive with wind from a purely economic perspective. Given its advantage as a seasonal complement to wind, however, it is clear though that prospects for CSP merit further attention.

Photovoltaic cells (PVs) offer an opportunity for direct conversion of solar energy to electricity. They can be mounted in interconnected modules (panels) and installed either on rooftops, on walls of buildings, or on ground level sites oriented to take best advantage of available sunlight (this means a southern exposure in the northern hemisphere). They can be deployed either to provide stand-alone power or to deliver power to the commercial electric grid. Worldwide PV installation is reported to have reached a record level of 2.8 GW peak capacity in 2007 with three countries (Germany, Japan, and the U.S.) accounting for 89% of the total.[6]

The efficiency with which solar energy is converted to electricity by PV systems presently on the market ranges from about 12 to 18% with prospects for significant future improvement. (Sun Power, a San Jose, California company, is currently advertising PV panels for residences with an efficiency of 18.1%.[7]) The cost per watt for installed PV in the U.S. in 2006 averaged between $7.50 and $9.50. While costs have been coming down, it seems likely that large-scale growth in solar power at least in the near future will emphasize CSP rather than PV. There may be important opportunities, however, for commercial institutions with significant demand for power and with convenient sites for placements of PV arrays to invest in PV given the savings that could be realized by avoiding costs for delivery from the commercial grid (the cost for delivery of power to my house in Cambridge accounts for almost half the total expense).

What would it take to provide 250 GW of solar-generated electricity capacity as projected for the U.S. in 2030 by Greenblatt (2008)? If we assume that energy from the sun is delivered to the surface at an average rate of 200 Wm^{-2} and that this solar energy would be converted to electricity using CSP with an efficiency of 30%, the area required to capture this energy would be equal to about 4.17 $\times 10^9$ m^2, equivalent to 1610 square miles or about a million acres. To accomplish the same objective using PV, adopting the efficiency of 18.1% advertised for residential installations by the Sun Power Company, would require 6.91 $\times$ 10^9 m^2, 2668 square miles, or 1.66 million acres. To place these numbers in context: the area occupied by the State of Arizona is about 114,000 square miles; Rhode Island, the smallest State in the U.S., covers 1545 square miles.

We have seen that solar energy has the advantage that it can serve as a useful complement to wind. When wind energy is abundantly available (in winter at higher latitudes for example), solar resources are limited. In contrast, when wind resources are low, solar potential is high (in summer, particularly in regions where demand for electricity for air conditioning is greatest). An optimal strategy could favor a combination of both wind and solar. As noted above, projected costs for solar-derived power are presently higher than those for wind. It is difficult, however, to forecast future prices. The wind market is currently more mature than solar. The market for solar power, however, is currently enjoying rapid growth. It will be important to combine analysis of what is potentially available for both wind and solar with informed projections of what each would cost if we are to chart the economically most efficient path to an effective, low-carbon, electricity future.

17.6 ELECTRICITY FROM THE EARTH

Geothermal sources could provide an important additional source of future sustainable electrical energy (with the advantage that the supply in this case would be essentially constant or base load in contrast to sources derived from either wind or sun). Temperatures below the Earth's surface increase on average by about 20 to 30 K per km reflecting the flow of heat from the Earth's deep interior. The rate of increase of temperature with depth is even greater in regions that are tectonically active (the western U.S. for example). Drilling into the Earth would provide an opportunity to tap this potential source of energy. Cold water injected into a well could be heated by exposure to deep hot rock and converted to steam. The steam could return then to the surface in another well where it could be used to drive a turbine and produce electricity. A study by MIT scientists suggests that with relatively modest investments in research over the next 15 years it should be possible to tap geothermal sources in the U.S. for as much as 100 GW of electricity by 2050.[8] Greenblatt (2008) envisages a source of 80 GW by 2030 projecting a cost by that time for what is referred to as enhanced geothermal power of about $3.50 a watt.

17.7 ELECTRIFICATION OF THE TRANSPORTATION SECTOR

Opportunities for significant savings in demand for imported oil could be realized if the energy source for cars and light trucks in the U.S. were provided not just by alternative fuels, as described above, but by electricity as discussed briefly in Chapter 12. We extend the treatment here with a more detailed analysis of the potential merits of a switch from vehicles driven by internal combustion engines to a fleet dominated by plug-in hybrids.

The average length of a typical vehicle trip in the U.S. is about 10 miles according to the 2001 National Household Travel Survey. Of such trips, 17% are for travel to and from work, 3.2% for work-related business, 46.2% for family/personal business, 8% for travel to and from school and church, 25% for recreation, with the balance, 0.6%, for unspecified purposes. The typical car or light truck in the U.S. is driven about 12,000 miles per year, which corresponds to an average travel distance per vehicle per day of about 33 miles. Consider the possibility of replacing the current fleet of cars and light trucks with plug-in hybrids capable of an all-electric range of 60 miles. If less than 10% of travel by the average vehicle following the most recent recharge of the vehicle's batteries involved distances greater than 60 miles (a reasonable assumption), it should be possible to reduce current consumption of gasoline and diesel fuels by cars and light trucks by 90% or more. Total demand for oil would be reduced in this case by about 36% (cars and trucks account for approximately 40% of total oil use), cutting requirements for imported oil by more than 50%. At an average price of $66.05 a barrel for oil in 2006 (prices climbed to a record of $147 a barrel in July of 2008 before dropping back to $75 a barrel in late October of 2008, below $65 by the end of the month, in May 2009 in the mid-$50's), the bill for imported oil for the U.S. totaled $299 billion. At $100 a barrel, it would exceed $450 billion, more than 3% of GDP, close to 50% of the international balance of payments deficit in 2006.

As discussed in Chapter 12, driving using electric propulsion is much more efficient from an energy point of view than driving using an internal combustion engine. Less than 20% of the energy in gasoline or diesel fuel is used to turn the wheels of a vehicle. In contrast, more than 90% of the electrical energy stored in the battery of a plug-in is deployed for the same purpose. There are

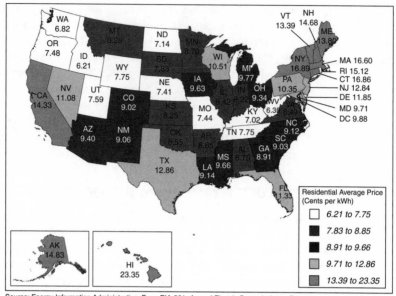

Source: Energy Information Administration, Farm EIA-861. *Annual Electric Power Industry Report.*

Figure 17.6 U.S. Electric Industry Residential Average Retail Price of Electricity by State, 2006 (cents per kWh).

important advantages also in terms of monetary savings for the motorist. As indicated in Box 12.1, at iretail prices for gasoline and electricity prevailing in Cambridge, Massachusetts in January 2006, the cost for electric propulsion would be lower than the cost for driving using gasoline by the equivalent of about $1.60 a gallon. Since the price for electricity in Cambridge is on the high end of the national average, the savings for the average U.S. motorist would be higher, $2.00 a gallon, or even more. A summary of U.S. retail prices for electricity on a state-by-state basis is presented in Fig. 17.6.[9] Savings in terms of U.S. emissions of CO_2 could amount to as much as 16% if the electricity deployed in the transportation sector were derived mainly from non-carbon sources of energy such as wind, solar, or nuclear (assuming a 36% reduction in overall consumption of oil).

Prospects for ethanol as a substitute for oil-based fuels in the transportation sector were discussed in Chapter 15. Opportunities to cut back on either fossil fuel use or on emissions of greenhouse gases using ethanol produced from corn were limited, we concluded, and there were other problems, the impact on food prices for example. Prospects for ethanol derived from sugar cane were more positive. We discussed also the possibility of a future market for ethanol produced from non-edible cellulose. The transition to an electrically driven transportation sector envisaged here, however, would largely eliminate the need for ethanol. Rather than expending energy, and financial resources, to convert cellulose to ethanol, why not burn the cellulosic biomass, convert its energy content to electricity, and use the resulting electricity rather than ethanol to drive the cars and trucks? Suppose that 100 units of fossil carbon were expended to produce 100 units of carbon as ethanol (probably a conservative assumption). Use of this ethanol as the energy source for an internal combustion vehicle would deliver the equivalent of 20 units of carbon energy to drive the vehicle with net emission of 100 units of carbon to the atmosphere as CO_2 (the carbon emitted from the fossil source used to produce the ethanol). Consider on the other hand burning 100 units of biomass carbon to produce 30 units of electricity to be used then with 90% efficiency to drive vehicles. In this case there would be zero net emission of CO_2 (the carbon used to produce the electricity would have been derived from the atmosphere by photosynthesis) and for 100 units of biomass carbon consumed, 27 units of energy would be deployed for vehicle propulsion—a net savings both in terms of demand for biomass and in emission of greenhouse gases.

A large-scale shift to electric propulsion in the transportation sector could offer important advantages also for overall improvement in the efficiency of the national electric power system. Operators of electric utilities face a persistent challenge to balance supply with demand (load) in real time. When demand peaks, they are obliged to turn on expensive sources of generating capacity (usually natural gas systems). Costs for this additional supply can exceed costs for base load supply by as much as a factor of 10 or even more (the stand-by systems must be maintained and supported financially even when they are not in use). Consider a future fleet of more than 100 million plug-in vehicles in the U.S. Imagine a choice of two household plugs to which these vehicles could be connected to draw power. One (a green plug) could provide the utility with discretion either to supply power to the vehicles' batteries or to draw power from them at times when the external demand was unusually high. The other (a red plug) would signal that the vehicle owner was in the market for power in real time and was prepared to pay whatever it would cost. The green plug would represent a component of what is known as a smart grid connection.[10] Under ideal circumstances, the vehicle owner would buy power when it was cheap and sell when it was expensive, taking advantage thus of an opportunity to profit from the investment in batteries!

Replacing 90% of current U.S. gasoline consumption with electricity would require an increase of about 23% in demand for electricity. If a significant fraction of this additional power were supplied off peak—at night for example when demand for electricity is normally low—the need for additional electric-generating capacity required to supply it would be relatively modest, potentially as little as 10% (McElroy, 2008).

17.8 ELECTRIC POWER SYSTEM

The primary objection to a large-scale switch to wind and/or solar is that the corresponding power sources are intrinsically variable and thus unreliable.

The wind does not blow all the time and clouds can obscure the sun on occasion, even in sunny climates.

Incorporating a mix of renewable energy supplies into the U.S. electricity distribution system, in part to service potential increased use of electricity in the transportation sector, will require an important investment in the nation's transmission infrastructure. Electricity in the U.S. is currently transferred through a system of three largely unconnected grid networks: the Eastern Interconnected System; the Western interconnected System; and the Texas Interconnected System. The backbone of each system consists of high voltage lines that can transport electricity efficiently over large distances (Chapter 11). Think of the transmission system as analogous to the network of roads that serves any particular region of the country. The high voltage backbone may be compared to the interstate highway system. Leading off and onto this highway system is a network of greater or lesser roads that allow motorists to find their way home or to the office or to any of a multitude of destinations. The analog of these secondary roads is the complex network of lines that connect multiple electricity-generating centers to multiple destinations where there is demand for this power. Accommodating the potential new sources of renewable electricity will require an updated road system. Ideally, it should be nationally connected (in contrast to the existing three unconnected grid systems) and traffic should be controlled by a policing system capable of redirecting traffic where necessary to avoid bottlenecks. The U.S. Department of Energy estimates that an additional 12,000 miles of transmission lines would be required to accommodate 130 GW of new electricity supply by 2030 at a cost of approximately $20 billion.[11] By way of comparison, it cost more than 20 times this amount to construct the interstate highway system. The overhead needed to connect the proposed renewable energy sources to the grid could amount to as little as 10% of the capital cost for the related generating facilities.

It is conceivable that incorporating large-scale wind and solar resources into the electric power system could reduce our dependence on imported oil in another way, complementing increased use of electricity by the transportation sector. Consider an extreme in which wind was projected to supply the bulk of U.S. electricity. There is no doubt, as indicated by the results presented above, that the potentially available wind resources could readily accommodate this demand. Significant excess electricity would be generated, however, under this circumstance, especially in winter. Rather than waste this resource, it would be important to find an economically valuable application for it. We suggested that it could be employed to provide a source of H_2. Hydrogen so formed could be converted to other valuable energy products, methanol for example, potentially using CO_2 recovered from the stack gases of either conventional or oxyfuel power plants. Or it could be used in a fuel cell or combusted to produce electricity on occasions when demand might temporarily outstrip supply. It may be noted in this context that the State of California has plans to install 150 to 200 hydrogen fueling stations throughout the state by 2010 (one every 20 miles on the state's major highways) to service a projected future fleet of hydrogen fuel cell powered vehicles.[12] Electrolysis of water fueled by electricity produced from either wind or sun could provide a carbon-free source for this hydrogen.

17.9 CONCLUDING REMARKS

The image of a potential low-carbon energy future presented here envisaged a world in which energy services would be provided to a much greater extent than today by electricity. The extra electricity, we proposed, could be generated largely by non-fossil sources. We chose in this presentation to emphasize wind and solar. We also mentioned geothermal sources, which along with nuclear power could also make important contributions.

We discussed the merits of a transition in the transportation system from oil and diesel to alternative transportation fuels, and, even more important, a switch from internal combustion driven cars and light trucks to plug-in hybrids. There are potential savings both in monetary terms for the motorist, in expense for imported oil, and in terms of reduced emissions of CO_2. A further benefit of the proposed transition to a significant plug-in hybrid vehicle fleet, as discussed, would be the opportunity to take advantage of the batteries of such vehicles as temporary storage reservoirs for electricity when it is cheap, available for resale to the grid when demand is high and power is expensive.

Both of the recent candidates for the U.S. Presidency, Senators McCain and Obama, took the threat of climate change seriously. The approaches to dealing with the problem they espoused were similar. Both were in favor of a so-called cap-and-trade strategy to restrict future emissions of greenhouse gases. The cap-and-trade approach would use a permitting system to place a limit on allowable emissions. Obama proposed to sell the permits using an auctioning process while McCain recommended distributing them initially gratis to coal, oil, and gas companies (primary producers) based on prior use. The assumption is that the trading mechanism would seek out the economically most efficient means to reduce emissions. It would amount essentially to a tax on carbon emissions that would be expected to encourage development of non-carbon based energy options (nuclear and the renewable options highlighted here), or alternatively to promote a market for carbon capture and sequestration (Chapter 14) that would allow continued use of fossil fuels. Market forces would determine then the least-cost approach to limiting emissions.

As discussed in Chapter 2, fluctuations in climate have had an important influence on the success or demise of civilizations repetitively over the course of human history. These climate variations, however, were exclusively natural in origin. The challenge we face today is fundamentally different. Our human society has evolved now as a powerful force for global change, rivaling nature. Former Vice-President Al Gore issued an eloquent call for action in a speech delivered in Washington D.C. on July 17, 2008. He concluded as follows:

> Our entire civilization depends upon us now embarking on a new journey of exploration and discovery. Our success depends on our willingness as a

people to undertake this journey and to complete it within 10 years. Once again, we have an opportunity to take a giant leap for humankind.

The choice is clear: we can accept the consequences—and they will be inequitably distributed—or we can make intelligent choices to change course. We have attempted in this volume to indicate that there are options at our disposal that can ensure the future progress for our global civilization while sustaining and maintaining the integrity of the global life support system. What happens next depends on how we react.

NOTES

1 Were the light bulb to mimic natural daylight, the temperature of the filament would have to exceed 5000°C (Chapter 3).

2 (http://www.pickensplan.com/theplan/, read October 14, 2008).

3 (http://www.eia.doe.gov/cneaf/solar.renewables/ilands/fig12.html, read October, 18, 2008).

4 (http://www1.eere.energy.gov/solar/csp.html, read October 19, 2008).

5 (http://www.nrel.gov/csp/1000mw_initiative.html, read October 19, 2008).

6 (http://en.wikipedia.org/wiki/Photovoltaic, read October 19, 2008).

7 (http://www.sunpowercorp.com/Products-and-Services/%7E/media/Downloads/for_products_services/spwr_225bk, read October 19, 2008).

8 (http://geothermal.inel.gov/publications/future_of_geothermal_energy.pdf, read October 23, 2008).

9 (http://www.eia.doe.gov/bookshelf/brochures/rep/, read October 22, 2008).

10 For a discussion of the smart grid concept (see for example http://en.wikipedia.org/wiki/Smart_grid, read October 21, 2008).

11 (http://www.windpoweringamerica.gov/pdfs/20_percent_wind_2.pdf, read October 24, 2008).

12 (http://www.hydrogenhighway.ca.gov/vision/vision.pdf, read October 23, 2008).

REFERENCES

Greenblatt, J. (2008). Clean Energy 2030: Google's Proposal for reducing U.S. dependence on fossil fuels. (http://knol.google.com/k/-/-/15x31uzlqeo5n/1#, read October 16, 2008).

Lu, X., Mc Elroy, M.B., and Kiviluoma, J. (2009). Global potential for wind-generated electricity, *Proceedings of the National Academy of Sciences*, **106**(27), 10933–10938.

McElroy, M.B. (2008). Saving money, oil, and the climate: using non-fossil energy sources to power our vehicles. *Harvard Magazine*. March–April, 2008.

Mufson, S. (2007). In Energy Conservation, Calif. Sees Light. *Washington Post*, February, 17, 2007 (http://www.washingtonpost.com/wp-dyn/content/article/2007/02/16/AR2007021602274_2.html, read October 17, 2008).

Olah, G.A., Goeppert, A., and Prakash, G.K.S. (2006). *Beyond Oil and Gas: The Methanol Economy*. Weinheim: Wiley-VCH.

Rienecker, M.M., Suarez, M.J., Todling, R., Bacmeister, J., Takacs, L., Liu, H.-C., Gu, W., Sienkiewicz, M., Koster, R.D., R. Gelaro, I. S., et al. (2007). The GEOS-5 data assimilation system-documentation of versions 5.0.1, 5.1.0, and 5.2.0. In *Technical Report Series on Global Modeling and Data Assimilation* (ed. Suarez, M.J.), NASA. Washington, D.C.

Index